THE BILE ACIDS

Chemistry, Physiology, and Metabolism

VOLUME 1: CHEMISTRY

Contributors to This Volume

W. H. Elliott
Department of Biochemistry
Saint Louis University
St. Louis, Missouri

P. Eneroth
Department of Chemistry
Karolinska Institute
Stockholm, Sweden

S. L. Hsia
Departments of Biochemistry and Dermatology
University of Miami School of Medicine
Miami, Florida

D. Kritchevsky
The Wistar Institute
Philadelphia, Pennsylvania

A. Kuksis
Banting and Best Department of Medical Research
University of Toronto
Toronto, Canada

J. T. Matschiner
Department of Biochemistry
Saint Louis University
St. Louis, Missouri

P. P. Nair
Biochemistry Research Division
Department of Medicine
Sinai Hospital of Baltimore
Baltimore, Maryland

R. Ryhage
Laboratory for Mass Spectrometry
Karolinska Institute
Stockholm, Sweden

J. Sjövall
Department of Chemistry
Karolinska Institute
Stockholm, Sweden

D. Small
Section of Biophysics
Department of Medicine
Boston University School of Medicine
Boston, Massachusetts

THE BILE ACIDS

Chemistry, Physiology, and Metabolism

VOLUME 1: CHEMISTRY

Edited by

Padmanabhan P. Nair

Biochemistry Research Division
Department of Medicine
Sinai Hospital of Baltimore, Inc.
Baltimore, Maryland

and

David Kritchevsky

The Wistar Institute
Philadelphia, Pennsylvania

PLENUM PRESS • NEW YORK–LONDON • 1971

Library of Congress Catalog Card Number 71-138520

SBN 306-37131-6

A Division of Plenum Publishing Corporation
227 West 17th Street, New York, N.Y. 10011

United Kingdom edition published by Plenum Press, London
A Division of Plenum Publishing Company, Ltd.
Davis House (4th Floor), 8 Scrubs Lane, Harlesden, NW10 6SE, London, England

Printed in the United States of America

Preface

The bile acids as principal end products of cholesterol metabolism occupy a focal position in our understanding of the role of steroids in biological systems. The biogenesis of bile acids from cholesterol in higher animals, and their functions in regulating sterol metabolism and in gastrointestinal physiology have been elucidated by the development of elegant methodological approaches during the last two decades. The molecular pleomorphism exhibited by the bile acids and bile alcohols in the animal kingdom is a classic example of their role in biochemical evolution.

The total story of the bile acids, their chemistry, their role in normal and abnormal physiological processes, and their significance in biochemical evolution has never been available in the form of a comprehensive treatise written in the words of those who have contributed to the development of our knowledge in this area. *The Bile Acids,* in two volumes, will serve to fill this void, and will also bring together information which will prove invaluable to both the biochemist and the medical scientist.

We wish to thank Mrs. Sally Wiseman and Mrs. Lillian Haas for their invaluable assistance with the editing of the manuscripts. This work was supported in part by grants AM-02131, General Research Support 5SO-1-FR-05479 (P.P.N.), HE-03299, HE-05209, and a National Heart Institute Research Career Award (D.K.), K6-HE-734, from the National Institutes of Health, United States Public Health Service.

P. P. N.
Baltimore, Maryland

D. K.
Philadelphia, Pennsylvania

Contents

Chapter 5
Extraction, Purification, and Chromatographic Analysis of Bile Acids in Biological Materials
by P. Eneroth and J. Sjövall

Chapter 6
Ion-Exchange Chromatography of Bile Acids
by A. Kuksis

Chapter 1

Chemistry of the Bile Acids*

David Kritchevsky[†‡]

The Wistar Institute of Anatomy and Biology
Philadelphia, Pennsylvania

and

Padmanabhan P. Nair[§]

Biochemistry Research Division
Sinai Hospital of Baltimore, Inc.
Baltimore, Maryland

I. HISTORICAL

Investigations into the nature of the compounds present in bile date back to the first decade of the nineteenth century (1, 2) and possibly earlier. The early investigators (1-3) found that by treating bile with lead acetate they could separate two distinct substances, which they designated "bile resin" and "picromel." Berzelius (2, 4, 5) obtained similar fractions, which he identified as "choleic acid" and "bilin." Demarcay (6) was the first to recognize the uniformity of the solid matter in bile. The presence of nitrogen and sulfur was noted by several early workers but they were probably thought to be integral parts of the isolated acids. Demarcay found that treatment of the bile solid with alkali yielded a nitrogen-free acid, which he first called

*Supported, in part, by grants AM-02131, HE-03299, HE-05209, and 5-SO-1-FR-05479 from the National Institutes of Health.

† Recipient of a Research Career Award (K6-HE 734) from the National Heart Institute.

‡ Wistar Professor of Biochemistry, Division of Animal Biology, School of Veterinary Medicine, University of Pennsylvania.

§ Director, Biochemistry Research Division, Department of Medicine, Sinai Hospital of Baltimore, Inc.

"cholic" acid, but since Gmelin (3) had applied this name to what we now know as glycocholic acid, Demarcay changed the name to "cholinic" acid. The name "cholic" acid was reapplied to this material toward the end of the nineteenth century. The isolation of nitrogen-free biliary acids was achieved by putrefaction (4, 7), but until the introduction of a specific hydrolytic enzyme by Nair (8, 9), the usual method for obtaining nitrogen-free bile acids was alkali treatment. This type of treatment prompted Strecker (10) to coin the term "cholalic" acid to indicate the nitrogen-free compound obtained from cholic acid (of Gmelin) by treatment with alkali. The first preparation of crystalline bile acids was achieved by Platner (11, 12) by the addition of ether to an alcoholic solution of dried bile.

Strecker (10, 13) set out to obtain a more accurate elementary analysis of the various biliary acids and through an analysis of several of their metal salts, he established the formulas $C_{26}H_{43}O_6N$ and $C_{24}H_{40}O_5$ for "cholic" and "cholalic" acid, respectively. The latter formula was confirmed by Mylius (14) a generation later. Strecker recognized that the difference between the two formulas (plus H_2O) was $C_2H_5O_2N$, the formula for glycine. He also saw the structural analogy with hippuric acid, which had recently been formulated as benzoyl-glycine. By analogy, he recognized taurine as the other moiety in the sulfur-containing acid and confirmed the postulated formula $C_{26}H_{45}O_7NS$ ("cholalic" acid plus taurine minus water). The taurine-conjugated acid was found mostly in the lead acetate filtrate (picromel). Strecker was of the opinion that bile from all species contained only these two conjugated acids and only differed in the proportions. Lehmann (15), in 1850, coined the names "glycocholic" and "taurocholic" acid.

With the elucidation of the elementary formula of cholic acid, a number of investigators began to contribute toward the nature of the functional groups. Hoppe-Seyler (16), for instance, showed that cholic acid was a monocarboxylic acid which yielded a monomethyl or monoethyl ester, that the three remaining oxygen atoms were present as hydroxyl groups and that, under proper conditions, it was possible to obtain the triacetate of methyl cholate (17). The elucidation of the structure of the bile acids was carried out by Wieland and his co-workers, beginning in 1912 (18).

Lachinov (19) isolated "choleic acid" from ox bile, where it was a companion of cholic acid. This substance was thought at first to be a true bile acid isomeric with desoxycholic acid,* which Mylius (14) had isolated from ox bile. Wieland and Sorge (20) finally showed that choleic acid was in fact a molecular compound of fatty acid (primarily palmitic) bound to eight

*The use of "desoxycholic" instead of "deoxycholic" is retained in this chapter as part of a general policy to adhere to nomenclature and terminology that will place the development of the subject matter in the correct historical perspective.

molecules of cholic acid. Desoxycholic acid has since been shown to form intermolecular compounds with a variety of acids, esters, alcohols, ethers, phenols, and hydrocarbons (21). These are now designated choleic acids. The number of molecules of desoxycholic acid that combine with a molecule of a given fatty acid is a function of the hydrocarbon part of the acid. There are two other common dihydroxy bile acids. Gundelach and Strecker (22) isolated an acid from hog bile that was isomeric with, but not identical to, desoxycholic acid. This compound was named "hyodesoxycholic" acid. Another isomer of desoxycholic acid was first isolated from goose bile by Heintz and Wislicenus (23), who called it "chenocholic" acid. Upon finding it to be isomeric with desoxycholic acid, Windaus renamed it "chenodesoxycholic" acid. This acid also occurs in human bile, as well as in the bile of a number of other animal species. Fischer (24) isolated a monohydroxy bile acid, lithocholic acid, from bovine gall stones. This acid is a minor component of human bile.

II. STRUCTURE OF BILE ACIDS

The most abundant naturally occurring bile acids in higher vertebrates are derivatives of cholanic acid (I) (Fig. 1), a 24-carbon-atom steroid possessing the characteristic cyclopentanophenanthrene nucleus. The structure and nomenclature of this class of compounds have been described in detail by Fieser and Fieser (25), Kritchevsky (26), and Van Belle (27).

There are two isomeric cholanic acids: one in which the plane of fusion of rings A/B is *cis*-oriented, whereas in the other it is *trans (allo)* (II).

The structural features of the two stereoisomeric forms are represented in the CPK molecular models shown in Figs. 2 and 3.

The elucidation of the structure of bile acids stems from the early observations of Adolf Windaus and of Heinrich Wieland that cholanic acid could be obtained from either cholesterol or cholic acid; they inferred from this that bile acids are steroidal in nature. Wieland and his group showed

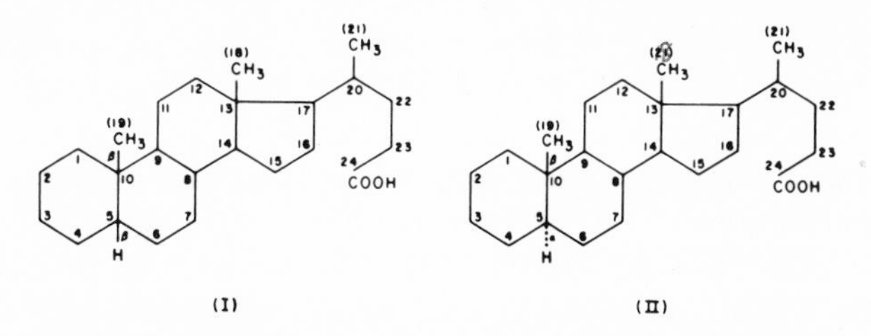

Fig. 1. Cholanic acid (5β-cholanoic acid) (I); allocholanic acid (5α-cholanoic acid) (II).

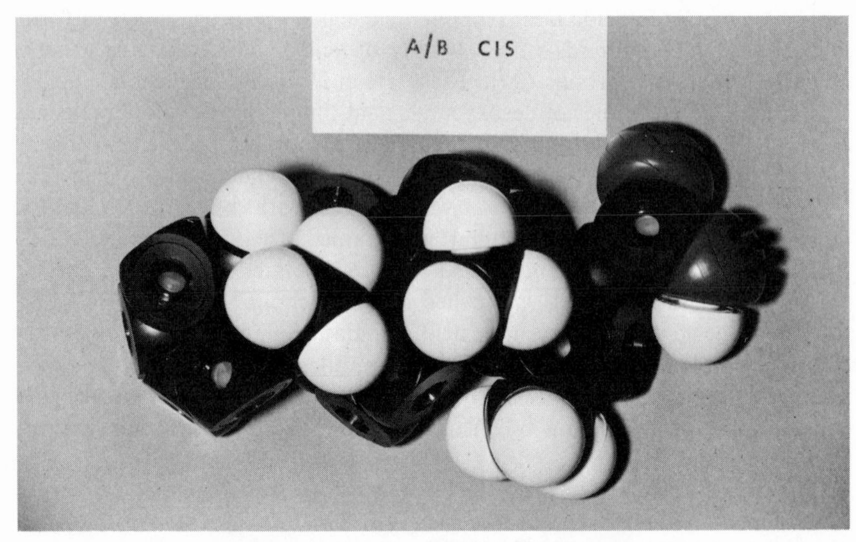

Fig. 2. Molecular model of cholanic acid.

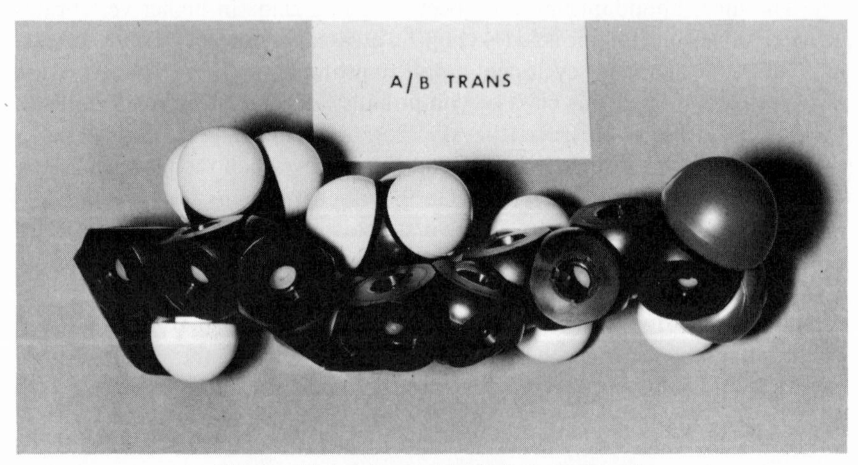

Fig. 3. Molecular model of allocholanic acid.

that cholic acid (III) could be dehydrated to yield a triene, which upon hydrogenation gave pure cholanic acid (I) (18) (Fig. 4). The careful oxidation of cholic acid gave a triketo compound (dehydrocholic acid) (IV) (28), which could be reduced to the same acid (29). Windaus (30) had speculated earlier that the sterols and bile acids were structurally related. He demonstrated

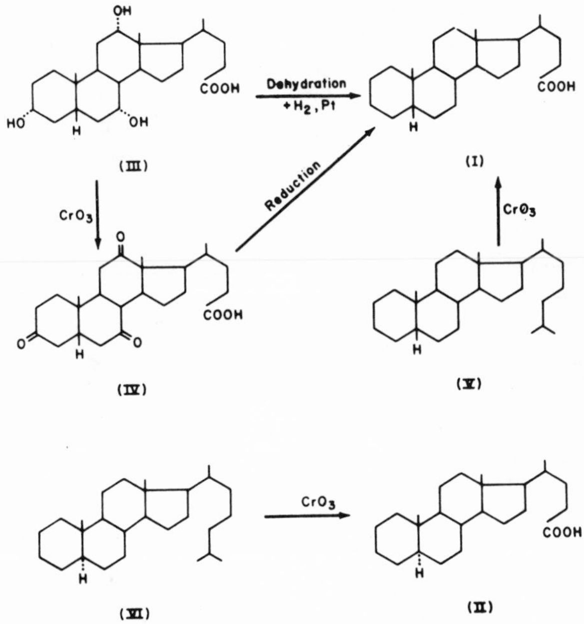

Fig. 4. Cholanic acid (I); allocholanic acid (II); cholic acid (III); dehydrocholic acid (IV); coprostane (V); cholestane (VI).

that cholestane (VI) (Fig. 4) upon oxidation yielded acetone plus an acid not quite identical with cholanic acid, but that coprostane (V), the corresponding isomeric hydrocarbon, under identical treatment, yielded cholanic acid (I) (31). It is known now that the acid obtained by the oxidation of cholestane (VI) is *allo* cholanic acid (II), the 5α epimer of cholanic acid (5β). The additional proof, if needed, was given by the resynthesis of coprostane from cholanic acid (32).

The dehydration of desoxycholic acid (33) yielded a diene mixture that could be hydrogenated to cholanic acid (34). Wieland and his co-workers also showed that both lithocholic and chenodesoxycholic acids were related to cholanic acid (35, 36). The positions and configurations of the hydroxyl groups were determined in the course of a series of oxidative cleavage studies, which actually bore upon the question of the ring structure.

The data that have emerged from these studies, which are fully described elsewhere (25, 37, 38), indicate the following structures for the common bile acids (Fig. 5). Sterically, the fusion of the A/B ring is *cis;* the B/C and C/D rings are *trans.* With reference to the methyl group at C_{10}, the hydrogen

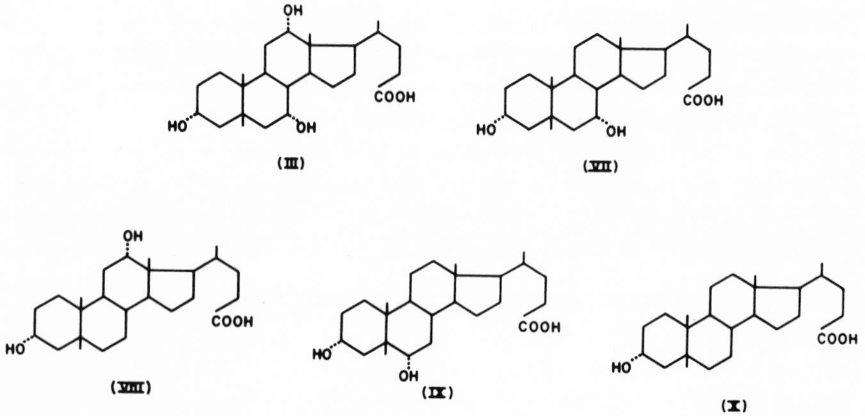

Fig. 5. Cholic acid (III); Chenodesoxycholic acid (VII); desoxycholic acid (VIII); hyodesoxycholic acid (IX); lithocholic acid (X).

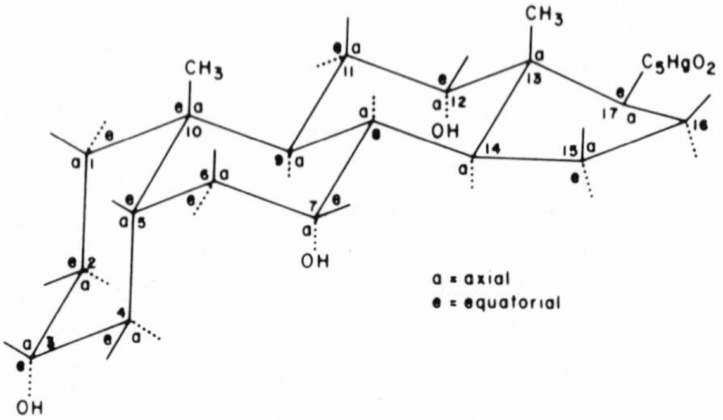

Fig. 6. Conformation of cholic acid.

atoms at C_5 and C_8 are *cis* and those at C_9 and C_{14} are *trans*. The methyl group at C_{13} is *cis* to that of C_{10}. The hydroxyl groups are in the α-configuration (behind the plane of the rings). A steric representation of cholic acid is shown in Fig. 6.

III. NATURALLY OCCURRING BILE ACIDS

A. Bile Acids with 24 Carbon Atoms

Bile acids that are formed as end products of cholesterol metabolism in the livers of higher vertebrates generally possess only 24 carbon atoms, in contrast to the 27 carbon atoms of cholesterol. These primary bile acids are derived from cholesterol by a series of hydroxylations of the sterol nucleus, followed by the loss of an isopropyl fragment from the side chain, as described in a later chapter by Danielsson.

Cholic (III) and chenodesoxycholic (VII) acids (Fig. 5), two of the common primary bile acids, undergo further chemical alterations by the action of microbial flora in the gut to give rise to several secondary bile acids (also possessing 24 carbon atoms), such as desoxycholic (VIII), hyodesoxycholic (IX), and lithocholic (X) acids (Fig. 5).

A few other compounds found in bile of specific animal species should be mentioned briefly. Hammarsten (39) isolated a substance from walrus and sea-lion bile that he showed to be a trihydroxy C_{24} acid. Windaus and Van Schoor (40) showed this to be an α-hydroxy acid, 3,7,23-trihydroxy-cholanic acid. This compound is called "β-phocaecholic" acid. "Pythocholic" acid ($C_{24}H_{40}O_5$) occurs in the bile of the family of snakes that includes boas and pythons (41, 42). The formula $3\alpha,12\alpha,16\alpha$-trihydroxycholanic acid has been suggested (43). "Ursodeoxycholic" acid (44) occurs in bear bile and is apparently $3\alpha,7\beta$-dihydroxycholanic acid. "Hyodeoxycholic" acid ($3\alpha,6\alpha$-dihydroxycholanic acid) has been isolated from hog bile.

B. Bile Acids with 27 and 28 Carbon Atoms

Some of the early investigations into the nature of bile acids in the bile of older vertebrates suggested the existence of acidic sterols more closely related to cholesterol than to cholic acid. Kurauti and Kazuno (45) isolated a bile acid from the bile of the frog, *Rana catesbiana,* which was later found to be an isomer of $3\alpha,7\alpha,12\alpha$-trihydroxycoprostanic acid (XI) (Fig. 7). This latter acid, possessing 27 carbon atoms, was isolated by Haslewood (46) in 1952 from the bile of the crocodile, *Alligator mississippiensis*. The inability to oxidize the side chain of C_{27} acids to C_{24} acids is regarded as a primitive characteristic in the biochemical evolution of the bile acids (for a fuller

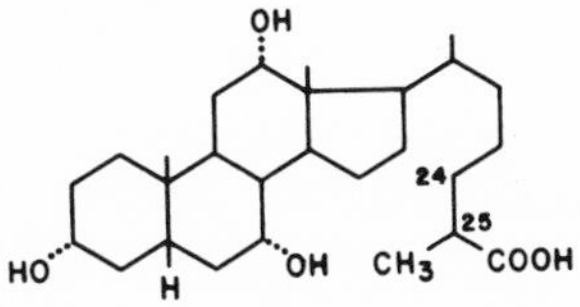

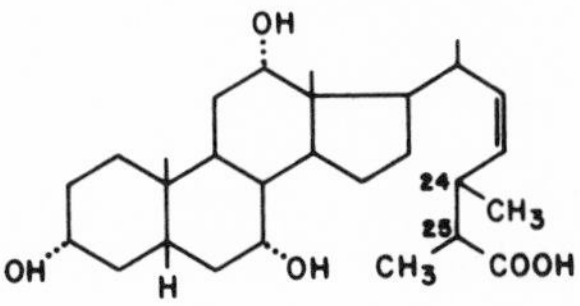

Fig. 7. $3\alpha,7\alpha,12\alpha$-trihydroxycoprostanic acid (XI); $3\alpha,7\alpha,12\alpha$-trihydroxy-24-methyl-Δ^{22}-coprostenic acid (trihydroxybufosterocholanic acid) (XII).

discussion see Chapter 2, this volume. This substance has been also isolated from human (47, 48) and baboon (49) bile and shown to be an intermediate in the biogenetic sequence between cholesterol and cholic acid in man (50).

"Trihydroxybufosterocholenic" acid (XII) (Fig. 7) ($3\alpha,7\alpha,12\alpha$-trihydroxy-24-methyl-Δ^{22}-coprostanic acid) is an example of a 28-carbon-atom bile acid first isolated from the bile of the toad, *Bufo vulgaris,* by Shimizu and Oda (51). All known C_{27} and C_{28} bile acids occur in nature as conjugates with the amino acid, taurine. There are several other acids in this group, which will be described in detail in subsequent chapters.

C. Conjugates of Bile Acids

Primary bile acids generally exist in bile in the form of conjugates with glycine or taurine through a peptide bond. These water-soluble conjugates, by virtue of their surface active properties, solubilize lipophilic substances in the gut, forming micelles in the process. The structures of two such compounds, glycocholic (XIII) and taurocholic (XIV) acids, are shown in Fig. 8.

A new series of conjugates were discovered by Palmer (52) in 1967 in which the 3α-hydroxyl group of glycolithocholic and taurolithocholic acids were esterified to give the corresponding 3α-sulfate esters. He administered lithocholic acid-24-^{14}C to human subjects 36 hr prior to surgery for gall bladder disease. Fractionation of the bile obtained at surgery revealed, in addition to labeled glyco- and taurolithocholic acids, two, more polar, metabolites. The identities of these metabolites were established by comparison

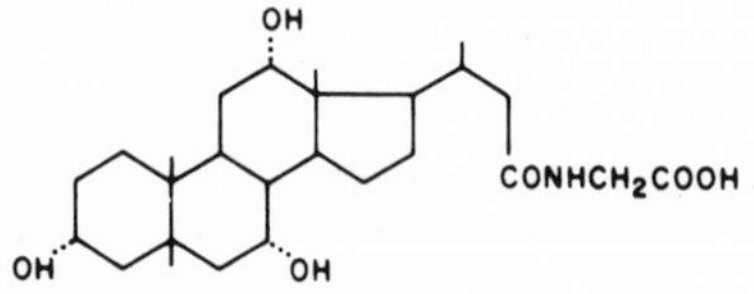

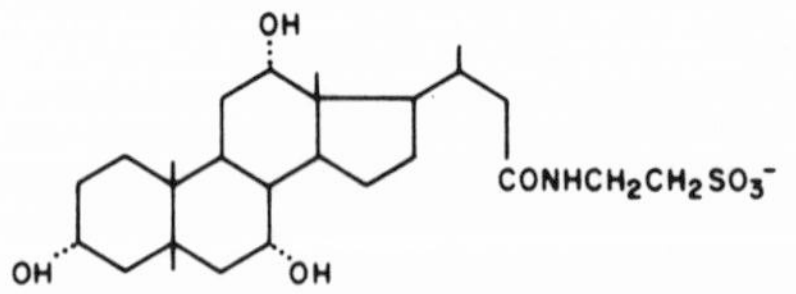

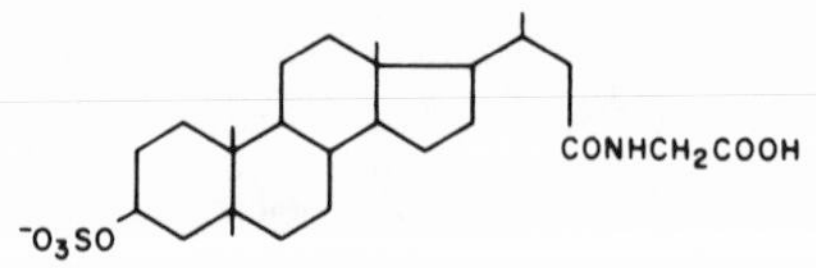

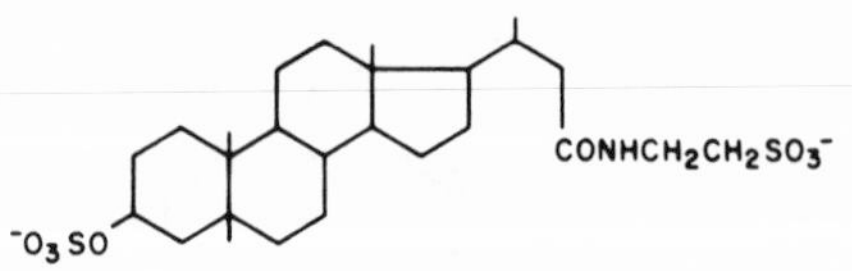

Fig. 8. Glycocholic acid (XIII); taurocholic acid (XIV); glycolithocholic sulfate (XV); taurolithocholic sulfate (XVI).

with synthetic glycolithocholic acid sulfate (XV) and taurolithocholic acid sulfate (XVI) (Figure 8).

D. The Bile Alcohols

The bile salts found in vertebrates classified below the level of the reptiles are unique in that they are mostly sulfate esters of polyhydric alcohols. The bile alcohols, such as scymnol (24,25-oxido-3α,7α,12α,26-tetrahydroxycoprostane), ranol (3α,7α,12α,24,26-pentahydroxy-27-norcholestane), and cyprinol (3α,7α,12α,26,27-pentahydroxycholestane), are discussed in a recent review by Hoshita and Kazuno (53) and in Chapter 2 by Matschiner (this volume).

IV. CHEMICAL REACTIONS

The bile acids undergo the reactions that might be expected at their various functional groups. The carboxyl group may be esterified, reduced, subjected to salt formation with metal ions, alkaloids, or organic bases and converted to the amide.

The hydroxyl group may be oxidized or esterified with a variety of acids. The hydroxyl groups at the different positions in the nucleus show variable reactivity toward oxidizing agents. The order of reactivity is $C_7 > C_{12} > C_3$ and $C_6 > C_3$. Oxidation may be carried out with CrO_3, K_2CrO_4, HNO_3, Cl_2, or Br_2 in appropriate media or *N*-bromosuccinimide.

The order of reactivity of the hydroxyl groups toward acetylation or hydrolysis is $C_3 > C_7 > C_{12}$. Thus methyl cholate when treated with acetyl chloride in pyridine will yield the 3-acetoxy-7,12-dihydroxy compound, then the 3,7-diacetoxy-12-hydroxy ester, and finally 3,7,12-triacetoxy methyl cholate as the severity of the reaction conditions increases. The reactivity at these positions *vis-à-vis* reduction or hydrogenation is also $C_3 > C_7 > C_{12}$.

A full discussion of the reactions involved in the chemistry of the bile acids may be found elsewhere in this volume. A summary of specific reaction conditions may be found in the books of Van Belle (27), Lettré (37), and Shoppee (38).

REFERENCES

1. L.J. Thenard, *Ann. Chim.* (I), **64**, 103 (1807).
2. J. Berzelius, *Ann. Chim.* **71**, 218 (1809).
3. L. Gmelin, *in* "Die Verdauung" (F. Tiedemann and L. Gmelin, eds.), Heidelberg (1826).
4. J. Berzelius, *Ann.* **33**, 139 (1840).

5. J. Berzelius, *Ann.* **43,** 1 (1842).
6. H. Demarcay, *Ann.* **27,** 270 (1838).
7. E. Von Gorup-Besanez, *Ann.* **59,** 129 (1846).
8. P.P. Nair, M. Gordon, and J. Reback, *J. Biol. Chem.* **242,** 7 (1967).
9. P.P. Nair, *in* "Bile Salt Metabolism" (L. Schiff, J.B. Carey, Jr., and J.M. Dietschy, eds.), p. 172, Charles C Thomas, Springfield, Ill. (1969).
10. A. Strecker, *Ann.* **67,** 1 (1848).
11. E.A. Platner, *Ann.* **51,** 105 (1844).
12. E.A. Platner, *J. Prakt. Chem.* (I), **40,** 129 (1846).
13. A. Strecker, *Ann.* **70,** 149 (1849).
14. F. Mylius, *Ber.* **19,** 374 (1886).
15. C.G. Lehmann, "Lehrbuch der Physiolische Chemie," Engelmann, Leipzig, 1850.
16. F. Hoppe-Seyler, *J. Prakt. Chem.* **89,** 257 (1863).
17. H. Wieland and W. Kapitel, *Z. Physiol. Chem.* **212,** 269 (1933).
18. H. Wieland and F.J. Weil, *Z. Physiol. Chem.* **80,** 287 (1912).
19. P. Lachinov, *Ber.* **15,** 713 (1882).
20. H. Wieland and H. Sorge, *Z. Physiol. Chem.* **97,** 1 (1916).
21. H. Rheinboldt, *Ann.* **451,** 256 (1926).
22. C. Gundelach and A. Strecker, *Ann.* **62,** 205 (1847).
23. W. Heintz and J. Wislicenus, *Poggendorff's Annalen der Physik* **108,** 547 (1859).
24. H. Fischer, *Z. Physiol. Chem.* **73,** 204 (1911).
25. L.F. Fieser and M. Fieser, "Steroids," Reinhold, New York (1959).
26. D. Kritchevsky, *in* "Comprehensive Biochemistry" (M. Florkin and E.H. Stotz, eds.), Vol. 10, p. 3, Elsevier, New York (1963).
27. H. Van Belle, "Cholesterol, Bile Acids and Atherosclerosis," p. 10, North-Holland, Amsterdam (1965).
28. O. Hammarsten, *Ber.* **14,** 71 (1881).
29. H. Wieland and E. Boersch, *Z. Physiol. Chem.* **106,** 190 (1919).
30. A. Windaus, *Arch. Pharm.* **246,** 117 (1908).
31. A. Windaus and K. Neukirchen, *Ber.* **52,** 1918 (1919).
32. H. Wieland and R. Jacobi, *Ber.* **59,** 2064 (1926).
33. H. Wieland and H. Sorge, *Z. Physiol. Chem.* **98,** 59 (1916).
34. H. Wieland, K. Kraus, H. Keller, and H. Ottawa, *Z. Physiol. Chem.* **241,** 47 (1936).
35. H. Wieland and P. Weyland, *Z. Physiol. Chem.* **110,** 136 (1920).
36. H. Wieland and G. Reverey, *Z. Physiol. Chem.* **140,** 186 (1924).
37. H. Lettré, R. Tschesche, and H. Fernholtz, "Über Sterine, Gallensäuren und verwandte Naturstoffe," Enke Verlag, Stuttgart (1954).
38. C.W. Shoppee, "Chemistry of the Steroids," p. 115, Butterworths, Washington (1964).
39. O. Hammarsten, *Z. Physiol. Chem.* **61,** 454 (1909).
40. A. Windaus and A. Van Schoor, *Z. Physiol. Chem.* **173,** 312 (1928).
41. G.A.D. Haslewood and V.M. Wootton, *Biochem. J.* **49,** 67 (1951).
42. G.A.D. Haslewood, *Biochem. J.* **49,** 718 (1951).
43. W. Klyne and W.M. Stokes, *J. Chem. Soc.* **1954,** 1979.
44. O. Hammarsten, *Z. Physiol. Chem.* **36,** 537 (1902).
45. Y. Kurauti and T. Kazuno, *Z. Physiol. Chem.* **262,** 53 (1939).
46. G.A.D. Haslewood, *Biochem. J.* **52,** 583 (1952).
47. E. Staple and J.L. Rabinowitz, *Biochim. Biophys. Acta* **59,** 735 (1962).
48. J.B. Carey, Jr., and G.A.D. Haslewood, *J. Biol. Chem.* **238,** 855 (1963).
49. P.P. Shah, E. Staple, I.L. Shapiro, and D. Kritchevsky, *Lipids* **4,** 82 (1969).
50. J.B. Carey, Jr., *J. Clin. Invest.* **43,** 1443 (1964).
51. T. Shimizu and T. Oda, *Z. Physiol. Chem.* **227,** 74 (1934).
52. R.H. Palmer, *Proc. Natl. Acad. Sci. U.S.* **58,** 1047 (1967).
53. T. Hoshita and I. Kazuno, *Adv. Lipid Res.* **6,** 207 (1968).

Chapter 2

Naturally Occurring Bile Acids and Alcohols and Their Origins

John. T. Matschiner*†

Department of Biochemistry
St. Louis University School of Medicine
St. Louis, Missouri

I. INTRODUCTION

The first scientific phase of bile acid investigation began in the middle of the nineteenth century when these compounds were first obtained in crystalline form. Earlier observations of bile and its properties are often cited but the efforts of the early writers, Hippocrates, Galen, Paracelsus, and others, were confined to the subjective and speculative considerations of their age. The first scientific phase, characterized by uncertainty and confusion concerning the constituents of bile, terminated around the end of the nineteenth century and gave rise to a productive period during which much information was obtained. The beginning and improvements in methods of isolation and characterization of bile acids during this period provided a strong background of structural knowledge concerning these compounds. Studies were stimulated by the recognized relationship between the bile acids and cholesterol, and, later, the steroid hormones. The contributions of Wieland and his collaborators to the structure of the bile acids began in 1912 and are reviewed in Fieser's monograph on the steroids (1).

The third and current scientific phase of bile acid investigation began about 1950. It has been characterized by new procedures for the detection and characterization of bile acids, by extensive metabolic studies, and by the identification of new and known bile acids where these are not the

---

*The author acknowledges support from N.I.H. Grants AM 11311 and AM 09909 during the time taken for the preparation of this chapter.

†Presently at Dept. of Biochemistry, University of Nebraska College of Medicine, Omaha, Nebraska.

principal components. A few new C_{24} acids have been identified; however greater progress has been made in clarifying the occurrence and structure of bile alcohols and acids with more than 24 carbon atoms.

Several reviews of the occurrence of bile acids and alcohols are available and have been used extensively in the survey for this chapter. Thorough reports by Haslewood (2–5), contain not only the extensive literature of the occurrence of the bile salts but also correlate these data with their probable significance in biochemical evolution. The early literature is reviewed in the companion volumes of Sobotka (6, 7). More recent coverage of the chemistry of the bile acids may be found in the books of Fieser and Fieser (1), Shoppee (8), and Van Belle (9). Bile alcohols and higher bile acids have been reviewed by Hoshita and Kazuno (10). An extensive tabulation of bile acids, their properties and derivatives, is given in Part 4 of Elsevier's Encyclopedia (11). Trivial names for common bile acids and alcohols are used throughout the text of this chapter. They are identified in the tabulation of acids and alcohols in Section II. D.

II. NATURALLY OCCURRING BILE SALTS

A. Early Observations

The problem confronting early investigators was to obtain, in reproducibly pure form, a sample of the principal components of bile. The earliest observations were those of Thenard (12), Berzelius (13), and Gmelin (14). These investigators distinguished two components of bile—an acid precipitated by lead acetate and a soluble fraction presumed to be neutral. On the basis of these observations, Berzelius and others believed that the main constituent of bile was not an acid or soap, as had been generally held, but a neutral substance, the soluble fraction, which decomposed readily into acidic products.* Several investigators obtained crystalline fractions from bile (17–19) and recognized the inaccuracy of Berzelius' concept, but it was Strecker (19) who convinced his colleagues that two acids existed in bile—a nitrogen-containing acid (glycine-conjugated) and one containing both nitrogen and sulfur (taurine-conjugated)—and that the neutral fraction of Berzelius corresponded to the salt of the latter acid.

The next important realization was that bile from different species

*The decomposition of bile referred to in the older literature was probably of bacterial origin. Berzelius (15) and Gorup-Besanez (16) reported the isolation of non-nitrogenous fractions from putrefied bile.

contained acids that differed not only in conjugation but also in free form. Strecker (20) and others, utilizing the crystallization techniques of Platner (17), examined the bile of several species and found two exceptions to Strecker's original view that only two acids occurred in bile (glycocholic and taurocholic). Pig bile yielded a new acid (glycohyodeoxycholic acid) which was carefully examined by Gundelach and Strecker (21) and recognized as a conjugate of glycine but different from glycocholic acid. Taurochenodeoxycholic acid, the principal acid of goose bile, was discovered by Marsson (22), and obtained as a free acid (barium salt) by Heintz and Wislicenus in 1859 (23).

Although alkaline hydrolysis of glycocholic and taurocholic acid was known to yield a crystalline, non-nitrogenous product (24), it was several decades before this procedure was applied to the verification and extension of the experiments that had yielded "different" bile acids. Thudichum (25), in 1881, observed an "isomer" of cholic acid in ox bile. Latschinoff (26), in 1885, isolated "choleic acid" from ox bile and determined its approximate ratio to cholic acid. Mylius (27, 28), in 1886, isolated "deoxycholic acid" from putrefied ox bile. The co-identity of these preparations was accepted for many years, but it was not until 1916 (29), when the molecular complexes of deoxycholic were described by Wieland and Sorge, that the apparent inconsistencies were resolved.

B. Isolation

By 1900, only deoxycholic acid and cholic acid had been obtained in pure crystalline form. In 1911, Fischer (30) isolated lithocholic acid from a gallstone in the course of his classical studies on the bile pigments. Hyodeoxycholic acid was purified by Windaus and Bohne in 1923 (31) and chenodeoxycholic acid was rediscovered almost simultaneously by Wieland and Reverey (32) and by Windaus *et al.* (33) in 1924.

A typical procedure for the isolation of the free bile acids included precipitation of the bile protein with alcohol and hydrolysis of the conjugated acids for long periods of time in from 5 to 10 percent aqueous sodium hydroxide. Different methods were then applied for the isolation of each individual acid from the resulting crude mixture of free bile acids. Cholic and deoxycholic acid could be separated from each other by the addition of concentrated barium chloride to an ammoniacal solution of the hydrolyzed acids, since most of the barium deoxycholate is precipitated with the fatty acids, whereas barium cholate remains in solution (34). The greater solubility of chenodeoxycholic acid in ether was used by Wieland for its isolation; lithocholic acid, the weakest of the acids, was obtained by fractional precipitation with acid. Several classical references are available that contain typical procedures used for the early isolation of both free and conjugated bile acids (7, 34–38).

As the result of the efforts of a number of investigators, a wealth of information accumulated concerning the occurrence of bile acids in a large number of species, but it must be emphasized that the procedures used until about 1950 allowed the isolation of only the predominant bile acids. Minor constituents were isolated from mother liquors when large amounts of bile were examined.

The greatest single advance in the field of bile acid investigation came with the application of chromatographic procedures to the separation and purification of the bile acids. Bergstrom and co-workers applied the reversed phase partition chromatographic procedure of Howard and Martin (39) to the separation of free (40, 41) and conjugated (42, 43) bile acids. Another chromatographic procedure grew from the studies of Ahrens and Craig (44), who evaluated several solvent systems for the separation of the constituents of beef bile by the counter-current distribution method. Mosbach and co-workers (45) applied one of these solvent systems to a partition chromatographic procedure for the separation of the free bile acids, using Celite as the supporting medium, aqueous acetic acid as the stationary phase, and the mixtures of petroleum ether and isopropyl ether as the eluting solvent. Investigators in Doisy's laboratory (46) used the same system, substituting benzene for isopropyl ether in the movable phase. Another early procedure for separating the bile acids was paper chromatography, introduced in 1952 (47, 48). It has been used extensively since that time. Thin-layer and gas–liquid chromatography were introduced separately for the separation of bile acids, but used conjointly they provide a valuable means of identification (49, 50). Mass spectrometry, which was applied to the characterization of bile acids as early as 1958 (51), may be coupled with gas–liquid chromatography, and it has been widely used in recent years. Used in conjunction with thin-layer chromatography these methods provide a means of thorough analysis of any source of naturally occurring bile acid (52). These procedures and others reviewed more thoroughly in these volumes form the technical basis for extensive studies of metabolism of the bile acids and for the analysis of bile, feces, and tissues of various species.

C. Structural Identification

Investigations of the structure of bile acids and sterols were undertaken separately but eventually became fused into the common problem of determining the structure of the steroid nucleus. Such a relationship was suspected by Windaus in 1908 (53) but was predicted even earlier by Latschinoff (54) in 1879. The first successful attempts at establishing this relationship were those of Windaus and Neukirchen in 1919 (55). Cholestane (C_{27}) was oxidized to yield an acid (C_{24}) similar but not identical to the parent acid of the bile acid group already obtained by Wieland and Weil (56). However, the oxidation of coprostane gave an acid identical in every respect to Wieland's cholanic acid.

Once the relationship of the problems of sterol and bile acid chemistry became known, the bile acids were important sources of structural information. The elaborate studies that centered around ketonic and ring-ruptured or bilianic steroid derivatives were carried out for the most part on the suitably oxygenated bile acids. In the course of these investigations, interconversion in the bile acid series demonstrated that different naturally occurring bile acids were structurally represented in most instances by hydroxylation at only three positions in the steroid nucleus. The important exception to this was hyodeoxycholic acid, which proved to be 3α,6α-dihydroxycholanic acid. Hyodeoxycholic acid enabled Windaus (31, 57) to establish the nature of the stereoisomerism at carbon 5 and provided a means of passing between the so-called normal or bile acid series (rings A/B *cis*) and the allo series (rings A/B *trans*). This was possible since oxidation of the hydroxyl group at carbon 6 in the normal series results in a ketone that may be readily epimerized to the more stable A/B *trans* ring system.

Several other "atypical" acids were eventually isolated from various species. Ursodeoxycholic acid, first isolated in crystalline form from bear bile in 1927 (58), was identified as the 7β-epimer of chenodeoxycholic acid. The so-called β-hyodeoxycholic acid (3β,6α), which Kimura obtained in small amounts from pig bile (59), was structurally identified in the course of a thorough investigation of the four possible 3,6-dihydroxycholanic acids (60). The lagodeoxycholic acids isolated from rabbit bile by Kishi (61) were not characterized until the recent studies of Danielsson *et al.* (62) identified one of these compounds as allodeoxycholic acid. The contention that one of them may have been the 12β-epimer of deoxycholic acid was placed in doubt by Koechlin and Reichstein (63), who prepared that acid and found that it did not exhibit the physical properties of the natural material.

A thorough review of the chemistry of many of the naturally occurring bile acids and related substances may be found in the monograph by Fieser and Fieser (1). A tabulation of known acids and alcohols, their properties, and a brief description of their origins is presented in the next section of this chapter. Identifications in some cases have not been confirmed but the burden of accuracy has been left with the original authors.

D. Tabulation of Known Acids and Alcohols

1. C_{24} Acids

Cholanic acid

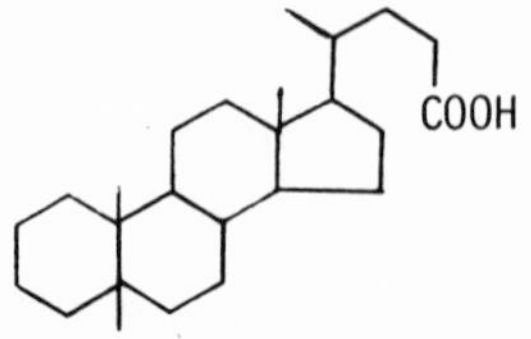

COOH

m.p. 164°C; α_D + 22° (1)

This acid, the parent compound of the hydroxylated C_{24} 5β acids, was detected as a trace component of human feces by Ali *et al.* in 1966 (50). It has not been detected elsewhere as a naturally occurring substance. Preparations of cholanic acid have been obtained by Clemmenson or Wolff–Kishner reduction of fully oxidized bile acids or by pyrolysis of hydroxy acids followed by hydrogenation of the resulting mixture (1).

Lithocholic acid

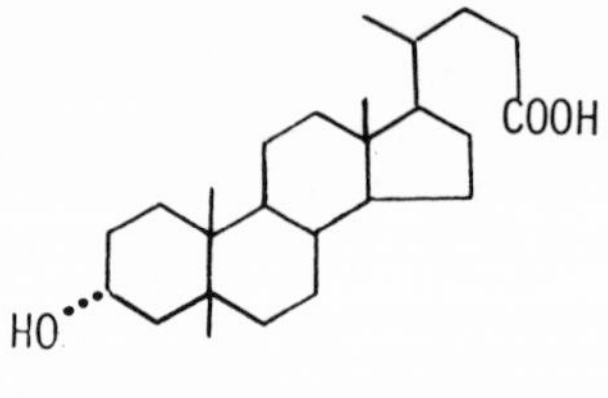

COOH

HO

m.p. 186°C; α_D + 35° (1)

Lithocholic acid was first isolated from a gallstone by Fischer in 1911 (30). It was later isolated from ox bile (1 g from 100 kg of bile) (64); from rabbit bile (0.4 g from 900 ml) (65); and subsequently from monkey, human, pig, and guinea pig bile (2, 66, 67). Lithocholic acid has been identified as one of the bile acids in human blood (68) and as a principal fecal bile acid. Mosettig *et al.* (69) isolated lithocholic acid from human stool and estimated its concentration to be 3 g/100 kg of fresh stool. Lithocholic acid is particularly insoluble, is not hydroxylated to an appreciable extent in man (70), and may be the cause of liver disease (4). In early studies lithocholic acid was not available in sufficient amounts from natural sources and was prepared from cholic acid. Lithocholic acid was particularly valuable in establishing the correspondence of the B/C ring structure between bile acids and cholesterol (71).

3β-Hydroxycholanic acid

COOH

HO

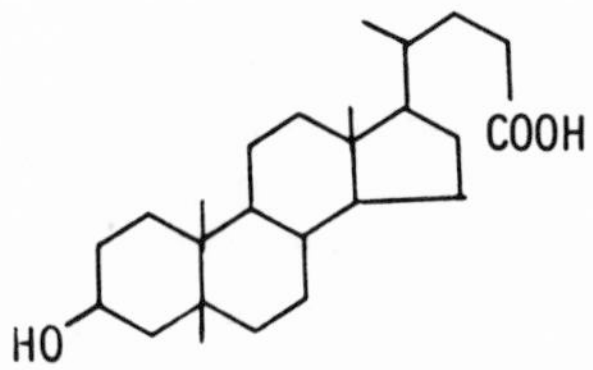

m.p. 177 °C; α_D + 20° (72)

This acid, the 3β-epimer of lithocholic acid, has been detected as a component of the feces of man and rabbit (32, 62, 73, 74). A compound named isolithocholic acid was isolated from chicken bile in an early study by Hosizima *et al.* (75). However, the properties of this acid do not coincide with those of 3β-hydroxycholanic acid. The 3β acid has been prepared from the ketone by hydrogenation under acidic conditions (72).

7β-Hydroxycholanic acid

COOH

OH

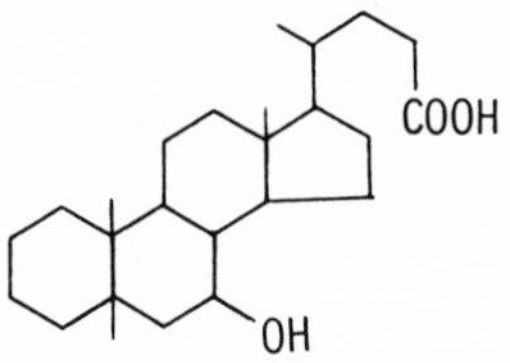

m.p. 133–134 °C; α_D + 56° (76)

This acid was detected by Ali *et al.* (50) as a trace component of human stool. It has not been detected elsewhere in nature. 7β-Hydroxycholanic acid may be prepared by reduction of the 7-keto acid with sodium and propanol (76).

Chenodeoxycholic acid

COOH

HO

OH

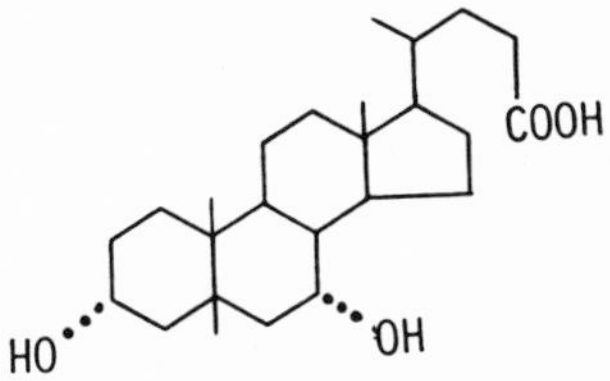

m.p. 143 °C; α_D + 11.5° (1)

Wieland (32) and Windaus (33) independently discovered chenodeoxycholic acid in different biles (man and goose) in 1924. The same acid had been detected previously in goose bile in 1848 by Marsson (22), who named it "chenocholinic" acid, and by Heintz and Wislicenus (23) in 1859, who gave it the name "chenocholic" acid. The present name was given by Windaus (33), who showed that the acid was isomeric with deoxycholic rather than cholic acid. Chenodeoxycholic acid was soon detected as a trace component of ox bile (77) and was isolated as a complex with 12-ketolithocholic acid called Weyland's acid (36). Chenodeoxycholic acid is a most soluble bile acid and was purified from other acids by this property. It occurs widely in fishes and birds. In birds, where it was first discovered, chenodeoxycholic acid is often present as the principal bile acid. Finally, chenodeoxycholic acid is widely distributed among mammalian species, making it as ubiquitous as any bile acid, with the possible exception of cholic acid (4). Chenodeoxycholic acid is a predominant component of rat bile (78). Wootton and Wiggins (79) were the first to obtain a true impression of the amount of chenodeoxycholic acid in human bile, where the chenodeoxycholic/cholic/deoxycholic ratio in fistula bile was found to be 32:60:8. The acid has also been identified in human blood (80, 81) and in the feces of dogs and man (50, 52, 82). The synthesis of chenodeoxycholic acid from cholic acid reported as early as 1932 by Kawai (83) has been accomplished in modified form by a number of investigators since that time (1).

Ursodeoxycholic acid

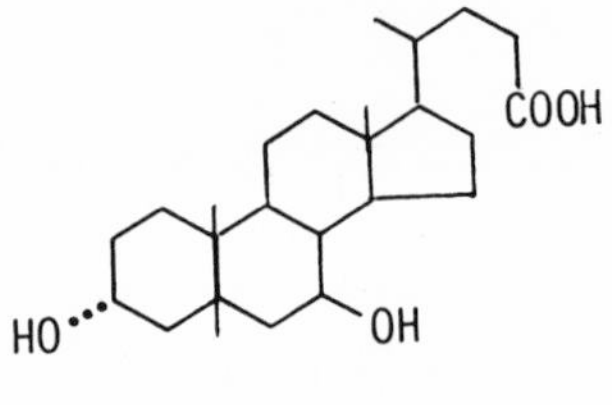

COOH

HO OH

m.p. 203 °C; α_D + 57° (84)

Ursodeoxycholic acid was detected very early by Hammarsten in polar bear bile (85). It was isolated in crystalline form by Shoda (58) and characterized later by Iwasaki (84). The acid, which is the 7β-epimer of chenodeoxycholic acid, may be prepared in good yield from 7-ketolithocholic acid by reduction with sodium in propanol according to Kanazawa *et al.* (86). Ursodeoxycholic acid was originally considered to be a unique constituent of bear bile but has since been detected as a minor constituent in the bile of several mammals including man (2). It is also present in human feces (52).

3β,7α-Dihydroxycholanic acid

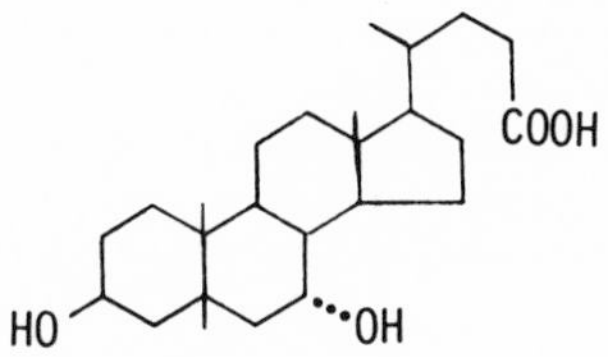

m.p. 193 °C; α_D + 8.5 (87)

This acid, the 3β-epimer of chenodeoxycholic acid, has been detected as a component of human feces (52). It has not been detected elsewhere in nature. It was prepared from chenodeoxycholic acid by Danielsson *et al.* (87).

Hyodeoxycholic acid

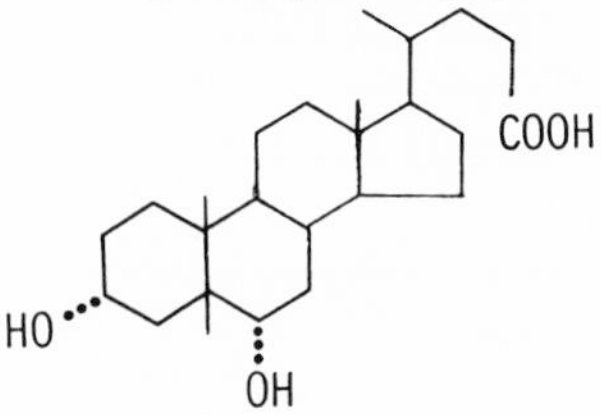

m.p. 197 °C; α_D + 8° (1)

Hyodeoxycholic acid was first obtained as its glycine conjugate by Gundelach and Strecker in 1847 (21). This unique conjugated acid from pig bile had become sufficiently well known by 1925 that a procedure for its preparation was included in Abderhalden's *Handbuch der Biologischen Arbeitsmethoden* (34). The free acid was studied by Windaus and Bohne in 1923 (31).

Hyodeoxycholic acid is not a primary acid though it is the principal constituent of bladder bile of the pig. The acid from which it is derived by bacterial action in the intestine is 3α,6α,7α-trihydroxycholanic (hyocholic) acid (88). 6-Hydroxylated acids were considered unique to the pig until similar acids were identified as minor constituents in rat fistula bile (89). Since that time hyodeoxycholic acid has been reported to occur in rat plasma and liver (90), in rat bile (91), and in rat feces (92, 93).

During studies of steroid structure, hyodeoxycholic acid became valuable for its ability to be interconverted between the two A/B stereochemical types. It was thus a source of supporting evidence for the C-5-isomers in the steroid family (1).

$3\alpha,6\beta$-Dihydroxycholanic acid

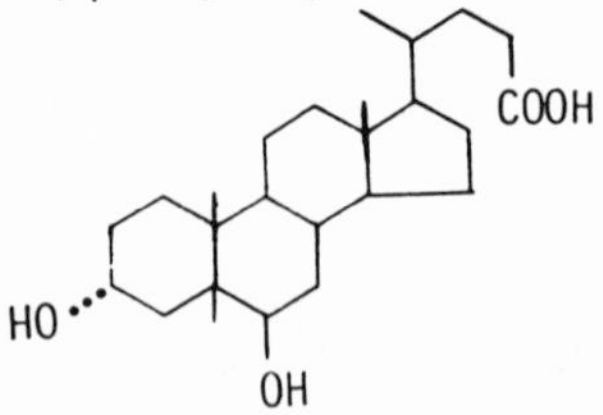

m.p. 208 °C; α_D + 37° (94)

This acid was isolated as a trace component of the mixed acids obtained from the bile of pig bladders (95). It has not been detected elsewhere as a naturally occurring compound. The acid may be prepared by $NaBH_4$ reduction of 6-ketolithocholic acid (96, 97).

$3\beta,6\alpha$-Dihydroxycholanic acid

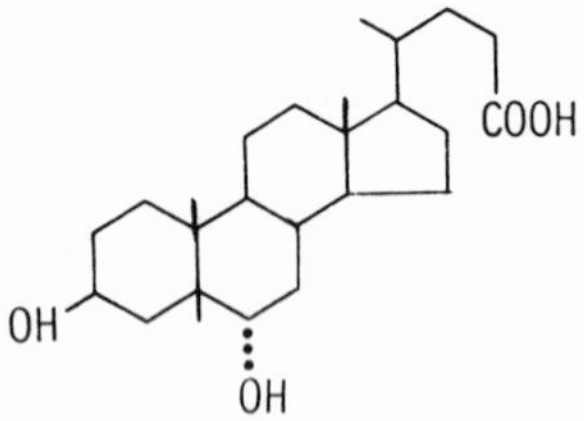

m.p. 190 °C; α_D + 5° (59)

$3\beta,6\alpha$-Dihydroxycholanic acid was isolated from pig bile by Kimura (59) who recognized the acid as a C-3-epimer of hyodeoxycholic acid. The conformation of the hydroxyl group at carbon 6 was finally established in 1947 (60). The 3β-hydroxy acid also occurs in gallstones of the pig, which characteristically contain large proportions of lithocholic acid (98, 99). $3\beta,6\alpha$-Dihydroxycholanic acid has not been detected in the bile of any other species. The acid may be synthesized by catalytic reduction of the 3-ketone under acidic conditions (100). The 3-keto acid is available by Oppenauer oxidation of hyodeoxycholic acid (101).

Deoxycholic acid

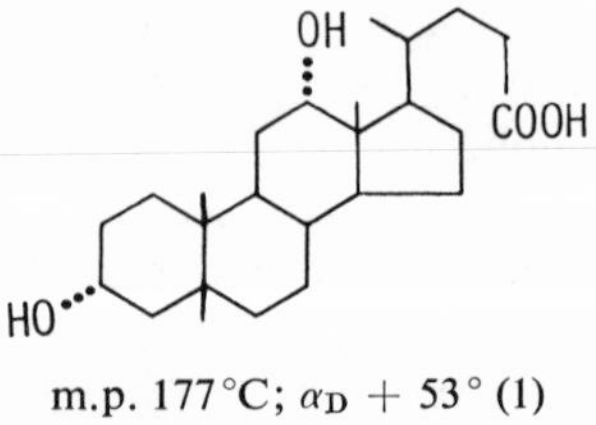

m.p. 177 °C; α_D + 53° (1)

This well-known bile acid was first reported by Latschinoff in 1885 (26). Latschinoff isolated the compound from ox bile, named it choleic acid, and recognized that it had a lower oxygen content than the trihydroxy acid, cholic acid. One year later Mylius (27, 28) reported an acid from putrefied ox bile which he named deoxycholic acid. Latschinoff (102) and later Lassar-Cohn (103, 104) presented evidence that choleic acid and deoxycholic acid were the same compound, but uncertainty continued until Wieland and Sorge (29) discovered the nature of the molecular complex that is choleic acid. Originally this name was applied to the stable mixture of eight molecules of deoxycholic acid and one molecule of fatty acid but later choleic acids were shown to form between deoxycholic acid and a variety of acholic compounds. The subject of choleic acids was thoroughly reviewed by Sobotka (7) and by Fieser and Fieser (1).

Deoxycholic acid is present in substantial amounts only in mammalian bile. It is a bacterial artifact produced from cholic acid during enterohepatic circulation and is not formed by the liver. The first observations of bacterial contribution to biliary composition were made in studies with deoxycholic acid (105, 106). The absence of this acid from rabbit fistula bile was observed as early as 1930 (107) but the importance of bacterial metabolites in the composition of bile was not appreciated until discovered by the Swedish workers (108). Several other bile acids have now been recognized as "secondary," resulting from similar alterations in the gut during enterohepatic circulation rather than from hepatic biosynthesis (4, 9).

The occurrence of deoxycholic acid in the bile of intact animals is dependent on the extent of its hydroxylation to cholic acid in the liver. In rabbits where a large amount of deoxycholic acid is formed in the intestine and little if any hydroxylation of deoxycholic acid occurs, this acid becomes predominant in the bile. In rats, where deoxycholic acid is also formed intestinally (108), the liver is capable of hydroxylating absorbed deoxycholic to cholic acid (109, 110) so that deoxycholic acid does not appear in bile in more than trace amounts. Deoxycholic acid is a principal component of human feces (111) and of the feces of other animals in which this acid is formed. Furthermore, intestinal stones or enteroliths are often rich in deoxycholic acid which precipitates as a choleic acid complex (4).

Deoxycholic acid was an important starting material for the early synthesis of cortisone and other cortical steroids. It was obtained principally from cholic acid by selective oxidation and subsequent Wolff–Kishner reduction. Cholic acid was obtained from beef bile which also contains substantial amounts of deoxycholic acid. The history of this period of steroid investigation is reviewed by Fieser and Fieser (1).

$3\alpha,12\beta$-Dihydroxycholanic acid

OH
COOH
HO

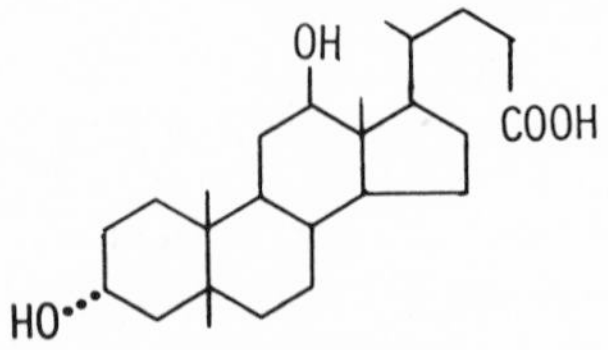

m.p. 186–188 °C; α_D + 38° (63)

This acid, the 12β-epimer of deoxycholic acid, has been detected as a fecal acid in man (50, 52). It has not been detected elsewhere as a naturally occurring compound. The acid was first prepared by Koechlin and Reichstein (63) by hydrogenation of the 12-keto acid in methanol with Raney nickel catalyst.

$3\beta,12\alpha$-Dihydroxycholanic acid

OH
COOH
HO

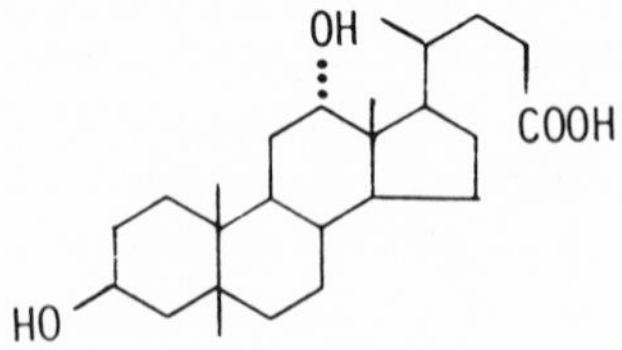

m.p. 176 °C; α_D + 47° (87)

The 3β-epimer of deoxycholic acid has been detected as a component of rabbit and human feces (50, 52, 62, 74, 87). It has not been detected elsewhere as a naturally occurring substance. The acid was prepared by reduction of the corresponding 3-keto methyl ester followed by hydrolysis (87). It has also been prepared in good yield by inversion of the 3-tosylate of methyl deoxycholate (112).

$3\beta,12\beta$-Dihydroxycholanic acid

OH
COOH
HO

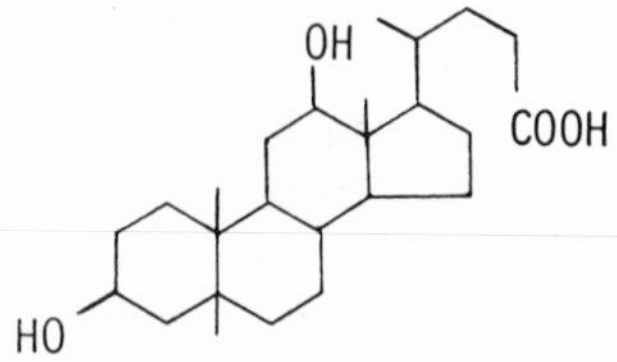

m.p. 204–206 °C; α_D + 27° (112)

This acid completes the citation of the four possible 3,12-dihydroxy-5β-cholanic acids as naturally occurring compounds. The $3\beta,12\beta$ acid and the two previous isomers have been detected only as fecal acids (3β,12β in man) (50, 52). Deoxycholic acid has been isolated from bile; however it is formed in the intestine and is also a principal fecal bile acid in several species (4). 3β,12β-Dihydroxycholanic acid has been prepared by Raney nickel reduction of the methyl ester of the corresponding 3-keto acid followed by hydrolysis (112).

Cholic acid

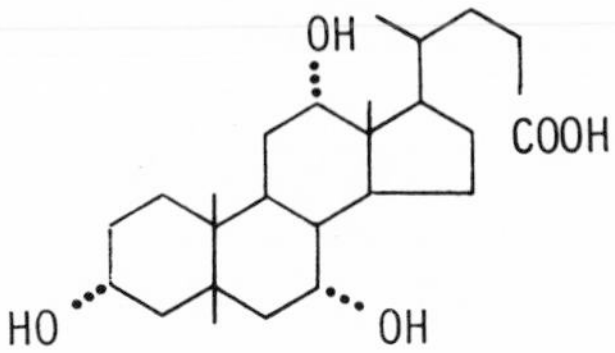

m.p. 199°C; α_D + 37° (1)

The history of cholic acid is intimately woven through the entire history of the isolation and chemistry of the bile acids. This most common acid was isolated in 1838 and 1843 but was first studied carefully by Strecker in 1848 (7). The name cholic acid given by Demarcay had already been used by Gmelin (14), probably for glycocholic acid, so that other names were substituted for the free bile acid. The name "cholalic" acid, which persisted for several decades, was given by Strecker (19). It referred to the treatment of bile with alkali in order to obtain the nitrogen-free bile acid. Strecker (19) obtained the correct $C_{24}H_{40}O_5$ empirical formula for cholic acid.

Little further was substantiated about cholic acid (or about bile acids generally) for nearly five decades. The name, cholic acid, had become well established, but trivial names based partly on imperfect characterization were common, so that consideration of nomenclature was a part of Wieland's first report on the bile acids in 1912 (56). A review of the evidence then available included that cholic acid was a trihydroxy, monocarboxylic acid and that two of the alcohols were secondary. The final presentation of the structure of cholic acid awaited the correct steroid formulation in 1932 (113, 114). By that time the structural relationship between the sterols and bile acids was well established. The preparation of cholanic acid from cholic acid had been reported in the paper by Wieland and Weil in 1912 (56). The preparation of cholanic acid from cholesterol (through coprostane) was reported in 1919 by Windaus and Neukirchen (55).

Cholic acid has been detected in the bile of most vertebrates. It also occurs in feces (115). A unique absence of this acid may occur in the bile of some members of the pig family. A trace of cholic acid was detected in a mixture of acids from domestic pig bile (116) but contamination could not be rigorously excluded. The acid was also detected in pig bile by GLC (117). In other studies cholic acid has not been detected in the bile of domestic and wild pigs (4). It has however been isolated from the bile of the warthog (99).

3α,7β,12α-Trihydroxycholanic acid

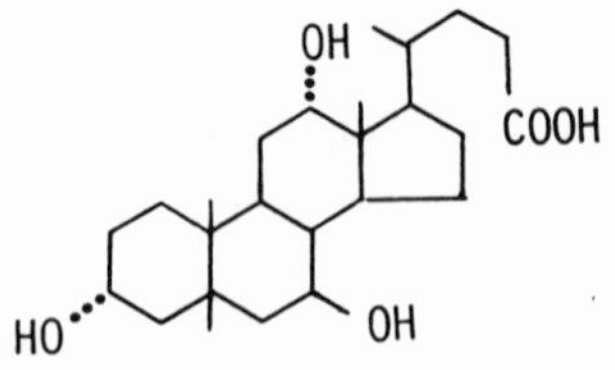

m.p. 127–129°C; α_D + 62° (118)

The 7β epimer of cholic acid has been detected in the feces of man, dog, and rat (82, 119–121). It has also been found in human bile (122) and in the bile of the rat as a bacterial metabolite of cholic acid (120). The acid may be synthesized from 7-ketodeoxycholic acid by reduction with sodium in propanol (118).

3β,7α,12α-Trihydroxycholanic acid

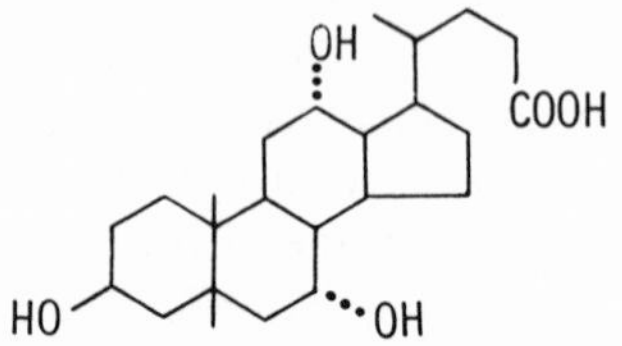

m.p. 202°C; α_D + 30° (87)

This acid, the 3β-epimer of cholic acid, has been detected as a component of the feces of man and dog (82,121). It may be synthesized from the corresponding 3-keto acid (87).

3β,7β,12α-Trihydroxycholanic acid

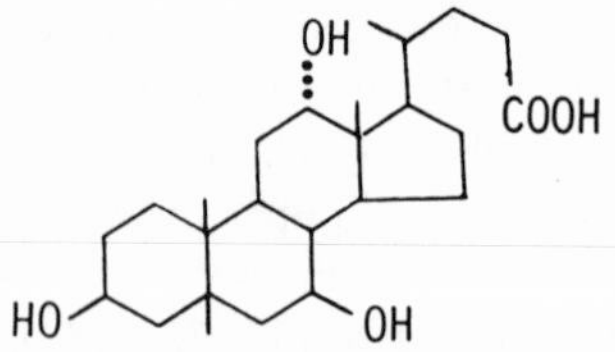

m.p. 154–157°C; α_D + 70° (121)

This trihydroxy acid has been detected as a component of human feces (121). It has not been detected elsewhere as a naturally occurring substance.

The acid was prepared by reduction of the 3,7-diketone with sodium in *n*-propanol (121).

Hyocholic acid

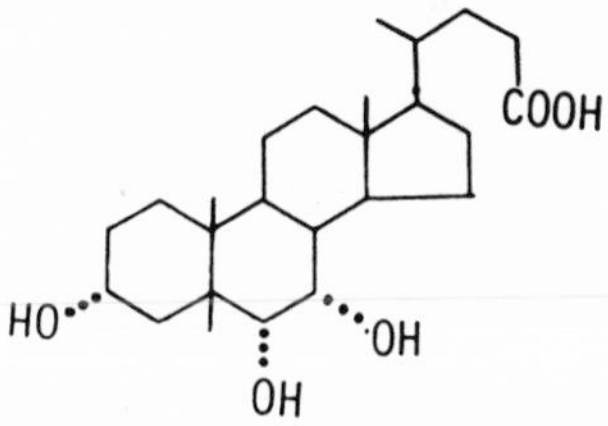

m.p. 189 °C; α_D + 5.5° (123)

Hyocholic acid was isolated from pig bile by Haslewood in 1954 (123, 124). It has not been detected in the bile of other species. Hyocholic acid was characterized by Haslewood (125), Ziegler (126,127) and by Hsia *et al.* (89) as 3α,6α,7α-trihydroxycholanic acid. The acid may be synthesized by $NaBH_4$ reduction of 3α,6α-dihydroxy-7-ketocholanic acid, which in turn is prepared from chenodeoxycholic acid (89).

α-Muricholic acid

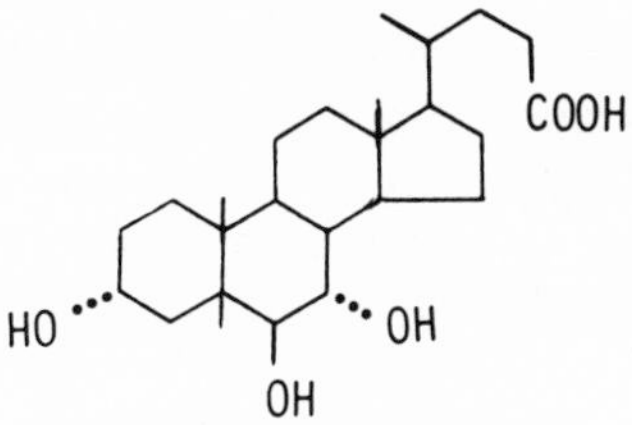

m.p. 200 °C; α_D + 38° (128)

This acid was isolated from rat bile by Matschiner *et al.* (46). It has not been isolated from the bile of other species. It was characterized by Hsia *et al.* (128). α-Muricholic acid may be synthesized from methyl 3α-acetoxy-Δ^6-cholenate either through epoxidation (128) or through the bromohydrin acetate (129).

β-Muricholic acid

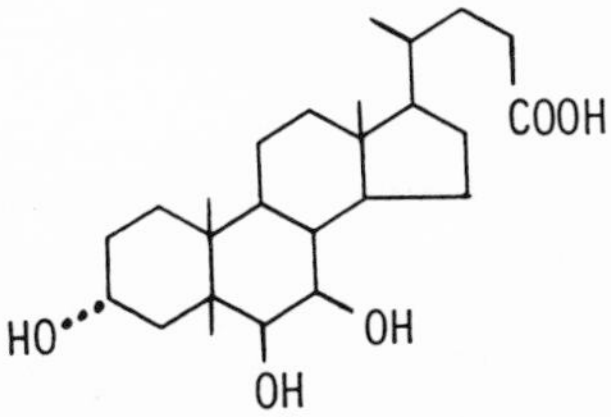

m.p. 226 °C; α_D + 62° (89)

β-Muricholic acid was isolated from rat bile along with α-muricholic acid (46) and was characterized by Hsia *et al.* (130). It becomes the principal bile acid in the urine of surgically jaundiced rats (131). β-Muricholic acid may be synthesized by Meerwein–Ponndorf reduction of a $3\alpha,6\alpha$-dihydroxy-7-ketocholanic acid (89).

Pythocholic acid

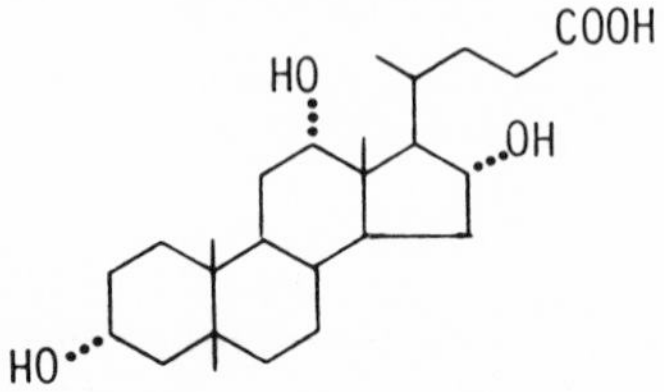

COOH
HO
OH
HO

m.p. 186–187 °C; α_D + 28° (methyl ester) (99, 132)

Pythocholic acid has been isolated from several species of snakes of the family Boidae (132). It was characterized as a 3-, 12-, 15- or 16-trihydroxy-cholanic acid by Haslewood and Wootton (132) and Haslewood (133). The assignment of the 16α-hydroxyl group is consistent with optical rotation data (134). The acid takes its name from the python where it is the principal bile acid. Despite its character as a unique trihydroxycholanic acid, pythocholic acid is probably not a primary acid, but rather formed by hydroxylation of deoxycholic acid returning from the gut (135).

Bitocholic acid

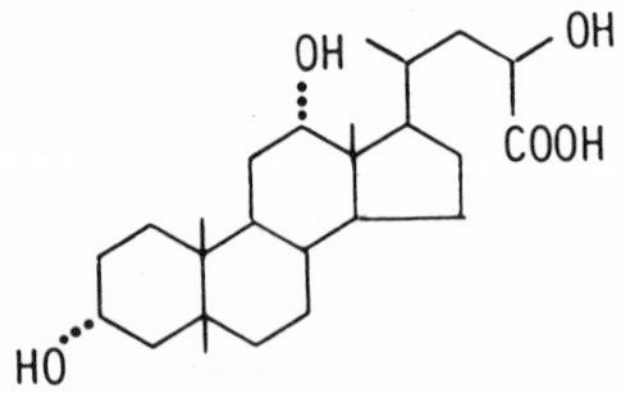

OH
OH
COOH
HO

amorphous; α_D + 48° (4)

This 23-hydroxylated derivative of deoxycholic acid has been detected as a component of the bile of certain snakes, especially of the family Viperidae. Specifically the acid has been isolated from *Bitis arietans, Bitis gabonica,* and Russell's viper (4, 136). In these snakes, bitocholic acid occurs along with the 23-hydroxylated derivative of cholic acid (see below).

Phocaecholic acid

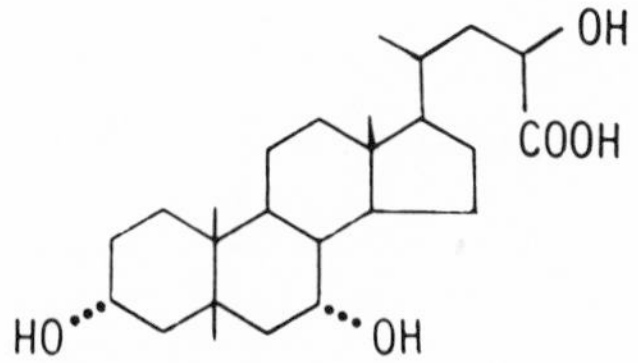

m.p. 222 °C; α_D + 11 ° (136)

The 23-hydroxy derivative of chenodeoxycholic acid was isolated very early by Hammarsten from seal and walrus bile (137, 138). It was characterized by Windaus and van Schoor in 1928 (139) and the structure confirmed by Bergström *et al.* (140) and Haslewood (136). Originally the acid described here as phocaecholic acid was referred to as β-phocaecholic acid. It has been found in all Pinnipedia that have been examined and not in other animals. The α-acid also isolated by Hammarsten was subsequently identified by Bergström *et al.* (140) as a tetrahydroxycholanic acid.

$3\alpha,7\alpha,12\alpha$,23-Tetrahydroxycholanic acid

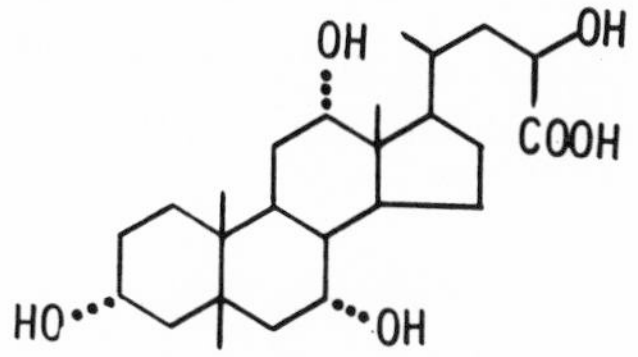

m.p. 152–154 °C (solvated); α_D + 36 ° (137)

This third bile acid containing the hydroxyl group at C-23 was isolated from seal bile and snake bile (136, 140). It is the only identified naturally occurring tetrahydroxycholanic acid. The acid had previously been isolated by Hammarsten (137, 138) who named it α-phocaecholic acid. The free acid has not been prepared in a form suitable for elemental analysis; however appropriate analysis may be made of the methyl ester tetraacetate (140).

3-Ketocholanic acid

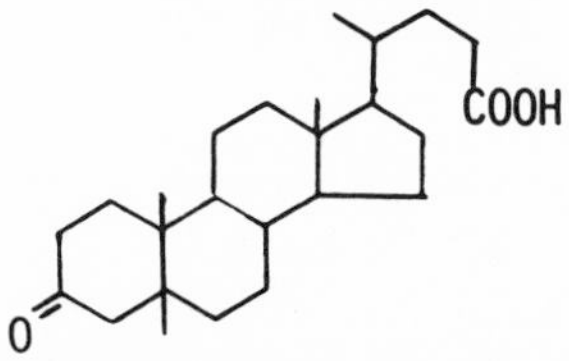

m.p. 140 °C; α_D + 33 ° (9)

3-Ketocholanic acid has been detected as a fecal bile acid in man (52) but has not been found in other sources. The acid may be prepared by CrO_3 oxidation of lithocholic acid (64, 141).

7α-Hydroxy-3-ketocholanic acid

COOH
O
OH

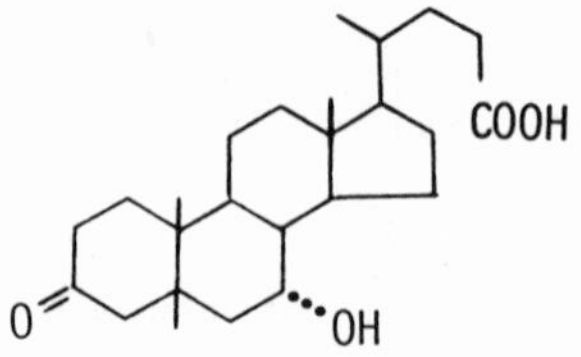

m.p. 129 °C (methyl ester); α_D + 20.5° (142)

This acid has been detected in human feces (52). It has not been found in other sources. The acid may be prepared by Oppenauer oxidation of methyl chenodeoxycholate (87) followed by hydrolysis of the product.

12α-Hydroxy-3-ketocholanic acid

OH
COOH
O

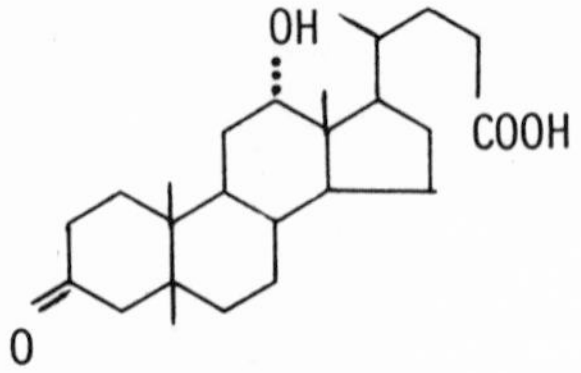

α_D + 50° (9)

This acid has been detected as a component of human feces (50, 52) but has not been found elsewhere in nature. The acid (m. p. 112–120 °C; α_D + 48°) was first prepared by Yamasaki and Kyogoku (143) by CrO_3 oxidation of the 12-monoacetate.

3α-Hydroxy-6-ketoallocholanic acid

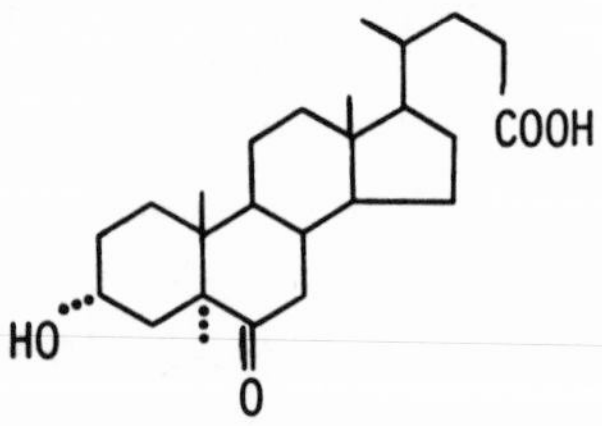

COOH
HO
O

m.p. 194 °C; α_D −9° (144)

This acid, first isolated by Fernholz (144), is considered to be an artifact of alkaline hydrolysis (i.e., 5β→5α), but clear evidence of the form in which it occurs in bile has not been presented. The acid has been found only in pig bile where it is presumed to be a secondary acid (88). 6-Ketoallolithocholic acid was also isolated as a metabolite of hyodeoxycholic acid in surgically jaundiced rats (95). Preparation of the acid is by the procedure of Wieland and Dane (145).

7-Ketolithocholic acid

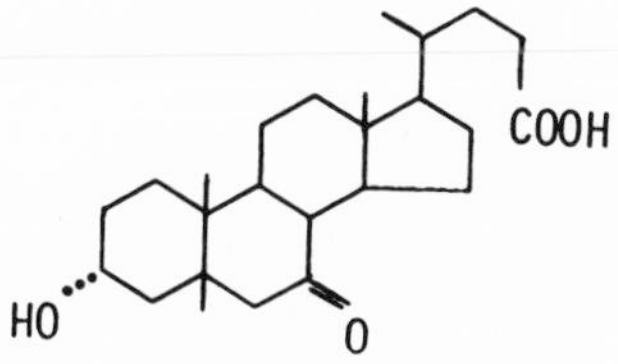

m.p. 203 °C; α_D −27° (84)

7-Ketolithocholic acid was isolated from guinea pig bile by Imai (146) and from the bile of the coypu by Brigl and Benedict (147). The acid from the coypu was characterized by Kazuno and Takuma (148) and by Haslewood in 1954 (149). 7-Ketolithocholic acid has been found in chicken bile and in the bile of a number of mammalian species (2). It also occurs in human feces (50, 52). The acid may be prepared by partial oxidation of chenodeoxycholic acid (150).

12-Ketolithocholic acid

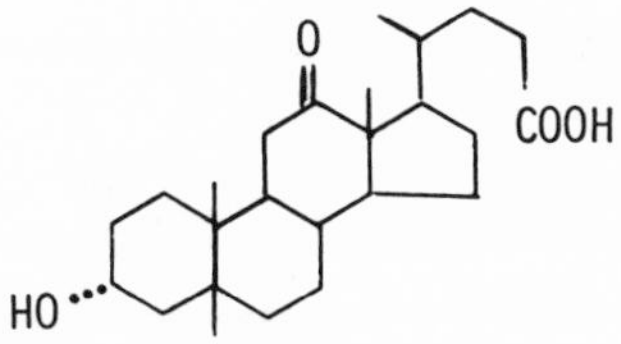

m.p. 165 °C; α_D + 87° (151)

This keto acid was first identified in beef bile by Wieland and Kishi (36). It had been obtained earlier as a molecular complex with chenodeoxycholic acid (152). 12-Ketolithocholic acid has been identified as a fecal acid in several species (4). The acid may be prepared by direct oxidation of deoxycholic acid (153) but better overall yields are obtained by protection of the hydroxyl group at carbon 3 (154).

3β-Hydroxy-12-ketocholanic acid

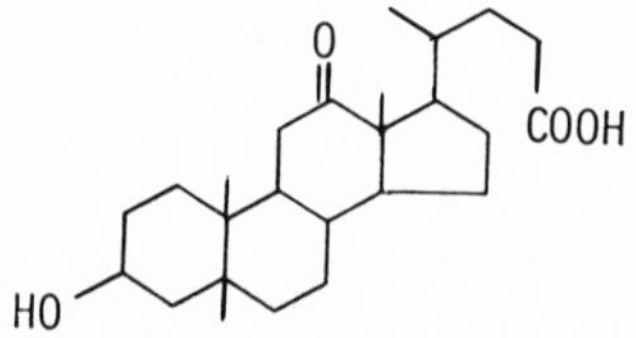

m.p. 218–220 °C; α_D + 91° (155)

This keto acid has been detected in the feces of man and rabbit (52, 62, 74) but has not been detected in other natural sources. It may be synthesized by reduction of dehydrodeoxycholic acid or its methyl ester followed by suitable chromatographic purification (74, 155).

3, 12-Diketocholanic acid

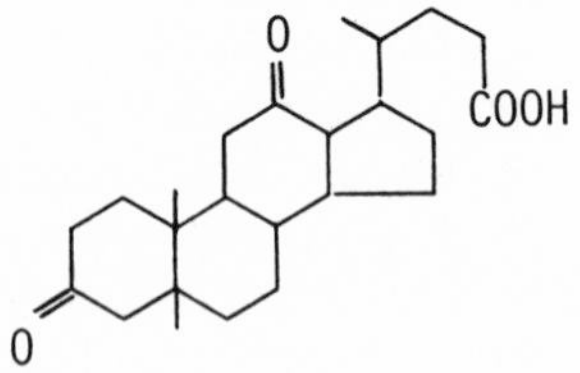

m.p. 187 °C; α_D + 92° (4)

Dehydrodeoxycholic acid has been detected as a fecal acid in man (52). It has not been detected in other sources. The acid may be obtained by oxidation of deoxycholic acid by CrO_3 in glacial acetic acid (156).

7α,12α-Dihydroxy-3-ketocholanic acid

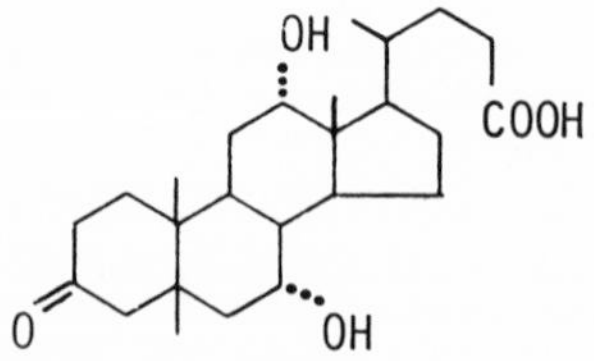

m.p. 121 °C; α_D + 38° (9)

This acid was isolated from cattle bile by Haslewood in 1946 (157) but has not been detected elsewhere. It may be synthesized by Oppenauer oxidation of cholic acid or from the methyl ester triacetate of cholic acid by selective hydrolysis and subsequent oxidation (158, 159).

7-Ketodeoxycholic acid

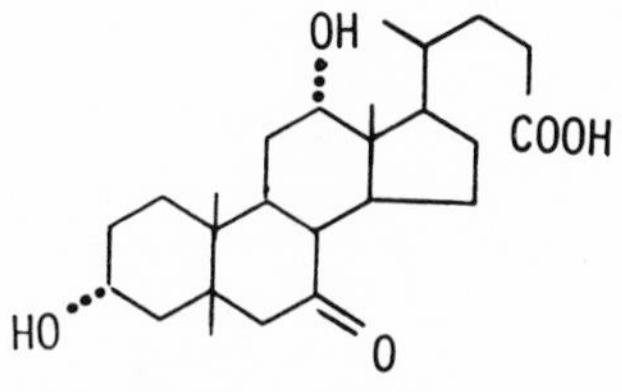

m.p. 200 °C; α_D + 1.5° (1)

3α,12α-Dihydroxy-7-ketocholanic acid was identified first by Haslewood in cattle bile (157). It is also present in the bile of other species [python (160), monkey (161), and rat (120)] and in the feces of man, dog, and rat (82, 119–121). The acid may be prepared by selective oxidation of cholic acid (162).

3α,7α-Dihydroxy-12-ketocholanic acid

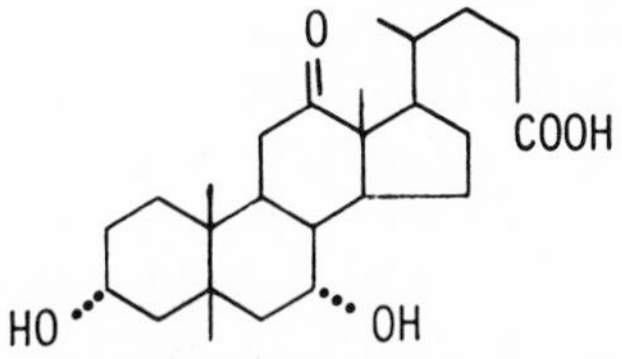

m.p. 219 °C; α_D + 81° (83)

This keto acid was detected in human feces (121) but is not known to occur elsewhere as a natural product. This acid and its esters have been valuable as intermediates in the preparation of chenodeoxycholic acid. The keto acid may be prepared by CrO_3 oxidation of the methyl ester diacetate of cholic acid (163) followed by rigorous hydrolysis. Mild hydrolysis yields the 7-monoacetate (m.p. 239 °C) which may also be reduced by the Wolff–Kishner procedure to the 3,7-dihydroxy acid.

3-Ketochola-4:6-dienic acid

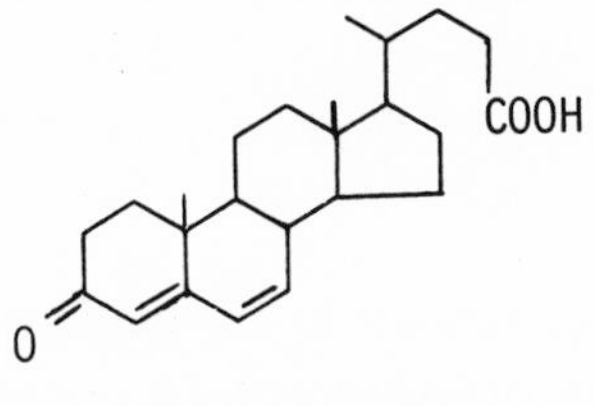

m.p. 150–152 °C (164)

This acid is the only unsaturated C_{24} bile acid to be recognized as a natural product. The acid was found in chicken bile and may be an artifact of preparation (164). The dienic acid shown above was prepared by Wiggins (164) from methyl 3β-hydroxychol-5-enate by oxidation with MnO_2 in boiling benzene.

Allocholic acid

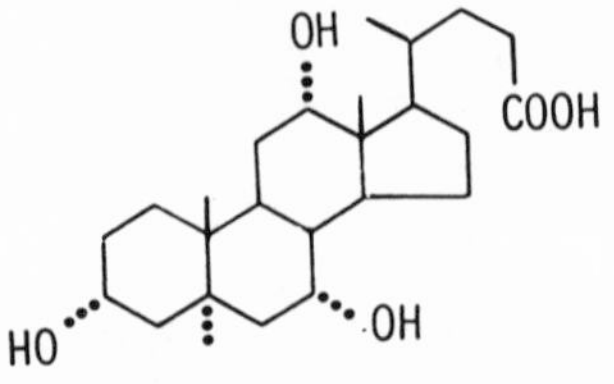

m.p. 240°C; α_D + 23° (165)

Until recent years, bile salts were considered to be exclusively 5β steroids. The first recognition of the occurrence of 5α or allo bile salts came from the studies of Anderson and Haslewood (165) who synthetized allocholic acid and identified it as a previously unidentified naturally occurring acid from fish bile. The earliest isolation of allocholic acid appears to have been that by Ohta from the bile of the Gigi fish (166). Since that time allocholic acid has been isolated from the bile of a large number of fishes, birds, and, to a smaller extent, mammals (4). It has also been detected in human feces (121). Allocholic acid may be prepared from cholic acid by allomerization of the 6,7-ketol in hot alkali (165) or by treatment of the 3-keto acid with Raney nickel in boiling isopropyl benzene (167).

Allodeoxycholic acid

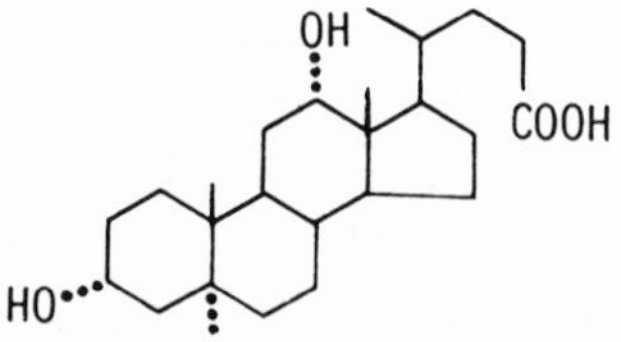

m.p. 214–215°C; α_D + 42° (62)

The natural occurrence of the 5α-epimer of deoxycholic acid was first demonstrated by Danielsson, Kallner, and Sjövall (62). The original isolation of this compound was probably by Kishi (61) who named the unidentified acid lagodeoxycholic acid from rabbit bile. In addition to its occurrence in rabbit bile, allodeoxycholic acid is present in rabbit feces (62) and accumulates as the glycine conjugate in gallstones of rabbits fed cholestanol (168). Allodeoxycholic acid may be synthesized from cholic acid by reactions similar to the preparation of allocholic acid (62, 168).

2. C_{27} *Acids*

3α,7α-Dihydroxycoprostanic acid

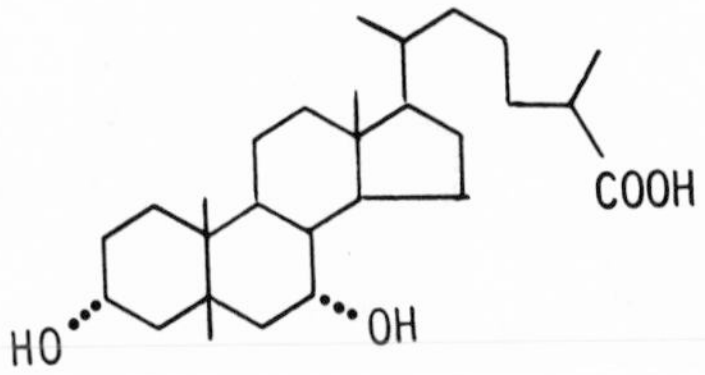

The only dihydroxycoprostanic acid thus far identified as a natural product was isolated from alligator bile by Dean and Whitehouse (169) and from human bile by Carey *et al.* (170). The acid was characterized by comparison with synthetic material (169, 171).

3α,7α,12α-Trihydroxycoprostanic acid

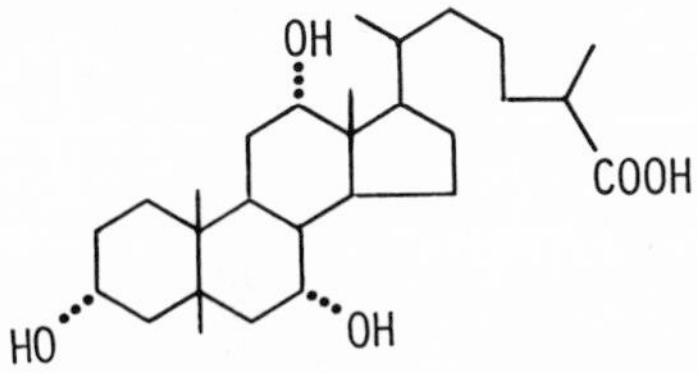

m.p. 180–182°C; α_D + 27° (25α) (172)
m.p. 194–196°C; α_D + 43° (25β) (172)

This trihydroxycoprostanic acid occurs in two epimeric forms (25α and 25β) in frog bile (173, 174) and in alligator bile (175, 176). The 25α acid has also been found to occur in human bile (177, 178) and in bile from the baboon (179). Both epimeric forms have been synthesized by Bridgwater (172). The 5α-epimer, 3α,7α,12α-trihydroxycholestanic acid, was obtained from iguana bile by Okuda, Horning, and Horning (180).

3α,7α,12α-Trihydroxy-25α-coprost-23-enic acid

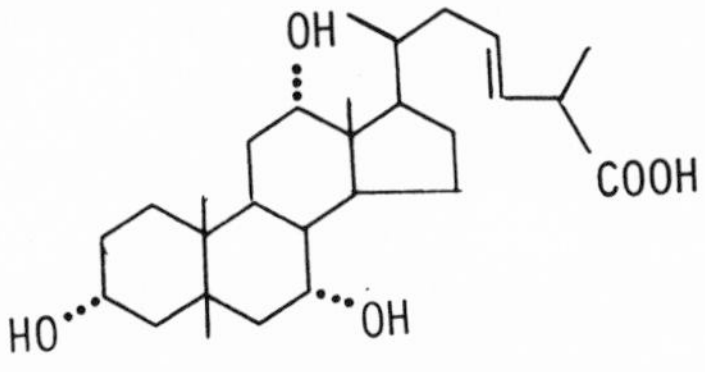

m.p. 179°C; α_D + 29° (181)

This unsaturated coprostanic acid was isolated from toad bile by Hayakawa (181). It was characterized by oxidation to norcholic acid and by hydrogenation to 3α,7α,12α-trihydroxy-25α-coprostanic acid (181, 182).

Varanic acid

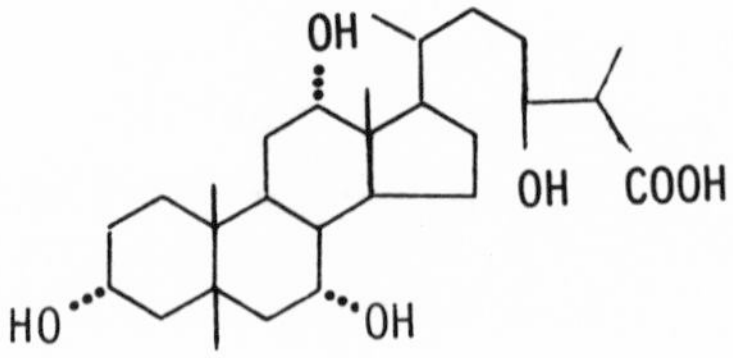

m.p. 120 °C (99)

Varanic acid was isolated from the bile of the monitor lizard by Haslewood and Wootton (99). An apparent stereoisomer occurs also in *Heloderma*. From comparative chromatographic and spectroscopic data with synthetic material Collings and Haslewood (183) have determined that the acid is very likely a C-24- or C-25-isomer of the 3,7,12,24-tetrahydroxy structure shown above. 3α,7α,12α,24-Tetrahydroxycoprostanic acid was synthesized by Inai *et al.* (184) and in radioactive form by Masui and Staple (185). The labeled acid was converted to cholic acid by rat liver (185).

3α,7α,12α, 22-Tetrahydroxycoprostanic acid

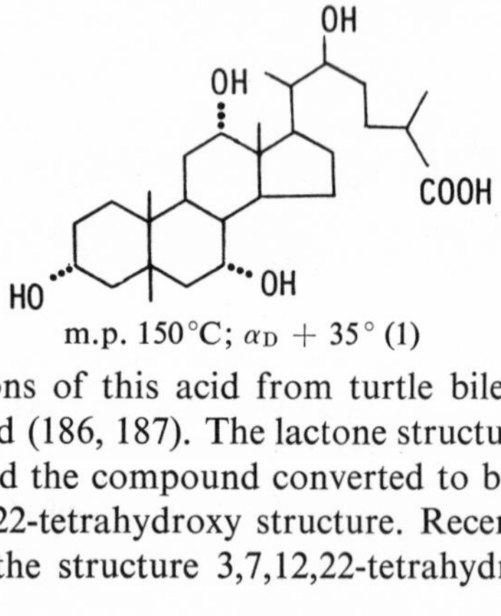

m.p. 150 °C; α_D + 35 ° (1)

Earlier preparations of this acid from turtle bile were called tetrahydroxysterocholanic acid (186, 187). The lactone structure obtained from this acid was confirmed and the compound converted to bisnorcholic acid (188) to support the 3,7,12,22-tetrahydroxy structure. Recent work by Amimoto *et al.* (189) confirms the structure 3,7,12,22-tetrahydroxycoprostanic acid.

3. C_{28} *Acid*

3α,7α,12α-Trihydroxycoprost-22-ene-24-carboxylic acid

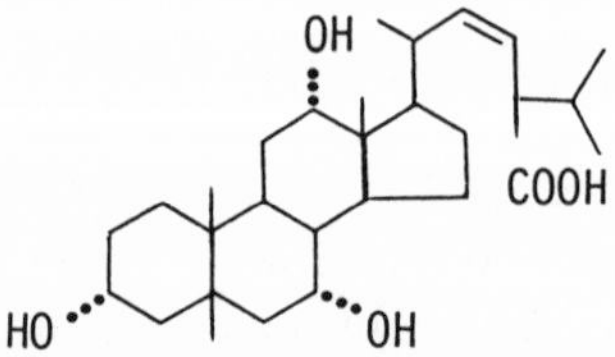

m.p. 160 °C; α_D −13 ° (1)

This is the only C_{28} steroid to be included as a bile acid or alcohol. It was isolated from toad bile by Shimizu and Oda (190) and named trihydroxybufosterocholenic acid. An isomeric acid has also been isolated (191). Trihydroxybufosterocholenic acid has been degraded by ozonization to bisnorcholic acid (192). The saturated derivative of this acid, $3\alpha,7\alpha,12\alpha$-trihydroxycoprostane-24-carboxylic acid, was characterized by partial synthesis (193). As a C_{28} acid, trihydroxybufosterocholenic acid presents a metabolic question. Cholesterol is not a precursor (194) though the livers of the animals from which the acid is isolated appear to contain mainly cholesterol (195). Hoshita *et al.* (193) proposed that the acid may arise from campesterol or from an intermediate formed during methylation at C-24.

4. C_{24} *Alcohol*

Petromyzonol

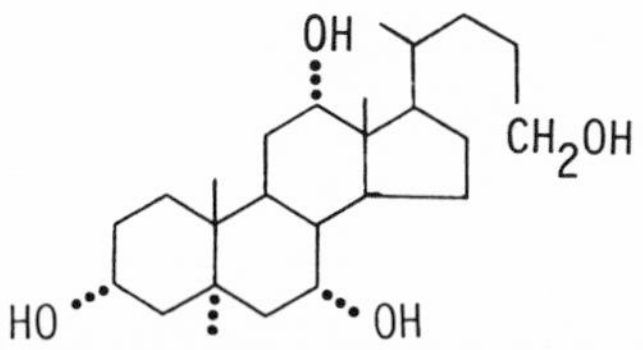

m.p. 240 °C (196); α_D + 28 ° (197)

Bile salts from larval lampreys have yielded a C_{24} alcohol characterized by comparison with the product obtained by $LiAlH_4$ reduction of the ester of allocholic acid (196). Petromyzonol is the only C_{24} alcohol thus far identified as a natural product.

5. C_{26} *Alcohols*

5α-Ranol

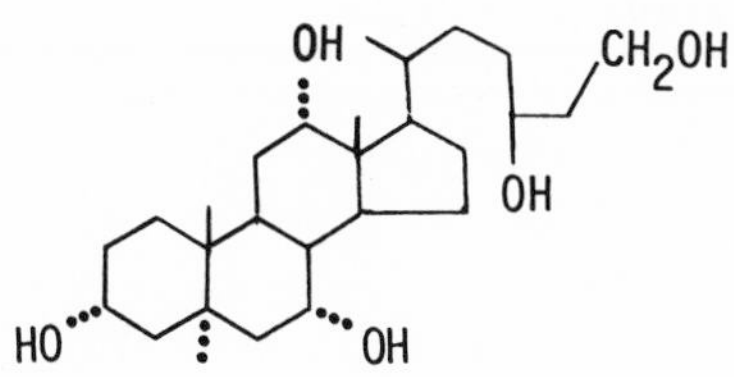

amorphous; α_D + 21 ° (198)

Ranol has been isolated from the frogs, *Rana temporaria* and *Rana catesbiana*. The first clear report of the compound presently identified as ranol is by Haslewood (199). The structure of ranol eluded investigators for several years. The free alcohol did not provide accurate elemental data and even the crystalline tetraacetate was believed to be a derivative of a C_{27} or C_{28}

steroid. By 1964 the unique structure of ranol was accepted (3, 198, 200). In the bile of *R. catesbiana* both 5α- and 5β-ranol are present. In this species, cholesterol-4-^{14}C and 3,7,12 trihydroxycoprostane have been observed to form 5β-ranol (201, 202). Betsuki (202) also identified radioactive 26-deoxy-5β-ranol and proposed that this compound is an intermediate in the formation of 5β-ranol. The conversion of cholesterol to ranol requires the loss of a single carbon from the C_{27} precursor. 5β-ranol is converted to cholic acid in the rat (203).

6. C_{27} *Alcohols*

Scymnol

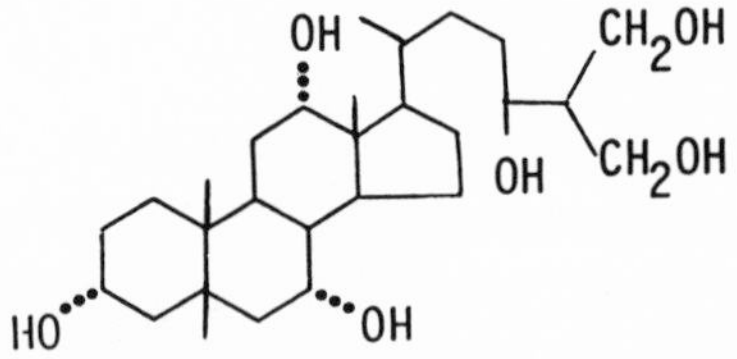

m.p. 123 °C (dihydrate); α_D + 34° (204)

The substance now called anhydroscymnol and containing the trimethylene oxide ring structure shown below was the single recognized bile alcohol at the time of Sobotka's monograph in 1938. This compound, originally designated in two forms, α and β, was isolated by Hammarsten after alkaline hydrolysis of shark bile in 1898 (205). Early work on the structure identified scymnol as an isomer of pentahydroxycoprostane (7) with oxygen atoms correctly positioned at carbons 7, 12, 24, and 26 or 27 but with incomplete information about the hydroxyl group at carbon 3 and the other oxygen atom. The correct structure in the side-chain (206, 207) was favored by Fieser and Fieser (1) on the basis that the oxide obtained after hydrolysis of the sulfate (208) probably possessed the four-membered ring structure

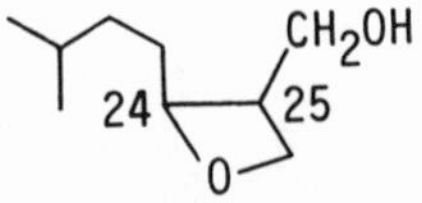

The prospect that the oxide ring might arise during hydrolysis of the native structure was suggested by Haslewood (209). The cholic acid nucleus was identified earlier (210, 211). The structure of the naturally occurring form of scymnol (shown above) has been verified by partial synthesis (204).

The early discovery of scymnol in a primitive chordate led to repeated mention of the probable evolutionary significance of this bile alcohol. Sobotka (7) called scymnol the phylogenetic precursor of the bile acids. Both Haslewood (3) and Fieser and Fieser (1) cite the suspicion of Windaus *et al.* (212)

that since elasmobranchii are one of the oldest vertebrate groups, the shark had not learned how to degrade C_{27} sterols to C_{24} bile acids and that scymnol was a stage of intermediary metabolism. This view was of course simplistic but was the prelude for extensive studies by Haslewood and his collaborators on the evolutionary aspects of bile acid metabolism.

Chimaerol

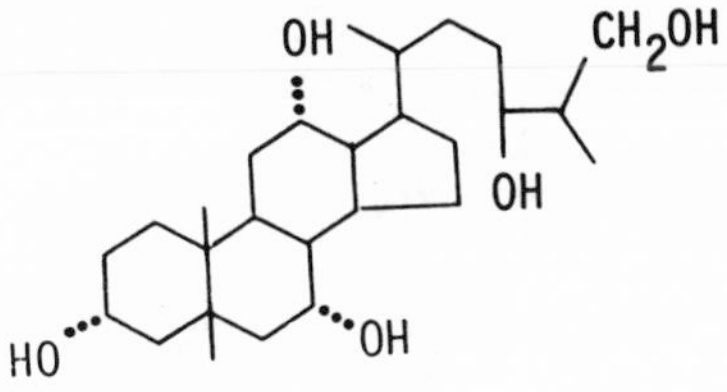

OH CH$_2$OH OH HO OH

m.p. 180–182 °C; α_D + 42° (213)

Chimaerol is 26-deoxyscymnol and may occur with scymnol (4). The alcohol has been found only in elasmobranch fishes, e.g. stingray, dogfish, and the rabbit fish (chimaera) from which the alcohol received its name (213–215). Chimaerol requires only C-terminal hydroxylation to from scymnol. The precise stereoisomer at C_{24} and C_{25}, which is chimaerol, has not been prepared but the above structure has been partially synthesized from cholic acid by Bridgwater *et al.* (204) and from anhydroscymnol by Cross (207).

Bufol

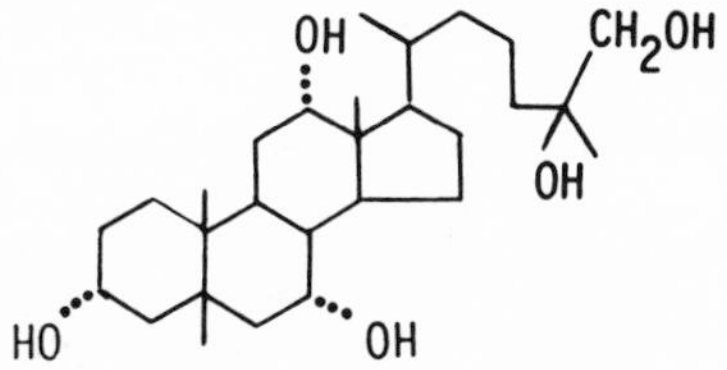

OH CH$_2$OH OH HO OH

m.p. 178 °C; α_D + 38° (216)

The 5β-epimer of this alcohol is the principal component of the bile of the toad (216). The compound isolated earlier by Kazuno (217) and named pentahydroxybufostane was probably an artifact of hydrolysis of 5β-bufol sulfate (194). The presence of a cholic acid nucleus was determined by Kazuno (217); the complete structure (pentahydric coprostane) was proposed by Okuda *et al.* (216). The structure has been verified by partial synthesis (183, 218).

The 5α-epimer of bufol was detected in the caudate amphibian *Diemictylus pyrrhogaster* (newt) (219, 220) and also in *Neoceratodus,* which is a bony fish relative of the African lungfish (*Protopterus)* (5). This incidence of 5α-bufol is correlated by Haslewood with the high incidence of other 5α steroids in the bile of bony fishes, salamanders, and lizards as a possible series of descendant species.

5α-Cyprinol

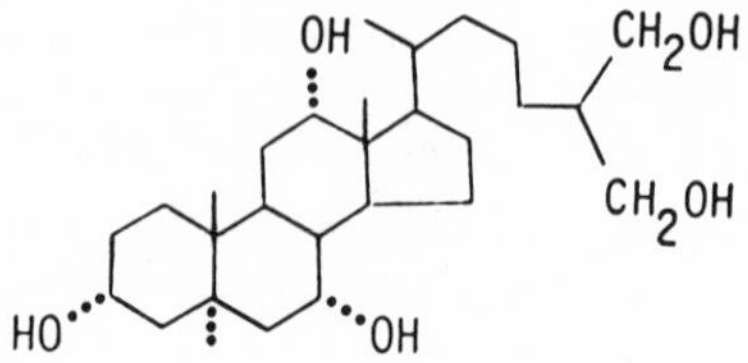

m.p. 244°C; α_D + 29° (4)

In the carp, chub, and related fishes, 5α-cyprinol is a principal bile alcohol (220). It also occurs in the lungfish and in the coelacanth, where it is accompanied by its 3β-epimer, latimerol. The characterization of 5α-cyprinol was accomplished by Hoshita *et al.* (221) and by Anderson *et al.* (222)

5β-Cyprinol

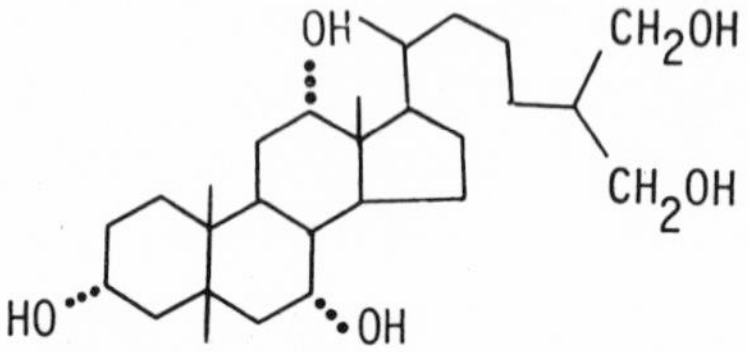

m.p. 175–177°C; α_D + 36° (223)

5β-Cyprinol has been obtained from the frog (*Rana nigromaculata*) (224), the eel (*Conger myriaster*) (225), the sturgeon, and the paddlefish (223). Hoshita, Kouchi, and Kazuno (226) prepared this compound before its isolation from natural sources. It has also recently been prepared by Haslewood and Tammar (223).

Latimerol

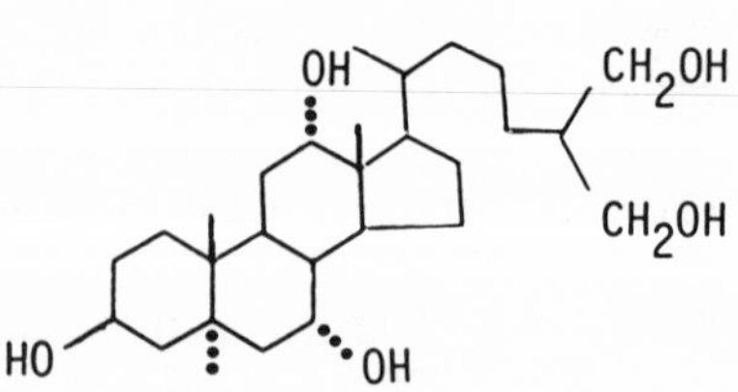

m.p. 236°C; α_D + 33° (227)

The ancient bony fish, the coelacanth, has bile that contains little if any bile acid and has as its principal bile alcohol the compound shown above, which is latimerol. Latimerol was isolated and characterized by Anderson and Haslewood (227). It occurs along with its 3α-epimer, 5α-cyprinol.

27-Deoxy-5α-cyprinol

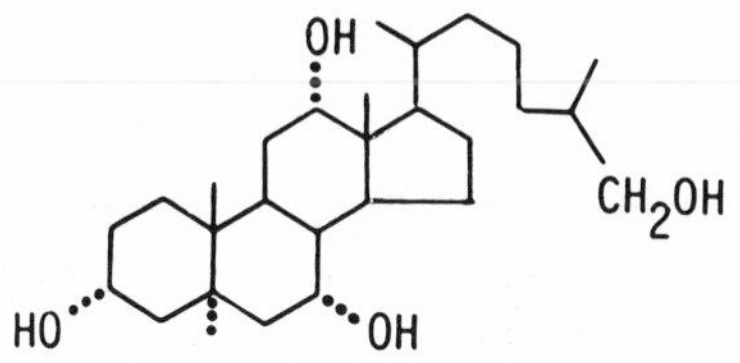

m.p. 232°C; α_D + 39° (228)

This alcohol has been isolated from the bile of the carp (228). It may be the alcohol isolated much earlier from the toad (*Bufo vulgaris japonica)* by Makino (229) and named tetrahydroxybufostane (194). 27-Deoxy-5α-cyprinol was prepared from anhydrocyprinol by Hoshita (230). Its 5β-epimer may be an expected intermediate in the formation of trihydroxycoprostanic acid, but the alcohol has not been widely found. Both the 5α- and 5β-epimers of 27-deoxycyprinol were formed in toads given ^{14}C-cholesterol (194).

Myxinol

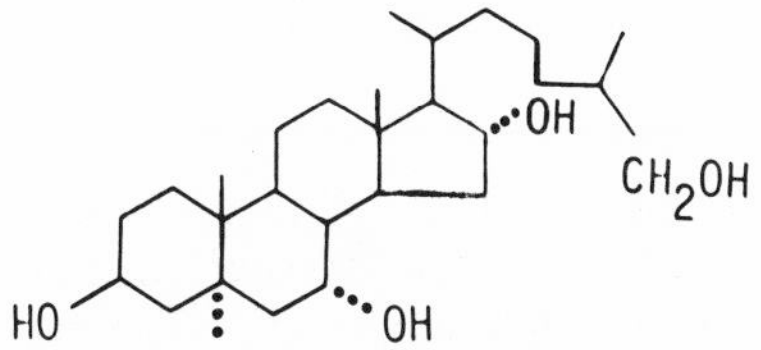

m.p. 206°C; α_D −15° (231)

Myxinol is a unique tetrahydroxy bile alcohol that occurs in the bile of the hagfish, a most primitive marine cyclostome. It has not been detected in the bile of other species. The structure of myxinol recently clarified by Anderson *et al.* (232) bears resemblance to other bile salts but offers no promise of metabolic relationship with other known structures. It is conjugated with two moles of sulfate. The hydroxyl group at carbon 3 is β; the hydroxyl at carbon 16 is reminiscent of pythocholic acid. Unlike pythocholic acid, myxinol has the A/B *trans*-conformation and contains other hydroxyl groups at carbons 3 and 7 rather than at carbons 3 and 12.

16-Deoxymyxinol

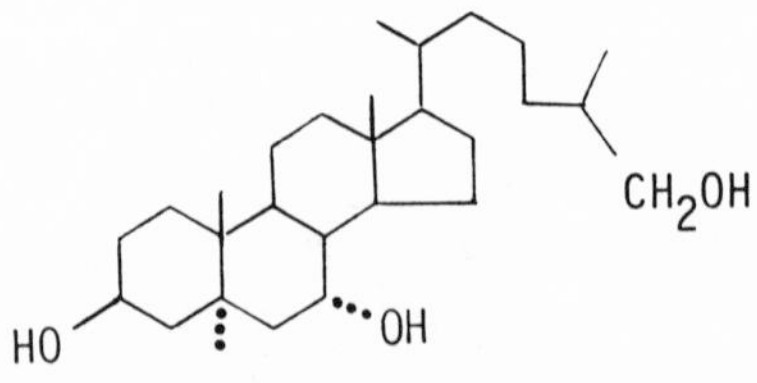

m.p. 219 °C; α_D + 13° (228)

A second bile alcohol from the bile of the hagfish is 16-deoxymyxinol, isolated and characterized by Anderson and Haslewood (233).

$3\alpha,7\alpha$-Dihydroxycoprostane

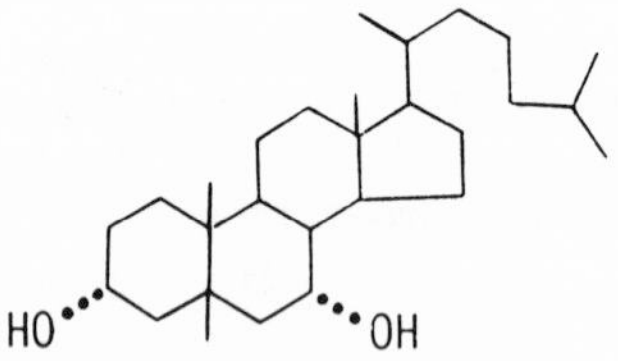

m.p. 84–86 °C (228)

Rabinowitz *et al.* (234) obtained this alcohol from human fistula bile. It was identified by comparison with synthetic material.

SUMMARY

In this chapter are tabulated 56 compounds accepted as bile acids and alcohols of natural occurrence. This does not include several that occur as isomers but are listed as a single compound. The list presented here could easily be modified by reexamination of the literature, particularly in the direction of compounds which have been identified as metabolites under conditions that promise their natural occurrence if only in trace amounts. Furthermore, the list will grow in subsequent years as new compounds are identified.

By inspection one may determine that had this chapter been written before 1950 very nearly one-third of the compounds recognized here would have been known. Taken by decades, no less than 26 bile acids and alcohols were detected as natural products for the first time since 1960. For the detection of trace components this is a tribute to those investigators who applied the effective techniques of TLC, GLC, mass spectrometry, singly and in combination. For the detection of higher bile acids and alcohols (where 14 out of the 18 compounds shown were reported since 1950) this is a tribute to those who have surmounted problems of chemical artifacts and challenging problems of characterization. There is little doubt also that those who contributed to the recent characterization and synthesis of allo acids opened the way for the identification of a number of bile alcohols.

Little has been said in this chapter concerning the evolutionary significance of the chemical nature of bile salt. For thorough discussions the reader is referred to references (3), (4), and (5). Consideration of the compounds listed here, however, shows the general structural progression C_{27} alcohols→C_{27} acids→C_{24} acids inherent in the development of any chemical concept for biological classification. Such a device must begin with primitive vertebrates since simpler forms of life do not possess discrete bile. The alcohol, scymnol, originally isolated and considered a primitive chemical structure, is preceded by several more primitive structures, particularly those with the 3β hydroxyl group and the A/B *trans* ring system (233). Bile alcohols have not been found in animals above the amphibian (5) but higher bile acids have been detected in primates. In later stages of development, the variety of hydroxyl and carbonyl derivatives of the bile salts now known, along with the wide extension of possibilities created by discovery of the 5α acids and alcohols complicate any effort to systematize biological forms with interpretable chemical structure.

ACKNOWLEDGMENT

The author is greatly indebted to Dr. G. A. D. Haslewood for reviewing the typescript of this chapter and making valuable suggestions for clarification and correction of the material presented.

REFERENCES

1. L. F. Fieser and M. Fieser, "Steroids," Reinhold, New York (1959).
2. G. A. D. Haslewood, *in* "Comparative Biochemistry" (M. Florkin and H. S. Mason, eds.), Vol. 3, part 3, p. 205, Academic Press, New York (1962).
3. G. A. D. Haslewood, *Biol. Rev.* **39,** 537 (1964).
4. G. A. D. Haslewood, "Bile Salts," Methuen, London (1967).
5. G. A. D. Haslewood, *J. Lipid Res.* **8,** 535 (1967).
6. H. Sobotka, "Physiological Chemistry of the Bile," Williams Wilkins Co., Baltimore (1937).
7. H. Sobotka, "The Chemistry of the Steroids," Williams Wilkins Co., Baltimore (1938).
8. C. W. Shoppee, "Chemistry of the Steroids" (2nd ed.), Butterworths, Washington (1964).
9. H. Van Belle, "Cholesterol, Bile Acids and Atherosclerosis," North-Holland Publishing Co., Amsterdam (1965).
10. T. Hoshita and T. Kazuno, *Adv. Lipid Res.* **6,** 207 (1968).

11. "Elsevier's Encyclopedia of Organic Chemistry" (F. Radt, ed.), Series 3, Vol. 4–Supplement, Steroids Part 4, Springer-Verlag, Berlin (1962).
12. L. J. Thenard, *Annales de Chimie (et Physique) (I)* **64,** 103 (1807); *cited in* Sobotka (7).
13. J. Berzelius, *Annales de Chimie (et Physique) (I)* **71,** 218 (1809); *cited in* Sobotka (7).
14. L. Gmelin, *in* "Die Verdauung nach Versuchen," Heidelberg and Leipzig (1826); *cited in* Sobotka (7).
15. J. Berzelius, *Ann.* **33,** 139 (1840).
16. E. von Gorup-Besanez, *Ann.* **59,** 129 (1846).
17. E. A. Platner, *Ann.* **51,** 105 (1844).
18. F. Verdeil, *Ann.* **59,** 311 (1846).
19. A. Strecker, *Ann.* **67,** 1 (1848).
20. A. Strecker, *Ann.* **70,** 149 (1849).
21. C. Gundelach and A. Strecker, *Ann.* **62,** 205 (1847).
22. T. Marsson, *Ann.* **72,** 317 (1849).
23. W. Heintz and J. Wislicenus, *Ann. Phys. Chem.* **108,** 547 (1859); cited by B. Moore *in* "Textbook of Physiology" (E. A. Sharpey-Shafer, ed.), Vol. 1, Macmillan, New York (1898).
24. H. Dermarcay, *Ann.* **27,** 270 (1838).
25. J. L. Thudicum, *Ann. Chem. Med.* **2,** 251 (1881); *cited in* Sobotka (7).
26. P. Latschinoff, *Ber.* **18,** 3039 (1885).
27. F. Mylius, *Ber.* **19,** 369 (1886).
28. F. Mylius, *Ber.* **19,** 2000 (1886).
29. H. Wieland and H. Sorge, *Z. Physiol. Chem.* **97,** 1 (1916).
30. H. Fischer, *Z. Physiol. Chem.* **73,** 204 (1911).
31. A. Windaus and A. Bohne, *Ann.* **433,** 278 (1923).
32. H. Wieland and G. Reverey, *Z. Physiol. Chem.* **140,** 186 (1924).
33. H. Windaus, A. Bohne, and E. Schwarzkopf, *Z. Physiol. Chem.* **140,** 177 (1924).
34. O. Hammarsten, *in* "Handbuch der Biologischen Arbeitsmethoden" (E. Abderhalden, ed.), Vol. 1, part 6, p. 211, Urban and Schwarzenberg, Berlin–Vienna (1925).
35. S. M. White, *Biochem. J.* **23,** 1165 (1929).
36. H. Wieland and S. Kishi, *Z. Physiol. Chem.* **214,** 47 (1933).
37. F. Breusch, *Z. Physiol. Chem.* **227,** 242 (1934).
38. H. Wieland and W. Siebert, *Z. Physiol. Chem.* **262,** 1 (1939).
39. G. A. Howard and A. J. P. Martin, *Biochem. J.* **46,** 532 (1950).
40. S. Bergström and J. Sjövall, *Acta Chem. Scand.* **5,** 1267 (1951).
41. J. Sjövall, *Acta Physiol. Scand.* **29,** 232 (1953).
42. S. Bergström and A. Norman, *Proc. Soc. Exp. Biol. Med.* **83,** 71 (1953).
43. A. Norman, *Acta Chem. Scand.* **7,** 1413 (1953).
44. E. H. Ahrens, Jr., and L. C. Craig, *J. Biol. Chem.* **195,** 763 (1952).
45. E. H. Mosbach, C. Zomzely, and F. E. Kendall, *Arch. Biochem. Biophys.* **48,** 95 (1954).
46. J. T. Matschiner, T. A. Mahowald, W. H. Elliott, E. A. Doisy, Jr., S. L. Hsia, and E. A. Doisy, *J. Biol. Chem.* **225,** 771 (1957).
47. J. Sjövall, *Acta Chem. Scand.* **6,** 1552 (1952).
48. D. Kritchevsky and M. R. Kirk, *J. Am. Chem. Soc.* **74,** 4713 (1952).
49. S. M. Grundy, E. H. Ahrens, Jr., and T. A. Miettinen, *J. Lipid Res.* **6,** 397 (1965).
50. S. S. Ali, A. Kuksis, and J. M. R. Beveridge, *Can. J. Biochem.* **44,** 957 (1966).
51. S. Bergström, R. Ryhage, and E. Stenhagen, *Acta Chem. Scand.* **12,** 1349 (1958).
52. P. Eneroth, B. Gordon, R. Ryhage, and J. Sjövall, *J. Lipid Res.* **7,** 511 (1966).
53. A. Windaus, *Arch. Pharm.* **246,** 117 (1908); *cited in* Fieser and Fieser (1).
54. P. Latschinoff, *Ber.* **12,** 1518 (1879).
55. A. Windaus and K. Neukirchen, *Ber.* **52,** 1915 (1919).
56. H. Wieland and F. J. Weil, *Z. Physiol. Chem.* **80,** 287 (1912).

57. A. Windaus, *Ann.* **447,** 233 (1926).
58. M. Shoda, *J. Biochem.* **7,** 505 (1927).
59. T. Kimura, *Z. Physiol. Chem.* **248,** 280 (1937).
60. L. F. Fieser and M. Fieser, "Natural Products Related to Phenanthrene," Reinhold, New York (1949).
61. S. Kishi, *Z. Physiol. Chem.* **238,** 210 (1936).
62. H. Danielsson, A. Kallner, and J. Sjövall, *J. Biol. Chem.* **238,** 3846 (1963).
63. B. Koechlin and T. Reichstein, *Helv. Chim. Acta* **25,** 918 (1942).
64. H. Wieland and P. Weyland, *Z. Physiol. Chem.* **110,** 123 (1920).
65. S. Kishi, *Z. Physiol. Chem.* **238,** 210 (1936).
66. H. Danielsson and K. Einarsson, *Acta. Chem. Scand.* **18,** 732 (1964).
67. L. J. Schoenfield and J. Sjövall, *Acta Chem. Scand.* **20,** 1297 (1966).
68. J. B. Carey, Jr., and G. Williams, *Science* **150,** 620 (1965).
69. E. Mosettig, E. Heftmann, Y. Sato, and E. Weiss, *Science* **128,** 1433 (1958).
70. J. B. Carey, Jr., and G. Williams, *J. Clin. Invest.* **42,** 450 (1963).
71. H. Wieland, E. Dane, and E. Scholz, *Z. Physiol. Chem.* **211,** 261 (1932).
72. L. F. Fieser and R. Ettorre, *J. Am. Chem. Soc.* **75,** 1700 (1953).
73. E. Heftmann, E. Weiss, H. K. Miller, and E. Mosettig, *Arch. Biochem. Biophys.* **84,** 324 (1959).
74. H. Danielsson, P. Eneroth, K. Hellström, S. Lindstedt, and J. Sjövall, *J. Biol. Chem.* **238,** 2299 (1963).
75. T. Hosizima, H. Takata, Z. Uraki, and S. Sibuya, *J. Biochem.* **12,** 393 (1931).
76. L. W. Wells, dissertation, St. Louis University, 1964, p. 68.
77. H. Wieland and R. Jacobi, *Z. Physiol. Chem.* **148,** 232 (1925).
78. S. Bergström and J. Sjövall, *Acta Chem. Scand.* **8,** 611 (1954).
79. I. D. P. Wootton and H. S. Wiggins, *Biochem. J.* **55,** 292 (1953).
80. D. H. Sandberg, J. Sjövall, K. Sjövall, and D. A. Turner, *J. Lipid Res.* **6,** 182 (1965).
81. J. B. Carey, Jr., *Science* **123,** 892 (1956).
82. S. Hirofuji, *J. Biochem.* **58,** 27 (1965).
83. S. Kawai, *Z. Physiol. Chem.* **214,** 71 (1932).
84. T. Iwasaki, *Z. Physiol. Chem.* **244,** 181 (1936).
85. O. Hammarsten, *Z. Physiol. Chem.* **36,** 525 (1902).
86. T. Kanazawa, A. Shimazaki, T. Sato, and T. Hoshino, *Proc. Japan Acad.* **30,** 391 (1954); *cited in C. A.* **49,** 14785 (1955).
87. H. Danielsson, P. Eneroth, K. Hellström, and J. Sjövall, *J. Biol. Chem.* **237,** 3657 (1962).
88. S. Bergström, H. Danielsson, and A. Göransson, *Acta Chem. Scand.* **13,** 776 (1959).
89. S. L. Hsia, J. T. Matschiner, T. A. Mahowald, W. H. Elliott, E. A. Doisy, Jr., S. A. Thayer, and E. A. Doisy, *J. Biol. Chem.* **225,** 811 (1957).
90. T. Okishio and P. P. Nair, *Biochemistry* **5,** 3662 (1966).
91. T. H. Lin, R. Rubinstein, and W. L. Holmes, *J. Lipid Res.* **4,** 63 (1963).
92. M. Makita and W. W. Wells, *Anal. Biochem.* **5,** 523 (1963).
93. H. G. Roscoe and M. J. Fahrenbach, *Anal. Biochem.* **6,** 520 (1963).
94. W. H. Hoehn, J. Linsk, and R. B. Moffett, *J. Am. Chem. Soc.* **68,** 1855 (1946).
95. J. T. Matschiner, R. L. Ratliff, T. A. Mahowald, E. A. Doisy, Jr., W. H. Elliott, S. L. Hsia, and E. A. Doisy, *J. Biol. Chem.* **230,** 589 (1958).
96. R. L. Ratliff, J. T. Matschiner, E. A. Doisy, Jr., S. L. Hsia, S. A. Thayer, W. H. Elliott, and E. A. Doisy, *J. Biol. Chem.* **236,** 685 (1961).
97. D. N. Jones and G. H. R. Summers, *J. Chem. Soc.* **1959,** 2594.
98. R. Schoenheimer and C. G. Johnston, *J. Biol. Chem.* **120,** 499 (1937).
99. G. A. D. Haslewood and V. Wootton, *Biochem. J.* **47,** 584 (1950).
100. R. B. Moffett and W. M. Hoehn, *J. Am. Chem. Soc.* **69,** 1995 (1947).
101. T. F. Gallagher and J. R. Xenos, *J. Biol. Chem.* **165,** 365 (1946).
102. P. Latschinoff, *Ber.* **20,** 1043 (1887).

103. Lassar-Cohn, *Ber.* **26,** 146 (1893).
104. Lassar-Cohn, *Z. Physiol. Chem.* **17,** 607 (1893).
105. S. Lindstedt and J. Sjövall, *Acta Chem. Scand.* **11,** 421 (1957).
106. O. W. Portman, *Arch. Biochem. Biophys.* **78,** 125 (1958).
107. S. Okamura and T. Okamura, *Z. Physiol. Chem.* **188,** 11 (1930).
108. S. Bergström, H. Danielsson, and B. Samuelsson, *in* "Lipide Metabolism" (K. Bloch, ed.), p. 291, Wiley, New York (1960).
109. S. Bergström, M. Rottenberg, and J. Sjövall, *Z. Physiol. Chem.* **295,** 278 (1953).
110. T. A. Mahowald, J. T. Matschiner, S. L. Hsia, R. Richter, E. A. Doisy, Jr., W. H. Elliott, and E. A. Doisy, *J. Biol. Chem.* **225,** 781 (1957).
111. J. B. Carey, Jr., and C. J. Watson, *J. Biol. Chem.* **216,** 847 (1955).
112. F. C. Chang, N. F. Wood, and W. G. Holton, *J. Org. Chem.* **30,** 1718 (1965).
113. O. Rosenheim and H. King, *Chem. Ind.* **51,** 954 (1932).
114. H. Wieland and E. Dane, *Z. Physiol. Chem.* **210,** 268 (1932).
115. M. Jenke and F. Bandow, *Z. Physiol. Chem.* **249,** 16 (1937).
116. J. T. Matschiner, T. A. Mahowald, S. L. Hsia, E. A. Doisy, Jr., W. H. Elliott, and E. A. Doisy, *J. Biol. Chem.* **225,** 803 (1957).
117. A. Kuksis, *J. Am. Oil Chemists' Soc.* **42,** 276 (1965).
118. B. Samuelsson, *Acta Chem. Scand.* **14,** 17 (1960).
119. J. G. Hamilton, *Arch. Biochem. Biophys.* **101,** 7 (1963).
120. A. Norman and J. Sjövall, *J. Biol. Chem.* **233,** 872 (1958).
121. P. Eneroth, B. Gordon, and J. Sjövall, *J. Lipid Res.* **7,** 524 (1966).
122. J. Sjövall, *Acta Chem. Scand.* **13,** 711 (1959).
123. G. A. D. Haslewood, *Biochem. J.* **56,** xxxviii (1954).
124. G. A. D. Haslewood and J. Sjövall, *Biochem. J.* **57,** 126 (1954).
125. G. A. D. Haslewood, *Biochem. J.* **62,** 637 (1956).
126. P. Ziegler, *Can. J. Chem.* **34,** 523 (1956).
127. P. Ziegler, *Can. J. Chem.* **34,** 1528 (1956).
128. S. L. Hsia, J. T. Matschiner, T. A. Mahowald, W. H. Elliott, E. A. Doisy, Jr., S. A. Thayer, and E. A. Doisy, *J. Biol. Chem.* **226,** 667 (1958).
129. S. L. Hsia, W. H. Elliott, J. T. Matschiner, E. A. Doisy, Jr., S. A. Thayer, and E. A. Doisy, *J. Biol. Chem.* **233,** 1337 (1958).
130. S. L. Hsia, J. T. Matschiner, T. A. Mahowald, W. H. Elliott, E. A. Doisy, Jr., S. A. Thayer, and E. A. Doisy, *J. Biol. Chem.* **230,** 573 (1958).
131. T. A. Mahowald, J. T. Matschiner, S. L. Hsia, E. A. Doisy, Jr., W. H. Elliott, and E. A. Doisy, *J. Biol. Chem.* **225,** 795 (1957).
132. G. A. D. Haslewood and V. M. Wootton, *Biochem. J.* **49,** 67 (1951).
133. G. A. D. Haslewood, *Biochem. J.* **49,** 718 (1951).
134. W. Klyne and W. M. Stokes, *J. Chem. Soc.* **1954,** 1979.
135. S. Bergström, H. Danielsson, and T. Kazuno, *J. Biol. Chem.* **235,** 983 (1960).
136. G. A. D. Haslewood, *Biochem. J.* **78,** 352 (1961).
137. O. Hammarsten, *Z. Physiol. Chem.* **61,** 454 (1909).
138. O. Hammarsten, *Z. Physiol. Chem.* **68,** 109 (1910).
139. A. Windaus and A. van Schoor, *Z. Physiol. Chem.* **173,** 312 (1928).
140. S. Bergström, L. Krabisch, and U. G. Lindeberg, *Acta Soc. Med. Upsalien.* **64,** 160 (1959).
141. W. Borsche and F. Hallwass, *Ber.* **55,** 3324 (1923).
142. E. Hauser, E. Baumgartner, and K. Meyer, *Helv. Chim. Acta* **43,** 1595 (1960).
143. K. Yamasaki and K. Kyogoku, *Z. Physiol. Chem.* **233,** 29 (1935).
144. E. Fernholz, *Z. Physiol. Chem.* **232,** 202 (1935).
145. H. Wieland and E. Dane, *Z. Physiol. Chem.* **212,** 41 (1932).
146. I. Imai, *Z. Physiol. Chem.* **248,** 65 (1937).
147. P. Brigl and O. Benedict, *Z. Physiol. Chem.* **220,** 106 (1933).
148. T. Kazuno and T. Takuma, *J. Japan Biochem. Soc.* **19,** 14 (1947).

149. G. A. D. Haslewood, *Biochem. J.* **56,** 581 (1954).
150. L. F. Fieser and S. Rajagopalan, *J. Am. Chem. Soc.* **72,** 5530 (1950).
151. E. Schwenk, B. Riegel, R. B. Moffett, and E. Stahl, *J. Am. Chem. Soc.* **65,** 549 (1943).
152. P. Weyland, dissertation, Munich, 1920, p. 46; *cited in* Wieland and Kishi (36).
153. K. Kaziro and T. Shimada, *Z. Physiol. Chem.* **249,** 220 (1937).
154. B. F. McKenzie, V. R. Mattox, L. L. Engel, and E. C. Kendall, *J. Biol. Chem.* **173,** 271 (1948).
155. K. Kyogoku, *Z. Physiol. Chem.* **246,** 99 (1937).
156. H. Wieland and E. Boersch, *Z. Physiol. Chem.* **106,** 190 (1919).
157. G. A. D. Haslewood, *Biochem. J.* **40,** 52 (1946).
158. G. A. D. Haslewood, *Biochem. J.* **38,** 108 (1944).
159. T. S. Sihn, *J. Biochem.* **27,** 425 (1938).
160. M. Kuroda and H. Arata, *J. Biochem.* **39,** 225 (1952).
161. H. S. Wiggins and I. D. P. Wootton, *Biochem. J.* **70,** 349 (1958).
162. L. F. Fieser and S. Rajagopalan, *J. Am. Chem. Soc.* **71,** 3935 (1949).
163. H. Wieland and W. Kapitel, *Z. Physiol. Chem.* **212,** 269 (1932).
164. H. S. Wiggins, *Biochem. J.* **60,** ix (1955).
165. I. G. Anderson and G. A. D. Haslewood, *Biochem. J.* **85,** 236 (1962).
166. K. Ohta, *Z. Physiol. Chem.* **259,** 53 (1939).
167. M. N. Mitra and W. H. Elliott, *J. Org. Chem.* **33,** 175 (1968).
168. A. F. Hofmann and E. H. Mosbach, *J. Biol. Chem.* **239,** 2813 (1964).
169. P. D. G. Dean and M. W. Whitehouse, *Biochem. J.* **99,** 9P (1966).
170. J. B. Carey, Jr., I. D. Wilson, F. G. Zaki, and R. F. Hanson, *Medicine* **45,** 461 (1966).
171. P. D. G. Dean and R. T. Aplin, *Steroids* **8,** 565 (1966).
172. R. J. Bridgwater, *Biochem. J.* **64,** 593 (1956).
173. Y. Kurauti and T. Kazuno, *Z. Physiol. Chem.* **262,** 53 (1939).
174. H. Mabuti, *J. Biochem.* **33,** 117 (1941).
175. G. A. D. Haslewood, *Biochem. J.* **52,** 583 (1952).
176. P. P. Shah, E. Staple, and J. L. Rabinowitz, *Arch. Biochem. Biophys.* **123,** 427 (1968).
177. E. Staple and J. L. Rabinowitz, *Biochim. Biophys. Acta* **59,** 735 (1962).
178. J. B. Carey, Jr., and G. A. D. Haslewood, *J. Biol. Chem.* **238,** PC855 (1963).
179. P. P. Shah, E. Staple, I. L. Shapiro, and D. Kritchevsky, *Lipids* **4,** 82 (1969).
180. K. Okuda, M. G. Horning, and E. C. Horning, *Proc. 7th Intern. Congr. Biochem.*, Science Council of Japan, Tokyo, Vol. IV, p. 721 (1967).
181. S. Hayakawa, *Proc. Japan Acad.* **29,** 279 (1953).
182. S. Hayakawa, *Proc. Japan Acad.* **29,** 285 (1953).
183. B. G. Collings and G. A. D. Haslewood, *Biochem. J.* **99,** 50P (1966).
184. Y. Inai, Y. Tanaka, S. Betsuki, and T. Kazuno, *J. Biochem.* **56,** 591 (1964).
185. T. Masui and E. Staple, *J. Biol. Chem.* **241,** 3889 (1966).
186. K. Yamasaki and M. Yuuki, *Z. Physiol. Chem.* **24,** 173 (1936).
187. T. Kanemitu, *J. Biochem.* **35,** 155 (1942).
188. T. Kanemitu, *J. Biochem.* **35,** 173 (1942).
189. K. Amimoto, T. Hoshita, and T. Kazuno, *J. Biochem.* **57,** 565 (1965).
190. T. Shimizu and T. Oda, *Z. Physiol. Chem.* **227,** 74 (1934).
191. T. Shimizu and T. Kazuno, *Z. Physiol. Chem.* **239,** 67 (1936).
192. T. Shimizu and T. Kazuno, *Z. Physiol. Chem.* **239,** 74 (1936).
193. T. Hoshita, K. Okuda, and T. Kazuno, *J. Biochem.* **61,** 756 (1967).
194. T. Hoshita, T. Sasaki, Y. Tanaka, S. Betsuki, and T. Kazuno, *J. Biochem.* **57,** 751 (1965).
195. K. Morimoto, *Hiroshima J. Med. Sci.* **15,** 145 (1966).
196. G. A. D. Haslewood and L. Tökés, *Biochem. J.* **107,** 6P (1968).
197. G. A. D. Haslewood and L. Tökés, *Biochem. J.* **114,** 179 (1969).
198. G. A. D. Haslewood, *Biochem. J.* **90,** 309 (1964).
199. G. A. D. Haslewood, *Biochem. J.* **51,** 139 (1952).

200. T. Kazuno, T. Masui, T. Nakagawa, and K. Okuda, *J. Biochem.* **53,** 331 (1963).
201. T. Masui, *J. Biochem.* **49,** 211 (1961).
202. S. Betsuki, *J. Biochem.* **60,** 411 (1966).
203. H. Danielsson and T. Kazuno, *Acta Chem. Scand.* **18,** 1157 (1964).
204. R. J. Bridgwater, T. Briggs, and G. A. D. Haslewood, *Biochem. J.* **82,** 285 (1962).
205. O. Hammarsten, *Z. Physiol. Chem.* **24,** 322 (1898).
206. A. D. Cross, *Proc. Chem. Soc. (London)* **1960,** 344.
207. A. D. Cross, *J. Chem. Soc.* **1961,** 2817.
208. T. Briggs and G. A. D. Haslewood, *Biochem. J.* **79,** 5P (1961).
209. G. A. D. Haslewood, *Biochem. Soc. Symp. (Cambridge, Engl.)* **6,** 83 (1951).
210. H. Ashikari, *J. Biochem.* **29,** 319 (1939).
211. W. Bergmann and W. T. Pace, *J. Am. Chem. Soc.* **65,** 477 (1943).
212. A. Windaus, W. Bergmann, and G. König, *Z. Physiol. Chem.* **189,** 148 (1930).
213. R. J. Bridgwater, G. A. D. Haslewood, and J. R. Watt, *Biochem. J.* **87,** 28 (1963).
214. K. Okuda, S. Enomoto, K. Morimoto, and T. Kazuno, *J. Biochem.* **51,** 441 (1962).
215. M. Kouchi, *Hiroshima J. Med. Sci.* **13,** 341 (1964).
216. K. Okuda, T. Hoshita, and T. Kazuno, *J. Biochem.* **51,** 48 (1962).
217. T. Kazuno, *Z. Physiol. Chem.* **266,** 11 (1940).
218. T. Hoshita, *J. Biochem.* **52,** 176 (1962).
219. T. Hoshita, S. Nagayoshi, M. Kouchi, and T. Kazuno, *J. Biochem.* **56,** 177 (1964).
220. G. A. D. Haslewood, *Biochem. J.* **59,** xi (1955).
221. T. Hoshita, S. Nagayoshi, and T. Kazuno, *J. Biochem.* **54,** 369 (1963).
222. I. G. Anderson, T. Briggs, and G. A. D. Haslewood, *Biochem. J.* **90,** 303 (1964).
223. G. A. D. Haslewood and A. R. Tammar, *Biochem. J.* **108,** 263 (1968).
224. T. Kazuno, S. Betsuki, Y. Tanaka, and T. Hoshita, *J. Biochem.* **58,** 243 (1965).
225. T. Hoshita, M. Yukawa, and T. Kazuno, *Steroids* **4,** 569 (1964).
226. T. Hoshita, M. Kouchi, and T. Kazuno, *J. Biochem.* **53,** 291 (1963).
227. I. G. Anderson and G. A. D. Haslewood, *Biochem. J.* **93,** 34 (1964).
228. T. Hoshita, T. Sasaki, and T. Kazuno, *Steroids* **5,** 241 (1965).
229. H. Makino, *Z. Physiol. Chem.* **220,** 49 (1933).
230. T. Hoshita, *J. Biochem.* **52,** 125 (1962).
231. G. A. D. Haslewood, *Biochem. J.* **100,** 233 (1966).
232. I. G. Anderson, G. A. D. Haslewood, A. D. Cross, and L. Tökés, *Biochem. J.* **104,** 1061 (1967).
233. I. G. Anderson and G. A. D. Haslewood, *Biochem. J.* **112,** 763 (1969).
234. J. L. Rabinowitz, R. H. Herman, D. Weinstein, and E. Staple, *Arch. Biochem. Biophys.* **114,** 233 (1966).

Chapter 3

Allo Bile Acids*

William H. Elliott

Department of Biochemistry
Saint Louis University
St. Louis, Missouri

I. INTRODUCTION

The *allo* bile acids are the 5α-cholanoic acids, i.e., derivatives of cholanoic acid in which the hydrogen at C_5 is alpha-oriented rather than beta-oriented as in the large number of naturally occurring bile acids. The sequence of numbering in the cholane nucleus is the same as in the normal series. The

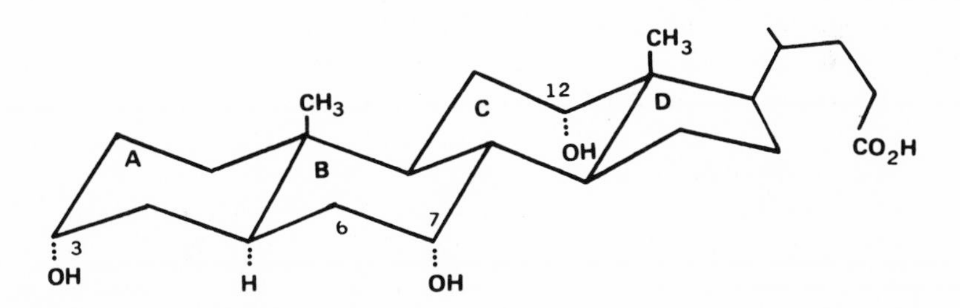

prefix allo (Greek *allos*, other) has been defined by *Chemical Abstracts* to designate the more stable of two geometrical isomers. The *trans* relationship of rings A and B existing in these derivatives is the more stable conformation

*This chapter represents Bile Acids XXIX in the series from this laboratory. The literature has been reviewed through June 1969. The author wishes to acknowledge support by the National Institutes of Health, Grant No. HE-07878, for certain aspects of studies reported in this manuscript.

[see Fieser and Fieser (1) for discussion of this matter]. The allo-acids presently known, in addition to the unsubstituted allocholanoic acid, are oxygenated at positions 3, 6, 7, and/or 12. Where the structure of the allo-acid is otherwise identical to a known 5β acid, the acid is frequently named by the trivial name of the 5β acid with the prefix allo; e.g., allocholic acid (I) (3α,7α,12α-trihydroxy-5α-cholanoic acid) is the 5α-derivative of cholic acid.

II. HISTORICAL BACKGROUND AND OCCURRENCE

During the period of the elucidation of structural formulas for sterols and steroids one can find repeated reference to two fundamental acids, cholanic and allocholanic acids. In the first of his papers on the constitution of bile acids (in 1912) Heinrich Wieland with Friedrich J. Weil (2) reported on the dehydration of cholic acid to a cholatrienic acid with subsequent catalytic reduction to a saturated acid, which was named cholanic acid, m.p. 157.5 °C, $[\alpha]_D$ + 20.3 °C. Seven years later Adolph Windaus and K. Neukirchen (3) reported results of studies designed to identify the C_{24} acid derived from the C_{27} hydrocarbon, cholestane. Upon oxidation of 10 g of cholestane they obtained 1.2 g of crude acidic product of m.p. 162 °C. On admixture of this material with cholanic acid a small but definite depression in melting point was observed. In 1935 Stoll *et al.* (4) reviewed the progress in purification of allocholanic acid. In 1923 Windaus and Bohne (5) reported a product of m.p. 162 °C. Ten years later Wieland *et al.* (6) reported an m.p. of 164–165 °C which could be raised to 167–168 °C by heating *in vacuo*; they showed that the two acids could form a substance with a congruent m.p. of 163.5 °C. Ruzicka (7) obtained a product of m.p. 169–170 °C from oxidation of cholestane with CrO_3; finally Stoll *et al.* (4) reported a sample of m.p. 172–173 °C, $[\alpha]_D$ + 22.5°, whose methyl ester melted at 93–94 °C. Thus, a span of 16 years elapsed from the time of isolation to the procurement of a pure specimen — a reflection of the purity of the starting materials and of the methodology of that period. This was a period of intense activity in the laboratories of Windaus with sterols and Wieland with bile acids culminating ultimately in a joint effort to arrive at a rational structural formula for these materials. Nobel Prizes were awarded to Wieland in 1927 and to Windaus in 1928 for their heroic efforts, despite an incomplete structure. The rapid developments in allied fields (the isolation of the estrogens, androgens, progestins, and adrenal cortical steroids) stimulated these studies, with revision in 1932 of the structures to the forms known today. The cholanic acids provided a firm foundation for many of these fundamental studies.

The studies of Windaus and Bohne (5) on hyodeoxycholic acid obtained from hog bile provided another entry into the allo-acid field. Oxidation of this dihydroxy acid provided a ketone which afforded allocholanic acid on Clemmensen reduction. Since cholanic acid was considered to be the parent of all naturally occurring bile acids known at that time, the isolation of the allo-acid was rather difficult to explain. In 1926 Windaus (8) offered the following explanation: the diketo acid (m.p. 162 °C) obtained from hyodeoxycholic acid was easily isomerized upon reflux with a mixture of acetic and hydrochloric acids or with alcoholic KOH to a new substance (m.p. 210 °C); Clemmensen reduction of the acid (m.p. 162 °C) proceeded via acid-catalyzed isomerization prior to reduction to the new acid (m.p. 210 °C) and thence to allocholanic acid. This new diketo acid (m.p. 210 °C, 3,6-diketo-allocholanoic acid) provided access to a number of dihydroxy or monohydroxymonoketo acids of the allo-series. Similar degradative studies have provided confirmatory evidence for the properties of the epimeric 3-hydroxy allocholanic acids obtained by Wieland, Ruzicka, Windaus, Stoll, Dalmer, Fernholz, and others.

Table I lists known allocholanic acids and some of their derivatives. Very few carbon–carbon unsaturated allocholanoic acids have been described (6, 9); several halogenated derivatives of the allo-acids and a number of 4,4,14α-trimethyl allocholanoic acids have been reported (10, 11) but they have not been included in this review. Of the monohydroxy derivatives, the "6-oxy" derivative reported by Wieland in 1932 (12) without assignment of configuration was prepared by catalytic reduction of the 6-keto acid, and is probably the axial 6β derivative (13). From an inspection of the data in the table it is clear that the subject of this chapter covers a span of 50 years with a renewed interest in the subject evidenced about 38 years after the initial paper of Windaus on allocholanic acid in 1919. Also it is evident that investigations on allo-acids flourished for about 18 years, then lagged, and were revived with new vigor again about 38 years later.

Although a few papers were published in the period 1937–1957, renewed interest in the allo bile acids may be attributed to studies of Haslewood and colleagues in England, and Kazuno and Yamasaki and their colleagues in Japan. In a continuing study on the comparative biochemistry of the bile salts, Haslewood revived the question of the composition of a bile acid obtained by Ohta (14) in 1939 from the bile of the Gigi fish, and later from other teleosts, to which was assigned the formula $C_{27}H_{46}O_6$ and the name "tetrahydroxynorsterocholanic acid." Ohta (1939) and Isaka (1940) (15) reported experiments that suggested that "Ohta's acid" was a 3α,6α,12α,X-tetrahydroxy acid. In 1951 Yamasaki (16) obtained the same acid from chicken bile and confirmed (14) the conversion of Ohta's acid to a mixture of cholanic and allocholanic acids. Based on these results Ohta had suggested that the

TABLE I. Properties of Allocholanic Acid and Derivatives

Substituent [a]	Melting point (°C)		$[\alpha]_D$		% Concentration; Solvent		M_D of ester		Reference	
	Acid	Ester	Acid	Ester	Acid	Ester	Calc.	Found	Acid	Ester
None	172–173	93	+22.2	+22.1	0.5;C	0.49;C	—	+ 82	71	71
	172–173	94	+22.5	—	1.0;D	—	—	—	4	4
	170		+22.2	—	1.8;C	—	—	—	6	—
3α	204–210	168–169	+25.5	+22.5	1.0;A	1.5;C	+ 87	+ 88	144	40, 71, 36
		167–168								
	204–208	169–170	—	+26.0	—	—	—	—	99, 145	146
	209	161–164	+29.0	+34	—	1.0;A	—	—	162	19, 162
	208–210	164	+25.4	+17.7	0.16;A	—	—	—	6	6
3α-AcO	—	147–148	—	+56	—	—	—	+242		145
	—	147–148	—	+21	—	—	+104	+ 91		162
3β	220	150–151	+23	+24.3	0.55;A	M	+ 80	+ 95	147	71
	218	151	—	+18.4	—	1.9;A	—	+ 72	6	6
	218	151–152	+17.2	—	—	—	—		148, 16	40
3β-AcO	—	154, 156	—	+14	—	0.33;A	+ 51	+ 60		147
7α	155–156	105–106	−1.0	+2.0	1.0;M	0.95;M	+ 23	+ 8	67	67
7β	—	91	—	—	—	—	+192	—		40
6	228	95	—	—	—	—	—	—	12	12
	240	101	—	—	—	—	—	—	149	149
12α	199	118–119	+42.2	+41.6	1.01;M	0.97;M	+175	+162	67	71
3-keto	183–184	113–114	+32.6	+40.5	0.98;M	0.99;M	+153	+157	71	71
	184	114	—	—	—	—	—	—	6	6, 41
6-keto	149–152	84	−3.5	—	—D	—	−33[b]	−13[b]	100	150
	151	—	—	—	—	—	—	—	12, 149	
7-keto	—	82.4	—	−41.6	—	1.0;M	−141	−161		40
3α,6β	247–248	—	—	—	—	—	—	—	151	—
	248–252	138	+10.7	—	0.99;D	—	+35[b]	+42[b]	97	145
3α-AcO-6β	—	192	—	+13	—	C	+54	+58		152
	—	190–193	—	+3.9	—	D	—	—		151, 99

TABLE I. (Continued)

Substituent [a]	Melting point (°C)		$[\alpha]_D$		% Concentration; Solvent		M_D of ester		Reference	
	Acid	Ester	Acid	Ester	Acid	Ester	Calc.	Found	Acid	Ester
3β,6α	233–234	—	—	—	—	—	—	—	151	
	228–229	—	+35	—	M	—	+133[b]	+137[b]	153	
3β,6β	273–274	184–186	—	—	—	—	+ 30	—	151	
	273–274	181	—	—	—	—	—	—	8	8
3β-6β-AcO	—	118	—	−11.5	—	—	−30	−51		152
3α,6α	227–228	165–166	—	—	—	—	—	—	151	151
3α,7α	245–246	125–126	+4	+6.4	0.81;M	1.0;M	+28	+26	40	68
	238–240	116–118	—	+7	—	0.9;M	—	—	39	66
3α-7α-AcO	—	142–143	—	−15.6	—	1.0;M	−56	−70		40
3α,7β	223–224	—	—	—	—	—	—	—	88	
3β,7α	273–274.5	159–160	—	+14	—	1.1;M	—	—	131	154
	—	160–161	—	+ 5.4	—	1.0;M	+21	+22		40
3β-7α-AcO	—	97–98	—	−10	—	1.0;M	−63	−45		40
3α,12α	213	175	+37.4	—	1.57;A	—			32	32
	214–215	174–176	+42	+35.6	0.75;A	1.0;M			34	34
	210	172	—	—	—	—			35	35
	219–220.5	179–180	—	+35	—	1.0;M			36	36
	215–216	177–178	—	+43.3	—	0.48;M	+180	+176	71	71
3α,12α-$(HCO_2)_2$	—	122–123	—	+61	—	1.0;M	—	—		36
3β,12α	233–234	137–138	+37.8	+41.0	0.66;M	0.99;M	+173	+166	155	34
	228	140							34	155
	231–232	137–138	—	+35.2	—	0.33;M			71	71
7α,12α	236–237	170–172	+22.0	+21.2	1.0;M	0.98;M	+166	+ 86	67	67
	185–187	—	—	—	—	—	—	—	12	
3α-6-keto	189–190	—	—	—	—	—			151	
	194	135–137	0.0		1.0;A		−26	0	42	42

TABLE I. (Continued)

Substituent [a]	Melting point (°C)		$[\alpha]_D$		% Concentration; Solvent		M_D of ester		Reference	
	Acid	Ester	Acid	Ester	Acid	Ester	Calc.	Found	Acid	Ester
3α-AcO-6-keto	—	185–187	—	−8.5	—	—	−9	−38		93, 13
	—	185	—	−8	—	1.09;C				42
3β-6-keto	247	93–95	−11	—	0.05;A	—	−35[b]	−43[b]	156	94
	244–245	—	—	—	—	—			94	
	243	—	—	—	—	—			151	151
3β-AcO-6-keto	—	156–158	—	−14	—	1.04;M	−62	−62		94
3-keto-6α	199–199.5	122–123	—	—	—	—			151	151
	195–197	90–97	+ 4	+55	1.06;P	1.04	+208	+222	94	94
3α-7-keto		160–161	—	—	—	—	—	—	—	40
3β-7-keto	185–187	147–149	—	−41.6	—	1.0;M	−143	−168	89	40
3-keto-7α	—	137–138	—	+16	—	1.1;M				66
	—	143–144	—	+18.5	—	1.0;M	+ 94	+ 75		40
3-keto-7α-AcO	—	126–127	—	+12	—	1.0;M	+ 10	+ 54		40
3-keto-12α	—	134–136	—	+56.8	—	1.0;M	+246	+229		34
	—	144–145	—	+51.7	—	1.0;M				71
3-keto-12α-AcO	—	143.5	—	—	—	—				87
7-keto-12α	192–193	164–165	−19.1	−13.8	0.83;M	1.01;M	−48	− 56	67	67
	199	158–159.5	—	—	—	—			157	157
12-keto-7α	187	128–129	+53.8	+41.2	0.71;M	0.29;M	+293 +291[b]	+166 +210[b]	67	67
3,6-diketo	208–209	149–151.5	—	—	—	—			151	151
	206–208	146–148	—	—	—	—			97	97
	209–210	148	± 0	± 0	C				8	8
	—	147–149	—	+0.5	—	—C	+40	+ 2	89, 159	158, 151, 159
3,7-diketo	—	160–161	—	−21.6	—	1.0;M	−70	−87		71
	185–187	158–159	—	—	—	—			89	40

TABLE I. (Continued)

Substituent [a]	Melting point (°C)		$[\alpha]_D$		%Concentration; Solvent		M_D of ester		Reference	
	Acid	Ester	Acid	Ester	Acid	Ester	Calc.	Found	Acid	Ester
3,12-diketo	222–223	136–138	+87.2	—	0.92;A	—			32	32
	230–232	157–160							36	36
	225	162–163	—	+88	—	0.55;M	+423	+354	155	71
	225	—	—	—	—	—			35	
7,12-diketo	194	143–144	+11	+9.4	0.94;M	0.99;M	+129	+38	67	67
	187.5-189	122–123	—	—	—	—			157	157
3, 6, 12-triketo	239–241	221–222	+32	—	—D	—	+308[b]	+129[b]	160	160
	248–249	212–214	+60	+25	0.40;M	0.49;M	+308[b]	+241[b]	161	161
3, 7, 12-triketo	229–232	164–165[a]	+28	—	1.5;A	—	+198[b]	+113[b]	19	19
	232–233	198–199	+19.8	+29.1	0.36;M	0.98;C	+200	+121	67	67
	230–233	226–228	—	—	—	—			28	28
3α,7α,12α	239–241	ca.225	+23	—	1.2;A	—	+119[b]	+94[b]	19	19
	250–251	225–226	+27.8	+26.7	0.75;M	0.84;M	+121	+113	67	67
	238–240	225–226	—	+28	—	1.02;M			28	66
3β,7α,12α	—	186–187	—	+58	—	0.65;M		+245		66
	241–242	198–199	+25.2	+23.2	0.68;M	1.0;M	+114	+98	67	67
3α,7β-12α	252–253	157–161[a]	+61	—	1.2;A	—	+288[b]	+249[b]	19	19
	253–255	—	+57	—	0.32;M	—			28	
3α,6β,7β	256–257	142–144	+49.5	+47.9	0.42;A	0.62;A	+147	+202	90	90
3-keto-7α,12α	—	152–154	—	+45	—	0.88;M	+187	+189		66
	—	156–157	—	+37	—	1.0;M		+155		67
3-keto-7α, 12α-$(AcO)_2$	—	132–133	—	+45	—	0.98;C	+290	+227		67
7-keto-3α, 12α	—	179–180	—	−11.6	—	0.76;M	−43	−49		67
7-keto-3β,12α	—	188–189	—	−5.1	—	0.3;M	−50	−21		67
6-keto-3α,7β	234–236	—	+39.6	—	0.19;M	—	+82[b]	+161[b]	90	
	232–234	—	—	—	—	—			19	
	226–227	—	+35	—	1.2;A	—		+142[b]	154	

TABLE I. (Continued)

Substituent [a]	Melting point (°C)		$[\alpha]_D$		% Concentration; Solvent		M_D of ester		Reference	
	Acid	Ester	Acid	Ester	Acid	Ester	Calc.	Found	Acid	Ester
12-keto-3α,7β	247–248	—	+94	—	0.70;A	—	+465[b]	+382[b]	19	
3,7-diketo-12α	223–224	178	—	±0.5	—	1.0;M	+23	+2	71	71
6-keto-3α,7β,12α	221–223	162–163.5	—	—	—	—			54	54
	218–220	—	+50	—	1.50;A	—	+175[b]	+211[b]	19	
	223–225	—	—	—	—	—			28	
	222-224.5	161–163	—	—	—	—			36	36
3β,5α,6β	242	203–205	−10.3	−7.6	P	P	—	—	153	153
	258–259	213–214	−4.4	—	A	—	—	—	163	163
Glycoallo-deoxycholic acid	221–223	184–185	+27	—	1.0;M	—			36	36

[a] Substituent hydroxyl groups in all tables are designated by number and Greek letter: AcO=acetate; (HCO_2)=formate; BzO=benzoate; Ester=methyl ester; a=ethyl ester; b=free acid; A=ethanol, 95% or absolute; M=methanol; C=chloroform; D=dioxane; E=90% ethanol; P=pyridine. Molecular rotations (M_D) are calculated from Δ values (164); the Δ-value for the 7-keto group has been corrected to −223 for these calculations (67).

new acid was substituted at position 6 by a hydroxyl group, since the derived ketone could provide the allo derivative on treatment with alkali. However, on investigation of the infrared spectrum of samples of the acid and its methyl ester obtained from Kazuno and from Yamasaki, Haslewood and Wootton (17) reported a resemblance to methyl cholate, They (18) subsequently isolated Ohta's acid from the bile of the king penguin and concluded that it was probably isomeric with cholic acid. A synthesis of allocholic acid was achieved (19, 19a) and they were able to show that the methyl (or ethyl) esters of the purest available preparations of Ohta's acid were indistinguishable from mixtures of methyl (or ethyl) allocholate and methyl (or ethyl) cholate in a ratio of about 2:1. Allocholic acid has subsequently been isolated from the bile of the leopard seal (20), chicken (21), salamander (22), several species of fish, reptiles, and birds (23, 24, 25), and from several mammals including man (25, 26, 27). Allocholic acid has been shown to be the major acidic biliary metabolite of cholestanol in the rat (28, 29), and the gerbil (30). The bile of the green iguana contains the taurine conjugate of allocholic acid as the major biliary acid (31).

Allodeoxycholic acid is a normal component of rabbit bile. In 1936, Kishi (32) obtained 1 g of a new bile acid, "β-lagodeoxycholic acid," from the contents of 3000 rabbit gallbladders and suggested that it was an isomer of deoxycholic acid. Six years later Matumoto (33) prepared $3\beta,12\alpha$-dihydroxy-5α-cholanoic acid and oxidized it to a diketo acid comparable to that obtained by Kishi from β-lagodeoxycholic acid. Danielsson, Kallner, and Sjövall (34) reported the isolation of allodeoxycholic acid from rabbit feces and noted the similarity of properties of their material with those reported by Kishi. These observations were confirmed independently by Yukawa (35), and Hofmann and Mosbach (36). Yukawa prepared allodeoxycholic acid by a Huang-Minlon reduction of $3\alpha,7\beta,12\alpha$-trihydroxy-6-keto-5α-cholanoic acid, and suggested identity with β-lagodeoxycholic acid. Hofmann and Mosbach isolated the bile acids from gallstones of rabbits fed a diet of 1% cholestanol, and identified the major component as glycoallodeoxycholic acid by comparison with their synthetic material. The comparative data of Kishi, Yukawa, and Danielsson *et al.*, and Hofmann and Mosbach thus establish β-lagodeoxycholic acid as allodeoxycholic acid. Hofmann *et al.* (37) established that allodeoxycholic acid constitutes about 6% of the bile acids of rabbit bile, but that it does not induce cholelithiasis in the rabbits until the concentration is increased above 20%, as achieved in animals on a diet of 1% cholestanol. Hofmann and Mosbach have prepared glycoallodeoxycholic acid (Table I) and compared its micellar properties with the corresponding 5β derivative; the calcium salt of glycoallodeoxycholate is less soluble than its 5β-epimer and does not dissolve in a solution of sodium chloride as does its epimer (36). Allodeoxycholic acid is undoubtedly a secondary bile acid (37, 38) (see Danielsson, this series, Vol. 2).

Allochenodeoxycholic acid has been shown to be a biliary constituent of the giant salamander (39) and a metabolite of cholestanol (40) and of allolithocholic acid in the bile fistula rat (41). Haslewood (25) has reasoned that a search for allochenodeoxycholic acid in the bile of the germ-free chick might be successful, since allocholic acid is present in bile of these animals and chenodeoxycholic acid is the major bile acid in this species; however, he has not been able to establish its occurrence.

Allolithocholic acid has been shown recently to be a metabolic product of allochenodeoxycholic acid in the rat after intracecal administration (41).

A derivative of hyodeoxycholic acid, *3α-hydroxy-6-keto-5α-cholanoic acid,* was isolated by Fernholz (42) from hog bile. Although this acid has been repeatedly referred to as a component of hog bile, it is most probably an artifact produced from the 6-keto-5β acid by alkaline hydrolysis of the conjugates present in bile.

A probable intermediate in the biosynthesis of allocholic acid, *3α,7α, 12α-trihydroxy-5α-cholestan-26-oic acid,* has been found in the bile of the giant salamander (39). Amimoto (43) obtained allocholic acid and a second radioactive acidic metabolite from bile of the giant salamander after administration of 27-deoxy-5α-cyprinol (3α,7α,12α, 26-tetrahydroxy-5α-cholestane). The unknown acid was esterified, reduced with $LiAlH_4$, and the product identified as 27-deoxy-5α-cyprinol. Okuda *et al.* (44) isolated a few crystals (m.p. 227 °C) from the bile of iguana and obtained the same product on reduction of the ester with $LiAlH_4$.

Presumably the few naturally occurring allo-acids are found as conjugates of taurine or glycine in bile or gallstones, or as the free acids in feces. At present the only conjugated allo-acid that has been isolated, characterized, and synthesized is glycoallodeoxycholic acid (36), a component of rabbit gallstones induced by feeding a diet containing cholestanol.

III. SEPARATION, ISOLATION, CHROMATOGRARHY

A. Separation and Isolation

A study of the chemistry and metabolism of the allo-acids has been inhibited by the inherent problem of separating these acids effectively from their 5β-epimers. This is evident from the first experiments of Windaus with allocholanic acid, for a period of 16 years elapsed before the pure acid was obtained. With the advent of chromatography and newer methods of separation of closely similar molecules the purified allo-acids emerged as entities unto themselves. The efficiency of these methods was assessed readily with labeled compounds.

Partial separations of the substituted methyl allocholanoates have now been achieved by crystallization. Kallner (41, 45) has reported the removal of 82% of the radioactivity after three crystallizations of a mixture of methyl allolithocholate-^{3}H and methyl lithocholate; 90% of the tritium was removed by several crystallizations of a mixture of methyl lithocholate-^{3}H and allolithocholate. Similarly, more than 90% of the radioactivity was removed from a mixture of methyl deoxycholate-^{3}H and allodeoxycholate after three crystallizations from aqueous acetic acid or aqueous methanol. Methyl 3β,12α-dihydroxy-5α-cholanate was separated from methyl deoxycholate by crystallization from aqueous methanol. Thomas *et al.* (46) reported the separation of 3α,6β-dihydroxy-5α- or 3α,6α-dihydroxy-5β-cholanoic acid from 3α,6β-dihydroxy-5β-cholanoic acid by crystallization from aqueous acetone or a mixture of methanol, ether, and hexane.

On the other hand, methyl allocholate and methyl cholate are not separated by crystallization (19, 19a, 47). Karavolas *et al.* (28) achieved a separation of 91% of allocholic acid from cholic acid by repeated crystallization from benzene, acetone, aqueous methanol, and benzene–acetone. Final separation was effected by chromatography on Florisil.

B. Chromatography

Bile acids are generally obtained from their source after alkaline or acid hydrolysis of the conjugates and the free bile acids are separated according to their polarity by partition or silicic acid chromatography (48, 49) (see Chapter 5 for details). In acetic acid partition chromatography the 3α-hydroxy-allo acids are eluted in fractions just preceding their 5β-isomer (e.g., allocholic acid is eluted in Fraction 60–4, whereas cholic acid appears in Fraction 80–1 (28). The reversed phase partition system of Norman with Hostalene (polyethylene powder) as supporting material for the stationary phase has been used (41) with solvent systems F and C (39) for the separation of allo-acids (43, 50). Table II contains data reported for elution of methyl allocholanoates from various grades of alumina. The results are obtained from six papers, and are therefore not necessarily related, i.e., frequently the solvent mixture was used to elute a *group* of related compounds.

Because of the *trans* fusion of rings A and B the allo-acids are more strongly adsorbed and consequently exhibit a slower mobility than their corresponding 5β-epimers. This is particularly well illustrated in *thin*-layer chromatography on silica gel G (Table III) in a comparison of the mobilities of the methyl esters in varying proportions of acetone in benzene. With the exception of the 3α-ols, the 7,12-diones, and the 3-keto-12α-ols, the methyl 5β-cholanoates are more mobile (larger R_f) than their 5α-isomers. Because

TABLE II. Chromatography of Methyl 5α-Cholanoates on Alumina

Functional groups	Grade[a]	Solvent Percentage (in C_6H_6)	Reference
3α	II	10% EtOAc[c]	41
	I	9% EtOAc	36
3β	II	5% EtOAc	41
3-keto	a[b]	20% Hexane	71
3α-7α	III	15–20% EtOAc	66
3β,7α	III	20–25% EtOAc	66
3α-AcO-12α	III	20% EtOAc	36
3α,12α	IV	45% EtOAc	38
	III	17–20% EtOAc	66
3β,12α	III	15% EtOAc	66
3α,12α-$(HCO_2)_2$	III	10% Acetone	36
3,12-diketo	II	5% EtOAc	38
3α,7α,12α	IV	5% MeOH[d] in EtOAc	66
3β,7α,12α	IV	75% EtOAc	66
3-keto-7α	III	8% EtOAc	66
3-keto-12α	III	3, 5 or 7% EtOAc	66, 34, 45
3-keto-7α,12α	IV	30% EtOAc	65

[a] Grade=Brockman number.
[b] a=deactivated with 12% water.
[c] EtoAc=ethyl acetate.
[d] MeOH=methanol.

of the *axial* character of the 3α-ols in the allo-series the methyl esters of these alcohols are more mobile than their corresponding *equatorial* 3α-ols of the 5β series. Eneroth (51) has noted that the presence of a 7α-hydroxyl group (e.g., methyl allochenodeoxycholate or allocholate) markedly retards migration on silica gel G in comparison with the 5β-isomers. This observation is confirmed and extended to include the 3α,6β,7β-triol (Table III). Mobilities on silica gel G are compared in Fig. 1.

The solvent systems of Eneroth (51, 52) and Sasaki (53) have been used frequently for the separation of allo-acids or their esters from their 5β-epimers. Of the free acids system, S-11 was used for the 3,7-diols (39) and S-12 for the 3,12-diols (54, 55); S-12 was reported for the 3,7,12-triols (41), the 3-keto-7α-ol (56), and the 3-keto-12α-ol (56). Allocholic was separated from cholic acid in the S-7 system (39, 22, 57) or in the EAW-2 system (53, 22). From Sasaki's data R_f's can be calculated for allocholic acid as 0.59 (EAW-2) (53), 0.91 (BAW-2), and 0.60 (S-7) (43). Methyl esters of the allo-acids were separated from their 5β-epimers on the basis of their polarity; for the 3α-ols, system S-15 was used (41); for the 3,7-diols, S-12 (56, 41) or S-6 (51, 39); for the 3,12-diols, S-12 (41, 56); for allocholate, S-6 (41, 58); and for the 3-keto-12α-ol, S-11 (45). From Hoshita's data (58) R_f's of 0.26 for allocholate and 0.33 for cholate can be calculated from the S-6 system; a better separation was obtained by a second exposure to the solvent system after initial

TABLE III. Thin-Layer (TLC) and Gas–Liquid (GLC) Chromatography of Methyl Allocholanoates

Substituent	TLC on silica gel G[a] $R_f \times 100$; % acetone in benzene		Relative retention times (RR_t)[b]					
			QF-1,230°		OV-1,260°		OV-17,260°	
	5α	5β	OH	TMS	OH	TMS	OH	TMS
None	—	—	0.18	—	0.40	—	0.24	—
3α	28(5); 50(8)	24(5); 37(8)	0.50	0.89	0.72	0.72	0.58	0.92
			0.52[c]	—				
3β	16(5); 33(8)	34(5); 45(8)	0.55	1.20	0.72	0.82	0.60	1.20
			0.58[c]					
7α	48(5); 60(8)	59(5); 70(8)	0.40	0.62	0.64	0.57	0.51	0.57
7β	40(5); 60(8)	43(5); 61(8)	0.43	0.81	0.67	0.78	0.52	0.83
12α	54(5); 68(8)	60(5)	0.37	0.66	0.60	0.60	0.49	0.64
3-keto	29(2); 60(5)	30(2); 62(5)	1.06	—	0.76	—	0.71	—
			1.08[c]	—				
7-keto	40(2); 71(5)	73(5)	0.72	—	0.66	—	0.57	—
12-keto	45(2)	45(2)	0.60	—	0.63	—	0.51	—
3β,6β	40(40)	—	1.27	1.40	1.20	1.33	1.29	1.31
			1.64[f]					
3α,7α	35(40)	45(40)	1.22	1.00	1.18	0.84	1.27	0.92
			1.14[d]	—				
			1.17[c]	—				
3α-7α-AcO	56(20)	—	1.40	3.07				
3β,7α	36(40)	63(40)	1.29	1.14	1.18	1.08	1.34	1.00
			1.22[d]	—				
			1.24[c]	—				
3β-7α-AcO	48(20)	—	1.77	4.06				
3α-12α	20(20); 61(40)	15(20); 53(40)	1.07	0.93	1.12	1.18	1.19	1.22
			1.07[d]	0.91[d]				
			1.07[c]	—				
3α,12α-$(HCO_2)_2$	—	—	1.30[e]	—				
3β,12α	22(20); 60(40)	66(40)	1.16	1.35	1.16	1.23	1.20	1.19
			1.15[c,d]	—				

TABLE III. (Continued)

Substituent	TLC on silica gel G[a] $R_f \times 100$; % acetone in benzene		Relative retention times (RR_t)[b]					
			QF-1,230°		OV-1,260°		OV-17,260°	
	5α	5β	OH	TMS	OH	TMS	OH	TMS
7α,12α	33(20); 82(40)	47(20); 90(40)	0.90	0.63	0.99	0.63	0.99	0.53
3-keto-7α	37(15); 42(20); 59(25)	46(15);50(20); 72(25)	2.32 2.23[d] 2.29[e]	4.12	1.27	1.08	1.60	1.60
3-keto-7α-AcO	34(8)	—	3.26					
3-keto-12α	49(20)	40(20)	2.23 2.14[c,d]	3.04	1.19	1.17	1.41	1.85
7-keto,3α	43(25); 35(20)	45(25); 39(20)	1.97	4.16	1.23	1.41	1.42	2.30
7-keto,3β	36(20)		2.12		1.26		1.41	
7-keto-12α	70(20)	74(20)	1.56	3.42	1.09	0.97	1.13	1.54
12-keto-7α	67(20)	—	1.40	2.43	1.03	0.90	1.12	1.18
3,7-diketo	70(20); 85(25)	74(20); 88(25)	3.96	—	1.17	—	1.52	—
3,12-diketo	77(20)	79(20)	3.52 3.18[e]	—	1.15	—	1.51	—
7,12-diketo	94(20)	94(20)	1.97	—	0.96	—	1.06	—
3,7,12-triketo	57(20); 92(50)	61(20); 95(50)	9.24 7.20[e]	—	1.64	—	2.98	—
3α,7α,12α	17(50)	23(50)	2.68 2.28[d] 2.37[e]	1.00	1.77	1.05	2.73	0.87
3β,7α,12α	23(50)	36(50)	2.78 2.44[d] 2.50[e]	1.18	1.88	1.01	2.71	0.79
3α,6β,7β	42(50)	52(50)	2.10	1.41	1.78	1.44	2.34	1.17
3-keto-7α,12α	6(20); 55(50)	9(20); 64(50)	6.10 4.49[d] 4.50[e]	5.07	1.96	1.09	3.12	1.54
7-keto-3α,12α	54(50)	51(50)	4.62	5.45	1.95	1.62	2.89	1.95

TABLE III. (Continued)

Substituent	TLC on silica gel G[a] $R_f \times 100$; % acetone in benzene		Relative retention times (RR_t)[b]					
			QF-1,230°		OV-1,260°		OV-17,260°	
	5α	5β	OH	TMS	OH	TMS	OH	TMS
7-keto-3β,12α	50(50)	—	4.50	7.00	1.85	1.96	2.87	2.46
6-keto-3α,7β	—	—	2.06		1.52		1.88	
3,7-diketo-12α	30(20)	33(20)	7.14	12.7	1.84	1.78	3.04	3.76

[a] In TLC on silica gel *G*, $R_f \times 100$ is followed by percentage of acetone in benzene in parentheses; 5α refers to allocholanates, and 5β to 5β cholanates. Data from (40, 67, 71, 85, 91, 170).

[b] Relative retention times for the columns labeled OH refer to methyl deoxycholate=1.00; for the columns labeled TMS (trimethylsilyl ethers) refer to the bis (TMS) ether of methyl deoxycholate=1.00. Absolute times of elution of methyl deoxycholate were: 3% QF-1, 29.0 min; 3% OV-1, 38.4 min; 3% OV-17, 44.0 min; methyl deoxycholate was injected simultaneously with each ester. Absolute times of elution of the bis (TMS) ether of methyl deoxycholate were: 3% QF-1, 10.0 min; 3% OV-1, 27.3 min; 3% OV-17, 14.2 min; this derivative was injected simultaneously with each ether (60).

[c] See (62).

[d] See (56).

[e] See (36).

[f] Relative retention time on 0.4% CNSi (63).

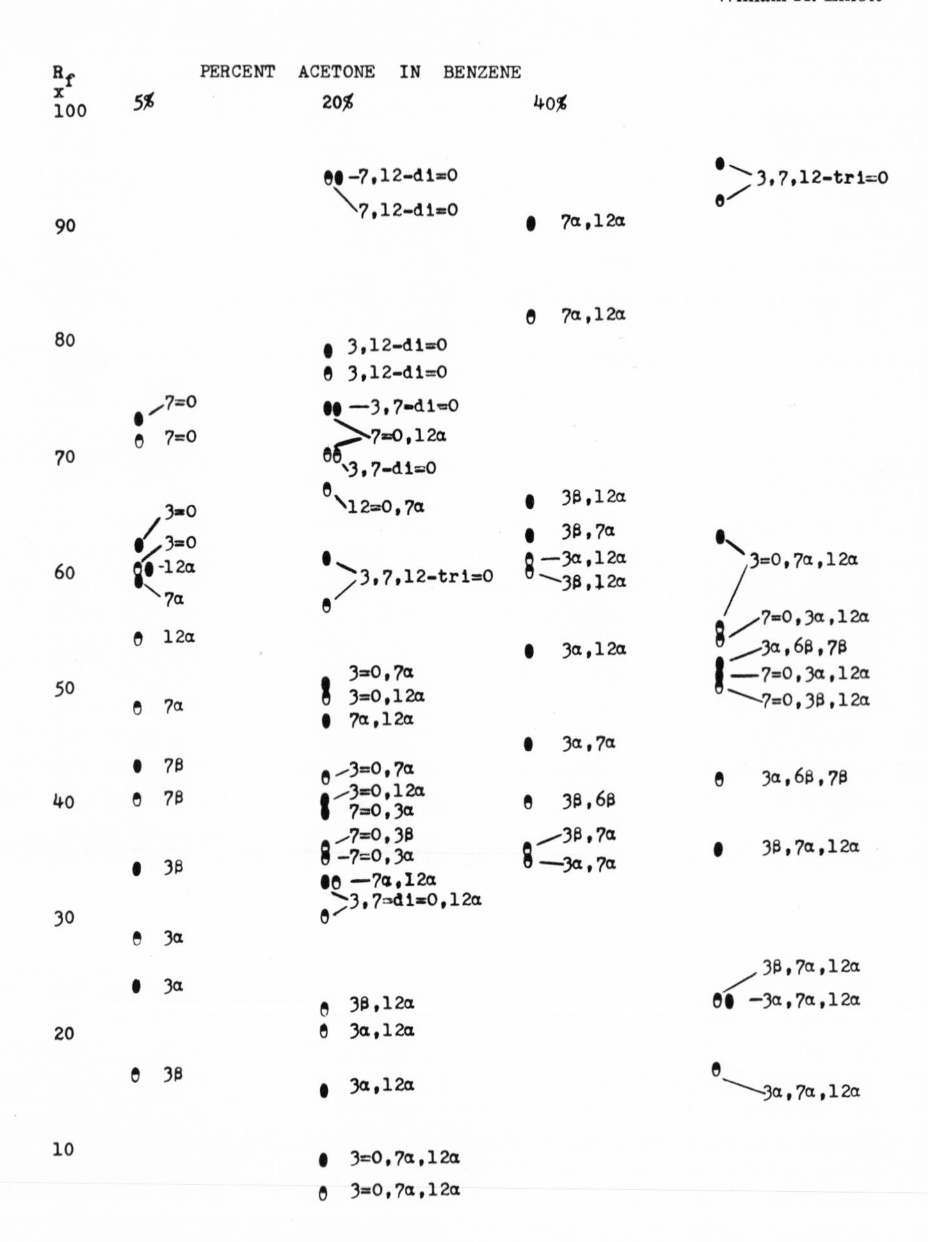

Fig. 1. Comparison of mobilities of methyl cholanoates on silica gel G. Solvent systems: 5%, 20%, 40%, or 50% acetone in benzene; R_f values are multiplied by 100. ●, 5β-cholanoate; ◓, 5α-cholanoate; 3=O, 3-keto; 3,7-di=O, 3,7-diketo; 3,7,12-tri=O, 3,7,12-triketo.

development and drying of the plate (58). Methyl $3\alpha,7\alpha,12\alpha$-trihydroxy-5α-cholestanoate was studied with system S-11 (50). Taurocholate was not well separated from tauroallocholate in any of the systems tested (53). From a comparison of mobility on alumina TLC of a series of 5α- and 5β-cholanoates Hodosan and Pop-Gocan (59) concluded that 6α-substituted compounds were more strongly adsorbed than the 3α compounds. Hofmann and Mosbach (36) reported a better separation of several methyl esters of allo-acids on Anasil B than silica gel G. They noted that adsorptivity was related to the presence of polar groups, since the mobilities of the epimeric methyl $3\alpha,12\alpha$-diacetoxy-cholanoates were the same in acetone–benzene systems. *Paper chromatography* (19, 39) and *glass–paper chromatography* have been used infrequently with the allo-acids.

Mobilities of methyl 5α-cholanoates in *gas–liquid chromatography* are given in Table III. Relative retention times, RR_t, are related to methyl deoxycholate as 1.00. In general, the esters are eluted in the expected order, i.e., the *axial* alcohol is eluted before the *equatorial* alcohol on the selective phases, QF-1 and OV-17. Since the 3α-ol is now *axial*, the order of elution from the column is reversed for the epimeric 3-ols in the allo-series from that observed in the 5β series. The slower mobility of the allocholanates is particularly evident in gas–liquid chromatography (Fig. 2); on the average, the allocholanates were eluted 1.22 times longer than their 5β-epimers from QF-1, 1.11 from OV-1, and 1.20 from OV-17 (60). Three exceptions to this generalization were noted: methyl 7-keto-12α-hydroxy-5α-cholanate was eluted more rapidly from OV-1, and methyl 3-keto-7α-hydroxy-5α- and methyl $3\alpha,6\beta,7\beta$-trihydroxy-5α-cholanate were eluted more rapidly from QF-1 than their respective 5β-epimers. Figure 2 shows the order of elution of 12 representative methyl cholanoates from the three phases QF-1, OV-1, and OV-17.

The relative retention times of the trimethylsilyl (TMS) ethers in Table III are related to the TMS ether of methyl deoxycholate as 1.00. The TMS ethers were prepared from a mixture of trimethylsilyl chloride, hexamethyldisilazane, and dry pyridine according to Makita and Wells (61) to assure the formation of the complete derivative (cf. Chapter 5 for a discussion of the preparation of partial TMS derivatives). These TMS ethers are generally eluted more rapidly from the column than the free alcohols; the TMS ethers of the *equatorial* 3β-ols of the allocholanates are generally retained on the column longer than the corresponding 3β ethers of the 5β-epimers (60). On the other hand, the TMS ethers of the allo-3α-ol, $3\alpha,7\alpha$-diol, $3\alpha,7\alpha,12\alpha$- and $3\alpha,6\beta,7\beta$-triols are eluted from each of the three phases before their respective 5β-cholanates. However, the TMS ether of methyl allodeoxycholate is eluted before the TMS ether of methyl deoxycholate on QF-1, but not on OV-1 or OV-17 (60). Hoshita *et al.* (39) observed that the TMS ethers of

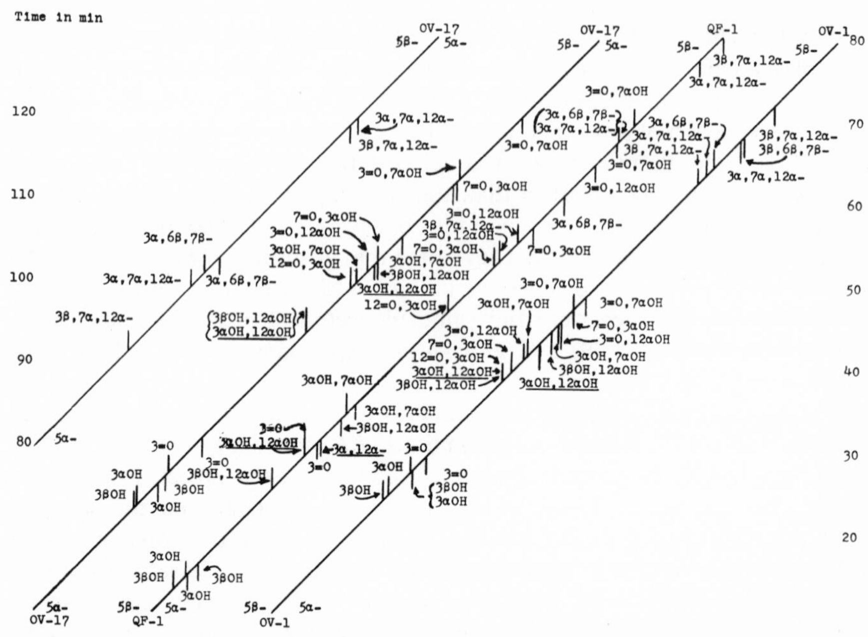

Fig. 2. Gas–liquid chromatography of methyl cholanoates on 3% OV-17, 3% QF-1, and 3% OV-1. Methyl 5β-cholanoates are plotted on the left side and methyl 5α-cholanoates are plotted on the right side of each diagonal line representing the phase on the column. Time of elution in minutes is read on the abscissa. The phases were mounted on Gas Chrom Q (100–120 mesh) (60).

methyl allocholate, allochenodeoxycholate, and allodeoxycholate each exhibited retention times on 1% SE-30 of 0.96 relative to that of the TMS ethers of the corresponding 5β-cholanates. Hofmann *et al.* (37) have reported retention times of 0.77 and 0.54 on 0.5% Hi Eff 8B for the TMS ethers of methyl allodeoxycholate and allocholate, respectively, relative to the TMS ether of methyl deoxycholate.

Retention times of a few of the trifluoracetates (TFA) of allocholanates have been reported, all relative to methyl deoxycholate. Eneroth and Sjövall (62) reported values of 1.18 and 1.09 for TFA esters of methyl 3β,6α-dihydroxy-5α-cholanoate and 3β,6β-dihydroxy-5α-cholanoate, respectively, on 0.5% QF-1, and 1.37 for the TFA ester of methyl allocholate on 3% QF-1 at 230°C. Tsuda *et al.* (63) reported relative retention times of 1.55 and 1.39 for the TFA esters of methyl 3β,6α-dihydroxy-5α-cholanoate and 3β,6β-dihydroxy-5α-cholanoate, respectively, on 0.5–2.3% SE-30 at 220°C.

IV. PHYSICAL PROPERTIES

The physical properties of known C_{24} allo-acids and several of their common derivatives are recorded in Table I. Although the melting points of a few of the methyl esters are reported below 100 °C, the free acids melt well above 100 °C and frequently above 200 °C.

The *trans* fusion of rings A and B in the allo-acids produces a more planar molecule than the 5β acid and contributes to the poorer detergency of glyco allodeoxycholate and consequent poorer solubility of the calcium salt (36). The Krafft point (critical micellar temperature) of several allo-acids has been determined and discussed (64). In contrast to the notorious character of deoxycholic acid to complex with a large variety of other substances, no evidence has been reported for the formation of "choleic acids" by allodeoxycholic acid.

A. Optical Activity

The *specific rotations* of the acid and methyl ester are given in Table I with the solvent and concentration; *molecular rotations* have been calculated for the methyl esters where specific rotations are available. Agreement between the calculated and found values is reasonably good for most substances. Although allolithocholic, allochenodeoxycholic, allodeoxycholic, and allocholic acids are less dextrorotatory than their corresponding 5β acids (65), the specific rotations of a number of the other allo-acids are either equivalent to or more dextrorotatory than the comparable 5β-epimer, thus precluding a general conclusion for this class of compounds. *Optical rotatory dispersions* of a few 3-keto-allo derivatives have been reported (34, 66, 67, 68, 40).

B. Infrared Spectrometry

Haslewood (25) has discussed the value of infrared spectroscopy for the characterization of bile acids and has indicated that bands at about 9.3, 9.6, 10.2, 10.5, and 10.95 μ are especially useful for detection of the cholic acid nucleus. Earlier he and Anderson (69) noted the importance of absorption bands at 10.4 and 11.2 μ for the identification of allocholic acid. He has pointed out the prominence of bands at about 9.2, 9.7, 9.9, 10.4, and 11.2 μ in the spectrum of allocholic acid. Those bands between 9.0 and 10.0 μ due to the C–O stretching frequency can be attributed to the *axial* hydroxyl groups at C_3, C_7, and C_{12}. This is apparent from the data in Table IV which show the absorption bands for a series of substituted allo-acids in the vicinity of the bands reported for allocholic acid. The band found in the vicinity of 9.9 μ can be attributed to the 3α-hydroxyl group; replacement with a 3β-hydroxyl group produces a hypsochromic shift.

The infrared spectrum of methyl allochenodeoxycholate is shown in Fig. 3. Infrared spectra or maxima for allocholic (19, 25, 70, 66), allodeoxy-

TABLE IV. Infrared Bands of Allo Bile Acids[a]

Compound	Wavelength (μ): 9.0 – 9.5 – 10.0 – 10.5 – 11.0 – 11.5
Allocholic acid	
Methyl allocholate	
3β,7α,12α-trihydroxy	
Methyl 3β,7α,12α-trihydroxy	
7α,12α-dihydroxy	
Methyl 7α,12α-dihydroxy	
3α,7α-dihydroxy	
Methyl 3α,7α-dihydroxy	
Methyl 3β,7α-diHO	
Methyl 3α,12α-diHO	
Methyl 3β,12α-diHO	
Methyl 3α-HO	
Methyl 3β-HO	
7α-HO	
Methyl 7α-HO	
12α-HO	
Methyl 12α-HO	
3α,7β,12α-triHO	
3α,6β,7β-triHO	
3=O	
Methyl 3=O	
Methyl 3=O,7α-OH	
Methyl 3=O,12α-OH	
Methyl 3=O,7α,12α-diOH	
Methyl 7=O	
Methyl 7=O,12α-OH	
Methyl 7=O,3β-OH	
Methyl 7=O,3α,12α-diOH	
Methyl 7=O,3β,12α-diOH	
Methyl 3,7-diketo	
Methyl 3,7-diketo-12α-OH	
Methyl 12=O,7α-OH	
12=O,7α-OH	
Methyl 7,12-diketo	

[a] Spectra were recorded in Nujol with a Perkin-Elmer Model 21 infrared spectrometer fitted with a sodium chloride prism. Relative intensities of the maxima are indicated as follows: ⁝, most intense; ¦, medium intensity; |, least intense.

cholic (34, 36, 71), allochenodeoxycholic (39, 40), allolithocholic (71), and their 3-keto and 3β-isomers (67, 71, 40) have been reported.

C. Mass Spectrometry

Mass spectra of the allo-acids and their derivatives generally resemble those of their 5β-epimers. Thus, allocholanoates (M^+) with free or protected hydroxyl groups provide fragment ions by ready loss of water (M-18), acetic acid (M-60), trifluoroacetic acid (M-144), or trimethylsilanol (M-90). The loss of the C_{17} side chain of the methyl esters affords the fragment ion (M-115); loss of a molecule of water from keto allo-acids, a common phenomenon in mass spectrometry is equally poor as a diagnostic aid. Although Bieman and Seibl (72) reported that *axial* terpenoid secondary alcohols lost water more readily than the epimeric *equatorial* alcohols, and Zaretskii *et al.* (73) concluded that the ratio of intensities of the mass peaks M-H_2O/M was greater for *axial* alcohols than their *equatorial* epimers in the progesterone series, Egger and Spiteller (74) concluded that the conformation about rings A/B played a more important role in mass spectrometry of steroids than the stereo-chemistry of the hydroxyl groups. This conclusion is supported by a comparative study of the mass spectra of 5β- and 5α-cholanoates (75). Among the monohydroxy bile acids studied (C_3, C_7, C_{12}) the 3α- and 3β-ols of the allo-series exhibit molecular ions (M^+) of larger intensity than their 5β-epimers, or of either series of the C_7 and C_{12} alcohols (75); accordingly, the ratio M-H_2O/M is considerably larger for the 3α- and 3β-ols of the 5β series than for the epimeric C_7 and C_{12} alcohols of either series. The ratio M-H_2O/M-(H_2O + 115) for the 12α-hydroxy derivatives is characteristically less than one in comparison to the C_3 and C_7 alcohols, but a significant difference is not found between these values for the 12α-ols of the 5β and 5α series. Preliminary investigations suggest that the ratio M-H_2O/M for the 3α,7α-diols of the allo-series is about twice that of the 5β series (75). For the 7α,12α-diols the ratio of M-18/M for the 5α series is about 10 times greater than the 5β series. A significant difference in this ratio does not appear to exist in comparison of the 3-, 7-, or 12-monoketo, or diketo derivatives. Among the monoketo monohydroxy derivatives this ratio is much larger for the 3-keto-12α-hydroxy-5α-epimer (a factor of 5-18), whereas for the 3-keto-7α-ol, the

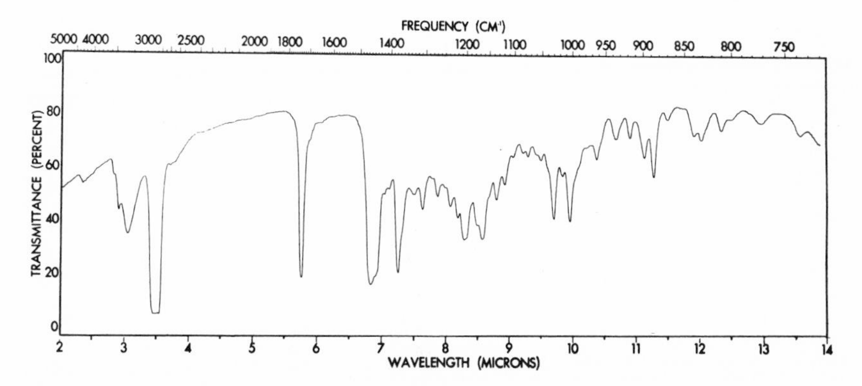

Fig. 3. Infrared spectrum of methyl allochenodeoxycholate in Nujol.

5α-epimer is larger by a factor less than 2. Although the 3-keto-7α, 12α-diol of the allo-series exhibits a larger ratio (2–5 times) than its 5β-epimer, the 5α-epimer of the 3,7-diketo-12α-ol more closely approximates that of its 5β-epimer (75); Kallner (65) noted that the intensities of the fragment ions, m/e 386 (M-H_2O) and 384 (M-2 H_2O) are larger in the spectra of methyl 3-keto-7α-hydroxy-5α-cholanoate and methyl 3-keto-7α,12α-dihydroxy-5α-cholanoate, respectively, than their 5β-epimers. A similar difference was not observed in the spectra of the 3-keto-12α-ols, suggesting a facile loss of the 7α-ol in the allo-series. The mass spectrum of methyl allochenodeoxycholate is compared with that of its 5β-isomer in Fig. 4.

A more intense peak in the ion fragment m/e 261 in the mass spectrum of the tris TMS ether of methyl allocholate (76, 44) and in the bis TMS ether

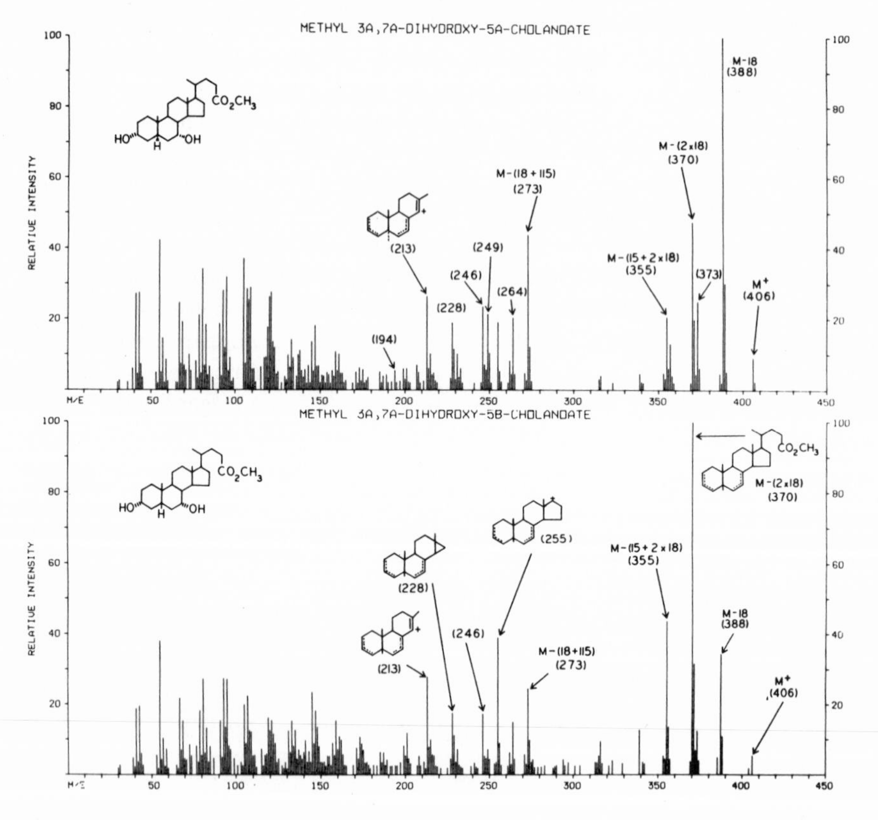

Fig. 4. Mass spectra of methyl allochenodeoxycholate (upper) and methyl chenodeoxycholate (lower). Spectra were obtained with the direct probe in an LKB 9000 mass spectrometer; ion source, 260°C; probe, 54°C; 3.5 kV accelerating potential; 60 μA, trap current; ionizing voltage, 70 eV.

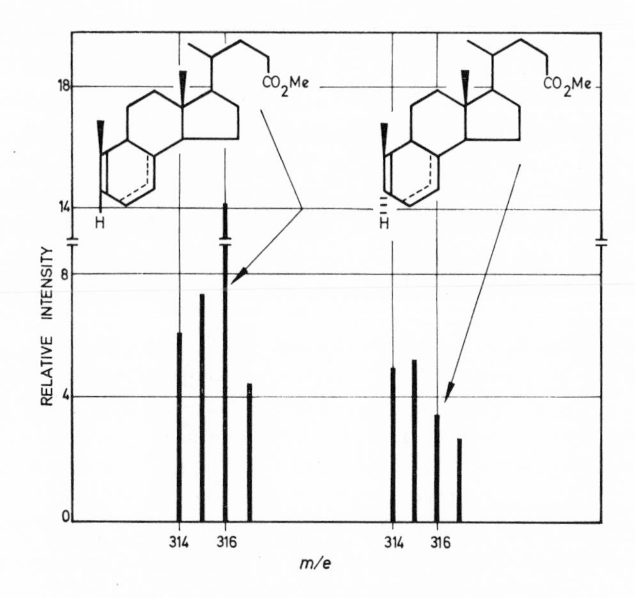

Fig. 5. Comparison of intensities of fragment ion m/e 316. Fragment ions derived from methyl 3-keto-7α-hydroxy-5β-cholanoate (left) and methyl 3-keto-7α-hydroxy-5α-cholanoate (right) as obtained from an LKB Model 9000 gas chromatograph mass spectrometer at 70 eV; 3.5 kV accelerating potential; 60 μA, trap current (68).

of the 3-keto-7α,12α-diol (67) than in the 5β derivatives has been reported. Sjövall (76) suggested that the fragment is represented by an ion containing the side chain, rings C and D, and C_7 with 2 double bonds in ring C. No significant differences are observed in the spectra of the trifluoroacetates of methyl cholate and allocholate (76).

Allen *et al.* (77) differentiate between the 3,6-diols of the 5β- and 5α-bile acids by comparison of the mass spectra of their derived bis (*O*-methyl) oximes. They report characteristic larger intensities of the molecular ions, of the ion M-46 (M-CH_3ONH), and m/e 179 for the 5α series, and smaller intensities for M-98, m/e 138, and 99 for the allo-derivatives.*

The cleavage of 3-keto steroids with loss of ring A was shown (78) to be more favored for the A/B *cis* (5β) steroids. This observation has been

*Since quantitative differences (intensities) of mass spectra vary with conditions and instruments of various designs (169), acceptance of the significance of the data reviewed here must await confirmation from other types of instrumentation. Most of the spectra referred to here were obtained with the LKB Model 9000 gas chromatograph mass spectrometer (66, 67, 75, 76, 77).

applied advantageously in the determination of structure of methyl 3-dehydro derivatives of allodeoxycholate (34), allochenodeoxycholate (66, 68, 40) and allocholate (66, 67). Figure 5 shows a comparison of the intensities of ions m/e 316 from methyl allochenodeoxycholate and its 5β-isomer.

V. CHEMICAL PROPERTIES AND DETERMINATION

Because of the change in conformation of the 3α-hydroxyl group in the allo-series certain differences in chemical reactivity may be anticipated. With the reagent potassium chromate:sodium acetate:acetic acid, 3α,7β,12α-trihydroxy-5α-cholanoic acid was oxidized principally at C_{12} to the monoketone (19); with methyl 3β,7α,12α-trihydroxy-5α-cholanoate, a mixture was obtained from which methyl 7-keto-3β,12α-dihydroxy-5α-cholanoate was separated by plc in about 23% yield (67). The major product from oxidation of methyl 7α,12α-dihydroxy-5α-cholanoate was the 7-keto-12α-ol, although the 7,12-dione and the 12-keto-7α-ol were also isolated (67). Oxidation of 3α,7β,12α-trihydroxy-5α-cholanoic acid, methyl allocholate or allochenodeoxycholate with *N*-bromo-succinimide (NBS) provided the 7-keto derivatives (19, 28, 67, 40), although the 3-keto-7α-ol was obtained in 14% yield from allochenodeoxycholate. The reduction of ketones in the allo-series proceeds as anticipated; reduction of the 3-ketones is discussed in *Preparation.*

The usual colorimetric assays applied to the normal (5β) bile acids have not been studied with allo-acids, probably because of the paucity of materials. The *Hammersten* test has been explored principally by Haslewood (25) who has reported a purple color given by allocholic (19) and cholic acids, their 3β-isomers (69) and their esters; 3α,7β,12α-trihydroxy-5α-cholanoic acid gives a yellow color (19). Undoubtedly gas–liquid chromatography will play an important role in quantitative analysis of the allo-acids in the future.

Digitonin has been used to precipitate the 3β-epimer of allocholic acid (69).

VI. PREPARATION OF ALLO-ACIDS

A. General Methods as Applied to Allocholic Acids

1. Haslewood's Synthesis from Cholic Acid

The initial studies of Anderson and Haslewood on the identification of Ohta's acid (79) sparked efforts to synthesize allocholic acid. Early attempts to reduce catalytically methyl 3α,12α-diacetoxy-7-ketochol-5-enoate to the

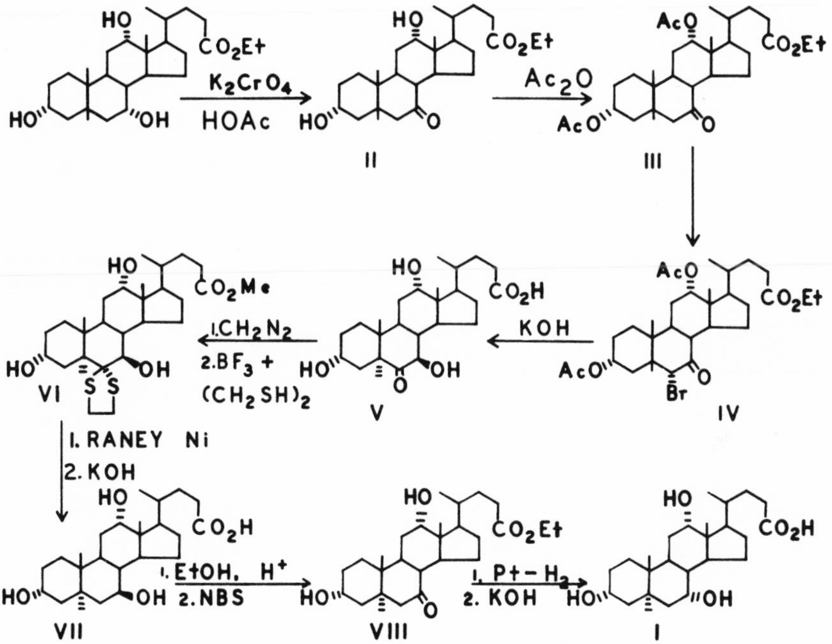

Fig. 6. Haslewood's preparation of allocholic acid (19).

5α derivative produced a mixture of 5β and 5α compounds in which the latter predominated. Separation of cholic acid from its 5α-epimer by crystallization was not feasible. Ample evidence exists for α-attack of the C_5 double bond to provide the allo-acid, e.g., catalytic reduction of 3β-acetoxy-7-ketocholest-5-ene provided the 5α-cholestane derivative (80). However, Anderson and Haslewood (19) successfully prepared allocholic acid as outlined in Fig. 6. Ethyl cholate was oxidized at C_7 and the product (II) acetylated at C_3 and C_{12}. The ketone (III) was brominated at C_6 and the product (IV) was allomerized to the important alloketol, (V). The C_6 carbonyl group of V was eliminated by desulfurization of the derived thioketal (VI) with Raney nickel to provide $3\alpha,7\beta,12\alpha$-trihydroxy-5α-cholanoic acid (VII). Oxidation of the ethyl ester of VII with NBS provided the ketone, (VIII), which was reduced catalytically to the desired 7α,ol; hydrolysis afforded allocholic acid, (I). The procedure was duplicated with minor variations by Karavolas *et al.* (28).

Because of the low yields attendant with this somewhat lengthy method, other procedures were investigated. Two general methods for the preparation of allo-acids have resulted.

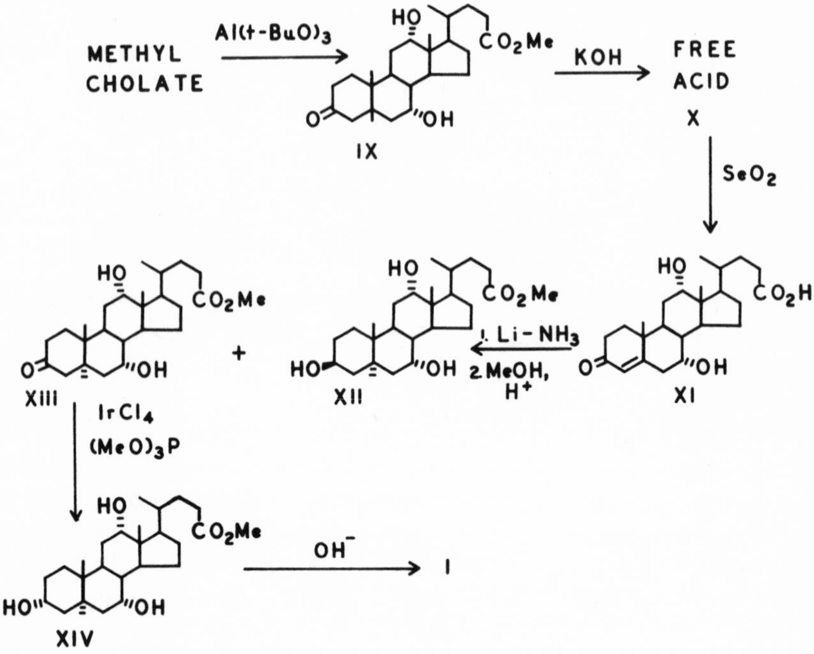

Fig. 7. Kallner's preparation of allocholic acid (66).

2. Reduction of 3-keto-Δ⁴-acids with Li–NH_3 and Trimethylphosphite-iridium (IV) Chloride

Kallner (66) studied the stereospecific formation of 5α steroids by the reduction of 3-keto-Δ^4-derivatives with Li in liquid ammonia (Fig. 7). Oppenauer oxidation of methyl cholate provided the 3-dehydro derivative IX which was hydrolyzed to the free acid (X) and the product dehydrogenated with SeO_2 to the dihydroxy-3-ketochol-4-enoic acid (XI). The unsaturated acid (XI) was reduced with Li in liquid ammonia, and the product methylated to provide a mixture from which XII was separated from the desired ketone (XIII) in yields of 18% and 30%, respectively. The ketone (XIII) was reduced with a mixture of iridium (IV) chloride and trimethylphosphite in aqueous isopropanol over a period of 72 hr to provide methyl allocholate (XIV) in 31% yield. Experiments of Haddad *et al.* (81) have shown high yields of the *axial* alcohols from substituted cyclohexanones. A recent paper of Browne and Kirk (82) suggests higher yields may be obtained by a prolonged reaction time (up to 5 days). In the preparation of methyl allodeoxycholate methyl 12α-hydroxy-3-keto-chol-4-enoate was prepared from the saturated ketone via the bromide according to Riegel and McIntosh (83). By this general procedure Kallner prepared methyl allocholate, allodeoxycholate, and allochenodeoxycholate and their 3β-epimers. The overall yields of these materials from the corresponding methyl 5β-cholanoates was less than 5%.

3. Transformation by Raney Nickel in Boiling p-Cymene

The method of Chakravarti, Chakravarti, and Mitra (84) for the transformation of 3-hydroxy-5β steroids to 3-keto-5α steroids has been utilized by Danielsson *et al.* (34) for the preparation of methyl 12α-hydroxy-3-keto-5α-cholanoate from the corresponding 5β derivative, and by Anderson and Haslewood (69) for the preparation of ethyl 3β,7α,12α-trihydroxy-5α-cholanoate from ethyl 7α,12α-dihydroxy-3-keto-5α-cholanoate in boiling cymene. The latter authors have commented on the poor and capricious yields obtained by this procedure (25).

The method has been studied more extensively (67, 71, 85) and in our hands it has been a useful procedure for the preparation of 3-keto-5α-acids from the esters of readily available acids. The following points are worthy of comment.

a. Raney Nickel. The catalyst (W-2) prepared carefully according to Mozingo (86), and preserved under ethanol, was washed three times by decantation with the solvent used in the reaction. The suspension of nickel in the solvent used was then added to the steroid such that a ratio of about 15–20 ml of solvent (e.g., *p*-cymene), 2 g of nickel, and 1 g of steroid was obtained. The temperature of the mixture was raised gradually to remove the last traces of ethanol, and a few milliliters of solvent were removed by distillation. Too little solvent results in excessive bumping and spattering of the catalyst on the side walls of the vessel; too large an amount of catalyst results in loss of product by adsorption on the catalyst.

b. The Reaction. Earlier studies were carried out for a period of 10 hr, but it has been shown that 1 hr is sufficient (85). With methyl lithocholate analysis by gas–liquid chromatography (Fig. 8) has shown that the mixture contained about 70% of the 3-keto-5α derivative, 20% of the 3-keto-5β-, and a small amount of the 3-keto-Δ^4 derivative; similar results were obtained with other 5β-cholanoates (85). Since a number of aromatic solvents or their saturated cyclohexyl derivatives can be substituted for *p*-cymene, the solvent does not undergo reduction in the reaction. The reaction proceeds equally well with the 3-hydroxy derivatives, so the 3-ketone need not be used as reactant. Oxygen must be present at C_3, however, for methyl 5β-cholanoate, 7α-hydroxy-, 7α,12α-dihydroxy-, and 12α-hydroxy-7-keto-5β-cholanoates were not converted to their corresponding allocholanoates.

c. Mechanism of the Reaction. A mechanism for the *diaxial* elimination of the 4α,5β hydrogens has been formulated (85) (Fig. 9). Support for this mechanism appears in a report by Inhoffen *et al.* (87) who showed that catalytic reduction of a dienolate such as E afforded 90% 5α-cholestanone

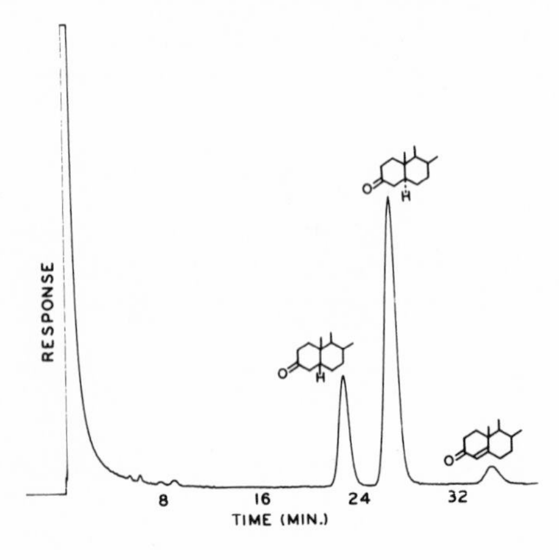

Fig. 8. Transformation of methyl lithocholate by Raney nickel. Gas chromatography of products from the reaction as determined with a Hewlett–Packard Model 402 gas chromatograph fitted with a hydrogen flame detector and a silanized glass U-shaped column (6 ft × 1/4 in o.d.) filled with 3% OV-17 on 100–120 mesh Gas Chrom Q under these conditions: flash heater, 280 °C; column, 260 °C; detector, 280 °C; helium, 40 psi at a flow rate of 80 cc/min (85).

and methyl 12α-acetoxy-3-keto-5α-cholanoate from the enol ethers of the respective 3-keto-Δ^4 derivative. With the 7-hydroxy cholanoates two additional reactions occur: (a) dehydrogenation at C_7 to provide 3,7-diones, or (b) complete loss of the C_7-OH, probably from the dienolate, E. Thus, methyl cholate provides 12α-hydroxy-3,7-diketo- and 12α-hydroxy-3-keto-5α-cholanoate in about equal quantities (Table V). Data in Table V show that the overall yield of desired 3-keto-allo derivative varies from 70 to 15%. The mixture of products, including the 5β derivative, can be separated by chromatography on alumina and preparative layer chromatography on silica gel G.

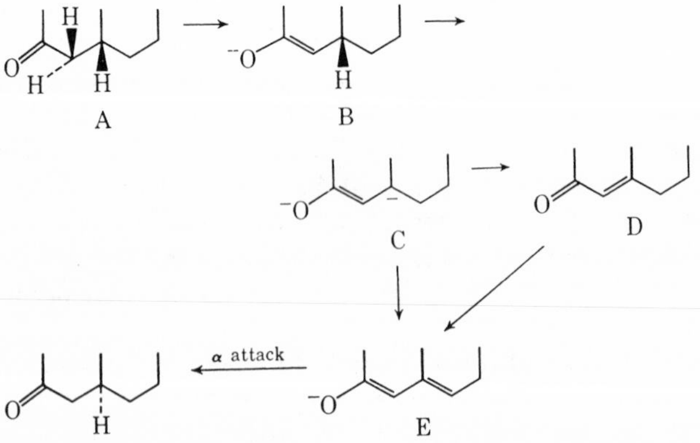

Fig. 9. Mechanism proposed for transformation of 5β steroid to 5α steroid with Raney nickel (85).

TABLE V. **Products of the Raney Nickel Allomerization in *p*-Cymene**

Reactant: Methyl 5β-cholanoate or Δ^4-cholenoate	Product: Methyl 5α-cholanoate	
Substituent	Substituent	Percentage yield
none	none	0
3α		76
3-keto	3-keto	82
3-keto-Δ^4-		76
7α	20% 7-keto-(5β) 80% unchanged	0
3α,12α		63
3-keto-12α	3-keto-12α	50
3-keto-12α-Δ^4		61
3,7α	3-keto-7α	15
	3,7-diketo	18
	3-keto	25
3α,7α,12α	3-keto-7α,12α	15
	3,7-diketo-12α	20
	3-keto-12α	20
3-keto-7α,12α	70% unchanged	0
7-keto-12α	unchanged	0
7α,12α	69% unchanged	—
	31% 7-keto-12α(5β)	—

d. Reduction of the 3-keto-5α Steroids. As predicted (1) reduction of the 3-keto-5α esters with $NaBH_4$ provided a better yield of the *equatorial* 3β-ol than the *axial* 3α-ol by a factor of 3-8. On the other hand, catalytic reduction provided more interesting results. Reduction of 3-keto-5α- or 12α-hydroxy-3-keto-5α-cholanoates in the presence of Adam's catalyst and a small amount of HCl afforded a ratio of 3β-ol to 3α-ol of about 2:1 (Table VI). However, with introduction of a 7α-hydroxyl group this ratio is changed in favor of the *axial* 3α-ol. Thus, about 53% of the 3α-ol is obtained from the 7α-hydroxy-3-keto-5α-cholanoate, and 79% of the 3α-ol from 7α,12α-dihydroxy-3-keto-5α-cholanoate. Hence, the overall yield of methyl allocholate from methyl cholate is about 11%. In catalytic reduction a small but detectable amount of the 3-deoxy derivative accompanies the desired products.

B. Other Allo-Acids

1. Allodeoxycholic Acid

Prior to the identification of β-lagodeoxycholic acid this acid was prepared by Yukawa (35) and by Hofmann and Mosbach (36) from the key intermediate, V, by the Huang-Minlon reduction. Although this reduction is

TABLE VI. Reduction of Methyl 3-Keto-5α-Cholanoates[a]

Additional substituent	Method	Yield 3α-ol	Yield 3β-ol	Ratio 3α:3β	Overall yield
none	Pt–H_2	22	44	0.51	17
12α	Pt–H_2	28	48	0.58	18
	$NaBH_4$	12	72	0.17	7
7α	Pt–H_2	53	36	1.47	8
	$NaBH_4$	17	83	0.20	2.5
7α-AcO[c]	Pt–H_2	46	42	1.10	N.A.[b]
	$NaBH_4$	11	89	0.12	N.A.
7α-BzO[d]	$NaBH_4$	15	85	0.18	N.A.
7α,12α	Pt–H_2	79	12	6.58	12
	$NaBH_4$	20	75	0.27	3

[a] References 131, 170. Overall yield refers to yield of methyl 3α-hydroxy-5α-cholanoate from methyl 3α-hydroxy-5β-cholanoate via Raney nickel transformation and reduction of the resultant 3-keto-5α-cholanoate.

[b] NA=not applicable, since the acetate and benzoate were prepared from the 3-keto-7-hydroxy-5α cholanoate before reduction.

[c] AcO=acetate.

[d] BzO=benzoate.

reported to provide olefins from the ketolic group, none was found in this instance; a yield of about 10% was reported for the final reaction. Danielsson *et al.* (34), Kallner (66), and Mitra and Elliott (67) prepared the methyl ester by the methods described under A, 2 and A, 3.

2. *Allochenodeoxycholic Acid*

Hoshita *et al.* (39) reported the synthesis of this acid from anhydro-5α-cyprinol (26,27-epoxy-3α,7α,12α-trihydroxy-5α-cholestane) (XV) as shown in Fig. 10. Kallner (66) and Ziller, Mitra, and Elliott (68,40) prepared the ester as described under A, 2 and A, 3 respectively.

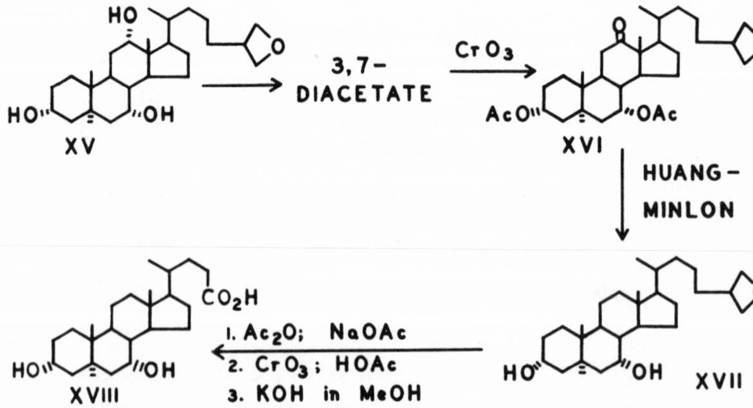

Fig. 10. Hoshita's synthesis of allochenodeoxycholic acid (39).

3. Alloursodeoxycholic Acid

Goto (88) obtained an allo-acid designated as 3α,7β-dihydroxy-5α-cholanoic acid in a manner somewhat analogous to the procedure of Anderson and Haslewood for the preparation of allocholic acid. Methyl 3α-cathyl-7,12-diketo-6α-bromo-5β-cholanoate was allomerized with alkali to provide three ketolic acids. One of these designated as 3α,7β-dihydroxy-6,12-diketo-5α-cholanoic acid, m.p. 235 °C, was reduced by the Wolff–Kishner procedure to provide equal quantities of allocholanic acid and alloursodeoxycholic acid, m.p. 223–224 °C.

*4. Other 3,7-Disubstituted-5α-cholanoates**

In a recent report Nakada *et al.* (89) detailed the preparation of 3,7-dioxo-5α-cholanoic acid (m.p. 185–187 °C) from 3β-hydroxy-7-keto-chol-5-enoic acid obtained from oxidation of 3β-hydroxy-chol-5-enoic acid with CrO_3 and subsequent reduction. Reduction of this material should provide the 3β-epimer of allochenodeoxycholic acid and its 7β-epimer.

5. 3,6-Dihydroxy-5α-cholanoates

Alkaline hydrolysis of fresh hog bile provides access to 3α-hydroxy-6-oxo-5α-cholanoic acid (about 3 g/1.5 liter of bile) (92, 93, 13). Figure 11 summarizes the routes to the four epimeric 3,6-diols from this acid. Kawanami (94) has studied the susceptibility of the hydroxyl groups of this series to NBS and concludes the following: 6β (a)$>$3β(e)$>$6α(e). Corbellini *et al.* (95, 96) and Ziegler (97) have investigated interconversions of isomers at C_3 through the mesylates and tosylates.

6. Monohydroxy 5α-cholanoates

Several of the monohydroxy allocholanoates listed in Table I were prepared by Wieland or Windaus. Removal by the usual methods of a carbonyl

---

*Anderson and Haslewood (19) explored the preparation of these derivatives from 3α-hydroxy-6-keto-5α-cholanoic acid in a manner analogous to that outlined in Fig. 6. The known 3α,7β-dihydroxy-6-keto-5α-cholanoic acid (90) was esterified, and the product converted to a crystalline ethylene thioketal. Treatment of the latter with Raney nickel provided allolithocholic acid and several fractions of unidentified material. From one of these fractions an acid (A) was obtained, m.p. 240–241 °C; $[\alpha]_D+60\pm1°$; from the succeeding fraction an acid (B) was converted to the methyl ester (C), m.p. 156–158 °C; $[\alpha]_D+58\pm1°$. The M_D for (C) (Found $+235$) agrees reasonably well with the calculated value for a 3β,7β-diol ($+190$) or a 3α,7β-diol ($+197$). In a similar sequence of reactions Ziller (91) obtained a methyl ester, m.p. 158-159 °C; RR_t 1.42 on 3% QF-1; free acid, m.p. 240–241 °C. Subsequent experiments have shown that the monohydroxy acids derived from this diol are the 3β- and 7β-ols, suggesting that this acid is the unreported 3β,7β-dihydroxy-5α-cholanoic acid (170).

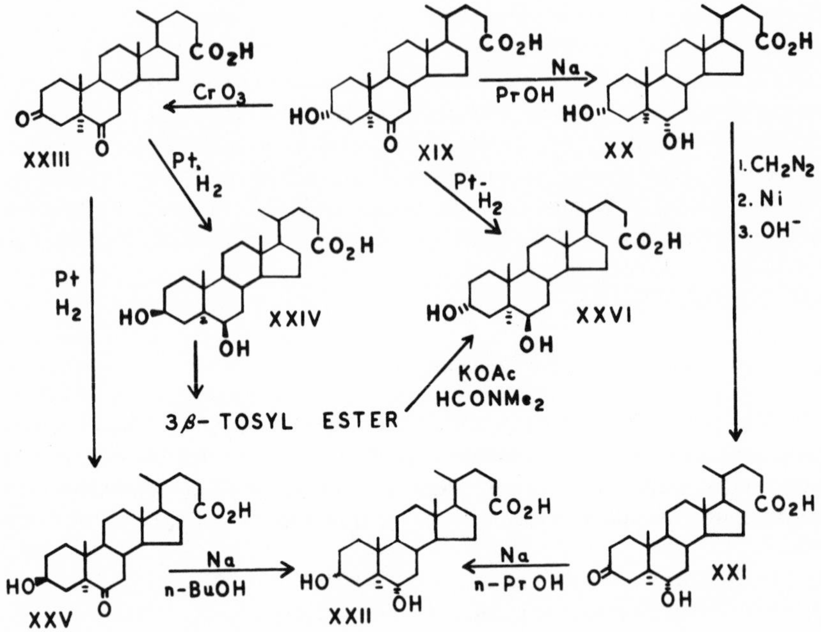

Fig. 11. Preparation of isomeric 3,6-dihydroxy allo bile acids.

group from monohydroxy monoketo compounds has afforded each of the derivatives. Methyl 3β-hydroxy-chol-5-enoate has been reduced catalytically to provide the 3β-hydroxy-5α-cholanoate (98, 99, 41). The allomerization of 6-keto cholanic acid (100) provided a product that could be reduced to the 6-hydroxy allo-acid (12). Kallner (41) has reduced methyl 3-keto-5α-cholanoate with $IrCl_4$ and trimethylphosphite to the 3α-ol in 57% yield with about 6% of the 3β-ol.

C. Conjugated Acids

Glycoallodeoxycholic acid was synthesized (36) by a modification (101) of the method of Norman (102).

VII. BIOSYNTHESIS OF ALLO BILE ACIDS

Three biosynthetic pathways for the elaboration of allo bile acids have been identified. Each appears to be an extension of the metabolism of a C_{27} sterol derived from acetate via mevalonate. The following sections delineate the derivation of allo-acids from cholestanol and cholesterol, from 5β bile acids, and from other 5α sterols. Whereas a comprehensive study of the pathway from cholesterol to the 5β bile acids has demonstrated the importance of liver microsomal mixed function oxidases in nuclear hydroxylations and the mitochondria-supernatant fractions in the oxidation of the side chain, comparable studies in the 5α series have only recently been initiated. The conclusions derived from such studies are likely to contribute significantly to an understanding of the mechanisms of conversion of sterols to bile acids, for several of the problems in the 5β series (i.e., ultimate reduction of the Δ^4 derivatives) are absent in the 5α series.

A. From Cholestanol and Cholesterol

The relationship of cholesterol to cholestanol has now been delineated. Early experiments of Schoenheimer *et al.* (103, 104) and Rosenfeld and Webster (105) resulted in a proposal that cholesterol was metabolized to Δ^4-cholestenone (XXIX), which was reduced to cholestanol or coprostanol (106); indeed, Baker and Greenberg (106) detected ^{14}C-cholestanol in rat feces after administration of radioactive acetate. Anker and Bloch (107) and Stokes *et al.* (108) found efficient conversion of labeled Δ^4-cholestenone to tissue cholestanol in rats. Cholesterol was shown (109) to be a precursor of cholestanol in the adrenals, liver, and intestine of guinea pigs. With cholesterol-4-^{14}C-4β-^{3}H Werbin *et al.* showed that the cholestanol isolated from the adrenals contained as much as 14% of the tritium at positions 5 and 6, whereas the cholestanol obtained from liver and intestine was virtually devoid of tritium at these positions (110). Based on these observations they proposed the following possible pathway:*

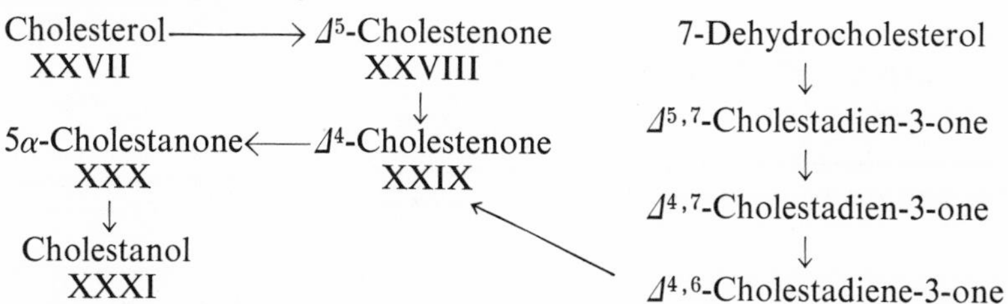

Support for a ketonic intermediate in this conversion was provided by Rosenfeld *et al.* (111, 112) who showed that cholestanol obtained from plasma, red

---

*Kandutsch (142) has also suggested the alternate pathway of biosynthesis of cholestanol from 7-dehydrocholesterol. Although unable to show the conversion of 7-dehydrocholesterol to the 3-keto derivatives, he has demonstrated the conversion of cholesta-4,7-dien-3-one and cholesta-4,6-dien-3-one to Δ^4-cholestenone with the microsomal fraction of mouse liver homogenates.

cells, and feces of man after administration of cholesterol-4-^{14}C-3α-^{3}H lost 80% of the tritium. The conversion of cholesterol to cholestenone (XXIX) was demonstrated (113) with the supernatant fraction of rat liver homogenates fortified with NADPH or NADH; fractions of beef adrenal tissue promoted the isomerization of XXVIII to XXIX (110). The formation of cholestanol from Δ^4-cholestenone (XXIX) was reported *in vivo* (107, 108, 114, 115) and *in vitro** (114); microsomal and mitochondrial preparations of rat liver were equally effective (116). Harold *et al.* (114) isolated cholestanol from rat liver homogenates incubated with 5α-cholestanone (XXX), ATP, and NAD$^+$.

Mosbach and colleagues have contributed extensively to a better understanding of the mechanism of biosynthesis of cholestanol. After intracardial administration of mevalonate-^{14}C to guinea pigs, radioactive cholestanol was isolated from liver, intestinal wall, and adrenals (117). With rat liver homogenates the incorporation of ^{14}C-mevalonate or ^{14}C-cholesterol into ^{14}C-cholestanol of the order of 0.05% in 4 hr (118).† The 5α-reductase of rat liver and adrenal localized (119) in the microsomal fraction, was shown to require NADPH, exhibited product inhibition, and was more active in preparations from female than male rats. The enzyme was considered distinct from the C_{19}- and C_{21}-5α reductases, since it was inhibited by C_{19}-3-keto-Δ^4 steroids.

The 3β hydroxysteroid reductase of rat liver which catalyzes the conversion of XXX to cholestanol (XXXI) was localized also in the microsomal fractions, and shown to provide the epimeric alcohols in a ratio of 10:1 (3β: 3α) in the presence of NADPH. The enzyme was not inhibited by cholestanol, but pronounced inhibition was noted with 7-keto- or 7α-hydroxycholestanol (XXXII) or Δ^5-cholestenone (XXVIII) (120). This enzyme differs from the C_{19}-steroid reductase, since the latter utilize NADH equally well and provides predominantly the 3α-ol.

The origin of allo bile acids from cholestanol is clearly established (121, 36, 28, 122, 40, 123). After separation of two unidentified biliary acids similar to, but not identical with, cholic and chenodeoxycholic acids, Harold *et al.*

---

*Harold *et al.* (114) also showed that the acidic biliary metabolites derived from Δ^4-cholestenone-C^{14} in the bile fistula rat appeared in fractions associated with trihydroxy and dihydroxy bile acids. Recrystallization with carrier cholic acid virtually eliminated the metabolite from this acid. By similar technique the ^{14}C-metabolite in the dihydroxy fractions was shown to differ from deoxycholic and chenodeoxycholic acids; however, in the latter case conversion to the diacetate was necessary to remove the ^{14}C.

† Support for the *in vivo* synthesis of cholestanol from acetate-1-C^{14} and mevalonate-2-C^{14} was also obtained from incubation with isolated arterial segments (143). Chobanian suggested the major pathway of synthesis in this tissue from these precursors does not involve cholesterol as an intermediate.

(124) suggested that these metabolites of cholestanol in the rat may be derivatives of allocholanic acid. Subsequently Karavolas and Elliott showed (121) that three principal radioactive metabolites were present in bile after intracardial administration of 5α-cholestan-3β-ol-4-^{14}C to rats with bile fistulas. Of the chromatographed ^{14}C, 11% was present in fractions associated with chenodeoxycholic acid, 70% with cholic acid, and 7% in more polar fractions. The major metabolite was separated from cholic acid by chromatography on Florisil and by crystallization, and was identified by isotopic dilution as allocholic acid (28). To obtain a larger amount of the metabolite associated with chenodeoxycholic acid, cholestanol-^{14}C was administered intraperitoneally to hyperthyroid rats with bile fistulas. After saponification and separation of the free bile acids (122), 38% of the chromatographed ^{14}C was associated with chenodeoxycholic acid, 48% with cholic acid, and 9% with more polar fractions. The first of these metabolites was separated from known dihydroxy acids by repeated chromatography of the methyl esters, and identified by isotopic dilution as allochenodeoxycholic acid (40). With the availability of synthetic allo-acids (67) the third metabolite was similarly identified as the 3β-isomer of allocholic acid 3β,7α,12α-trihydroxy-5α-cholanoic acid) (125). The presence of several 3β-hydroxy-allo acids in rat bile after intraperitoneal administration of 5α-cholestan-3β-ol-4-^{14}C-3α-^{3}H was confirmed by the identification of tritiated 3β,7α-dihydroxy-5α- and 3β,7α,12α-trihydroxy-5α-cholanoic acids (126).

Hofmann and Mosbach (36) have identified allodeoxycholic acid as the conjugate with glycine in gallstones from rabbits fed a diet of 1% cholestanol. Subsequent experiments have confirmed the origin of allodeoxycholic acid in this species from allocholic acid by intestinal dehydroxylation at C_7 (123); the intestinal anaerobe capable of this reaction in the rabbit has been characterized (38).

If the pathway of biosynthesis of allo bile acids from cholestanol is similar to that proposed for the 5β bile acids from cholesterol, 7α-hydroxylation should be the next step. The microsomal fraction of rat liver properly fortified with NADPH and oxygen was observed to hydroxylate cholestanol and cholesterol to the respective 7α-hydroxy derivatives at equivalent rates (127). With cholestanol the 7α-hydroxylase was inhibited by the product, or the 7β-ol or 7-keto sterol. The rate of hydroxylation could be enhanced by prior treatment of the rats with cholestyramine (a bile acid sequestrant) and/or phenobarbital. A C_5 double bond was not considered a requirement for enzymatic hydroxylation at C_7.

7α-Hydroxy-cholestanol can also be derived from the corresponding 7α-hydroxy-α,β-unsaturated ketone (XXXIV). Björkhem (128) showed that the microsomal reduction of 7α-hydroxycholest-4-en-3-one (XXXIV) to the 5α sterol (XXXV) proceeds via *cis* addition at C_4–C_5, as opposed to the *trans diaxial* addition in the formation of 5β sterols. With tritiated NADPH he demonstrated the selectivity of [4B-^{3}H]NADPH in addition of tritium to the 5α-position; presumably the hydrogen added at the 4α-position is derived from the medium. The addition of NADH to the microsomes catalyzed the reduction of the ketone (XXXV) to the 3β,7α-diol (XXXII).

Administration of 5α-cholestan-3β,7α-diol-4-^{14}C-3α-^{3}H to rats with bile fistulas has confirmed the conversion of this diol (XXXII) to allocholic, and allochenodeoxycholic acids and their 3β-isomers (126). The ratio of di- to trihydroxy allo-acids derived from the above diol was similar to that obtained from cholestanol (28). Since the ratio of ^{3}H to ^{14}C of the 3β-hydroxy allo-acids was the same as the administered diol, a 3-keto intermediate is not an obligatory prerequisite for oxidation of the side chain.

The C_{27} sterol, 7α,12α-dihydroxycholest-4-en-3-one (XXXVI), is an important intermediate in the pathway from cholesterol to cholic acid (129). Since allocholic acid is the major biliary acid of iguana bile, Hoshita *et al.* (31) incubated this sterol with microsomes of iguana liver and NADPH to ascertain whether 5α sterols would be formed. The principal product was 3α,7α,12α-trihydroxy-5α-cholestane (XXXVIII) although the enzyme system was partially inhibited by the intermediate ketone (XXXVII).*

12α-Hydroxylation is believed (130) to follow 7α-hydroxylation in the sequence to cholic acid. In a study of the mechanism of the *in vivo* conversion of tritiated 12α-hydroxycholest-4-en-3-one to bile acids, Einarsson (130) noted that deoxycholic and cholic acids were the principal metabolites in the bile fistula rat, but minor fractions behaved chromatographically like the corresponding allo-acids. In a similar experiment in the bile fistula rabbit a minor fraction behaved chromatographically like allodeoxycholic acid.

Recent investigations (131) have shown the *in vivo* conversion of tritiated samples of 5α-cholestane-3,26,-diol (XXXIX) and 5α-cholestane-3β,7α,26-triol (XL) to allocholic and allochenodeoxycholic acids and their 3β-isomers in the bile fistula rat. Small amounts of allolithocholic acid and its 3β-isomer were also detected from the above diol. A summary of the current status of knowledge in these metabolic transformations is given in Fig. 12.

B. From 5β Bile Acids

During studies on the metabolism of deoxycholic acid in the rabbit (34) a new acid was isolated from rabbit feces, whose methyl ester was eluted from alumina slightly before methyl deoxycholate, showed a mass spec-

---

*Unreported experiments (128) show that 7α-hydroxycholest-4-en-3-one (XXXIV) and 7α,12α-dihydroxycholest-4-en-3-one (XXXVI) are converted into the corresponding 3-keto-5α sterols and to the 5α-cholanic acids in the presence of NADPH and the microsomal fraction of rat liver (preferably female rats).

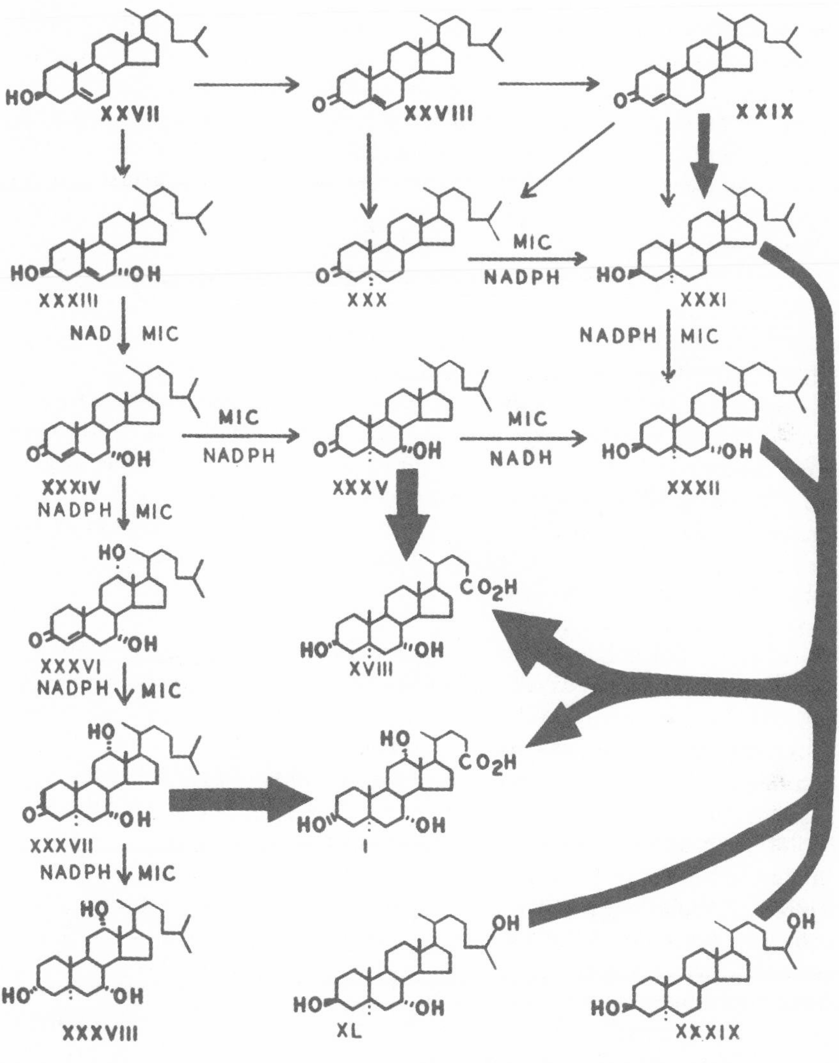

Fig. 12. Metabolic pathways to allo bile acids. ⬅ Bile fistula rat.

trum quite similar to methyl deoxycholate, and provided an oxidized product with a retention time on QF-1 of 1.18 relative to methyl 3,12-diketo-5β-cholanoate (similar to the relationship observed between 5α-cholest-3-one and 5β-cholestan-3-one). The unknown material was identified satisfactorily as allodeoxycholic acid, but the *in vivo* conversion of deoxycholic to allodeoxycholic acid in 70% yield remained unexplained. Kallner (55) continued the study and showed that radioactive allodeoxycholic acid was obtained from bile and feces after intracecal injection of radioactive deoxycholic acid, although no allo-acids were found in bile after intraperitoneal administration. On the other hand, administration of allodeoxycholic acid to the rat by either of these procedures provided allodeoxycholic and allocholic acids, thus demonstrating the conversion of the dihydroxy allo-acid to the trihydroxy allo-acid via liver enzymes, in a manner analogous to conversion of the 5β series.

Since the conversion of the 5β to the 5α acid appeared to be catalyzed by intestinal organisms, Kallner (45) utilized 3β-^{3}H-24-^{14}C-deoxycholic acid to ascertain whether a 3-ketone may be an intermediate. The doubly labeled acid was converted to the known bile acids with extensive loss of tritium after intracecal administration; although biliary deoxycholic and cholic acids retained 23 and 11%, respectively, of tritium, virtually no tritium was found in biliary or fecal allodeoxycholic, or fecal deoxycholic acids. Since these experiments suggested that deoxycholic acid is rapidly equilibrated in the intact rat with 12α-hydroxy-3-keto-5β-cholanoic acid, Kallner (45) administered tritiated 12α-hydroxy-3-keto-5β-cholanoic acid and its 5α-isomer intracecally to bile fistula rats; each of the keto acids provided deoxycholic and allodeoxycholic acids. The role of the 3-keto derivative in the conversion of 5β to 5α acids was amplified by the observation that 12α-hydroxy-3-keto-chol-4-enoic acid (XXXV) also served as an intermediate in the formation of deoxycholic and allodeoxycholic acids. After intracecal injection the distribution in the dihydroxy fraction of fecal bile acids was the following: deoxycholic and its 3-keto derivative were converted to allodeoxycholic acid in about 10–20% yield; allodeoxycholic acid and its 3-keto derivative were converted to deoxycholic acid in a yield of about 50%; 12α-hydroxy-3-keto-chol-4-enoic acid provided more deoxycholic acid (65%) than allodeoxycholic acid (35%). Finally, after intraperitoneal administration of 12α-hydroxy-3-ketochol-4-enoic acid, 55% of the deoxycholic and 45% of the allodeoxycholic acids were obtained from bile. Thus, liver enzymes are capable of reduction of the α,β-unsaturated keto acids to the 5β and the 5α series as has been shown for the comparable sterols (Fig. 12). Kallner (56) showed that incubation of 3-keto-Δ^{4}-C_{24} acids with the microsomal fraction and NADPH provided the corresponding 3-keto-5α-acids, i.e., 12α-hydroxy-3-ketochol-4-enoic acid or 7α-hydroxy-3-ketochol-4-enoic acid afforded 12α-hydroxy-3-keto-5α-cholanoic acid and 7α-hydroxy-3-keto-5α-cholanoic acid, respectively. However, incubation of 7α,12α-dihydroxychol-4-enoic acid with microsomes and NADPH or NADH failed to provide the corresponding 5α acid. Incubation of either of these three α,β-unsaturated keto acids with the 100,000g supernatant fraction provided the respective 5β acids. On the other hand, incubation of the 3-keto-5α-acids with the 100,000g supernatant fraction and NADPH or NADH was more efficient in formation of the 3α-

hydroxy allo-acids than similar incubation with fortified microsomal preparations (56).

Support for the concept of an unsaturated intermediate in the formation of allo-acids is provided by recent experiments of Yamasaki *et al.* (98, 89). After administration of 3-ketochol-4-enoic-24-^{14}C acid to rats and examination of the biliary metabolites, all four isomers of 3-hydroxycholanoic acid were identified; other di- and trihydroxy acids were not investigated. Of the four possible 3-hydroxy-isomers about twice as much lithocholate was present as each of the other isomers. Similar results were obtained following administration of 3β-acetoxychol-5-enoic-24-^{14}C acid; in addition, 3ξ,6ξ-dihydroxy-5α-cholanoic acids were obtained. Yamasaki *et al.* (89) propose that a 3β-dehydrogenase converts the 3β-hydroxy-Δ^5-cholenoic acid to the α,β-unsaturated ketone from which both 5β and 5α acids are derived, whereas hydroxylation of the above acid provides the diol from which only 5β acids are produced, somewhat analogous to the scheme of metabolism proposed by Mitropoulos and Myant (132) for the formation of chenodeoxycholic acid and the muricholic acids.

C. From Other 5*α* Sterols*

Allocholic acid occurs as a high proportion of the bile acids in certain lizards and to a lesser extent in various fishes and birds (25). A mechanism for the formation of the allo-acids from the 5α alcohols indigenous to these species has been investigated by Hoshita (58) and Amimoto (43). The bile salts in these animals are the sulfates of 5β and 5α bile alcohols (25). Amimoto (50) studied the metabolism of tritiated 5α-cyprinol (3α,7α,12α, 26,27-pentahydroxy-5α-cholestane-12β-^{3}H) in a giant salamander, and isolated two acidic biliary metabolites, allocholic acid, and an acid tentatively designated as 3α,7α,12α,26-tetrahydroxy-5α-cholestanoic acid. Since only 6% of the biliary radioactivity was accounted for as allocholic acid and the main portion of the biliary radioactivity was recovered in anhydro-5α-cyprinol (an artifact from 5α-cyprinol sulfate by alkaline hydrolysis), 5α-cyprinol is an inefficient precursor of allocholic acid; he suggested that 5α-cyprinol may be esterified to the sulfate but the ester is not readily metabolized. To test this hypothesis, randomly tritiated 27-deoxy-5α-cyprinol (3α,7α,12α,26-tetrahydroxy-5α-cholestane) was injected intraperitoneally into a giant salamander and the biliary constituents examined. Unfortunately only 7.2% of the tritium was recovered in bile; allocholic acid accounted for 28% of the biliary ^{3}H (a large portion of the ^{3}H of the substrate may have been concentrated in the

*For a thorough review of the bile alcohols see (168).

terminal end of the side chain). A small amount of ^{3}H was retained in an acid assumed to be 3α,7α,12α-trihydroxy-5α-cholestanoic acid by virtue of the identification of 27-deoxy-5α-cyprinol after reduction of the methyl ester with $LiAlH_4$.

To ascertain whether cholesterol is a precursor of 5α-cyprinol, Hoshita (133) administered ^{14}C-cholesterol intraperitoneally to carp, in which 5α-cyprinol is the principal biliary sterol (134). After 12 days bile was collected from the gallbladder, and radioactive 5α-cyprinol corresponding to 0.14% of the administered sterol was recovered from bile. To ascertain whether cholestanol or 7α-hydroxycholesterol was a precursor of 5α-cyprinol in this species Hoshita (58) administered these tritiated sterols intraperitoneally to carp and studied the distribution of ^{3}H in biliary constituents. From cholestanol-5,6-^{3}H, 1.1% of the biliary ^{3}H remained in the neutral fraction, of which 5α-cyprinol and 27-deoxy-5α-cyprinol were the major and minor constituents (~22:1). Allocholic acid was the only radioactive component identified in the acid fraction. Two weeks after similar injection of 7α-hydroxycholesterol-7β-^{3}H into carp, 17.6% of the radioactivity was recovered in bile, 86.5% of which was present in the neutral fraction and 6.3% in the acid fraction. 5α-Cyprinol and 27-deoxy-5α-cyprinol (~9:1) retained most of the tritium in the former fraction, whereas allocholic and cholic acids each retained ^{3}H (~1.5:1) in the acid fraction. Thus, carp liver enzymes convert 7α-hydroxycholesterol to 5α sterols and to the 5β- and 5α-C_{24} acids. Normally carp bile contains more cholic than allocholic acid.

Hoshita (135) subsequently showed that the major portion of the 4β hydrogen of cholesterol was lost in conversion to 5α bile alcohols in the carp, and suggested that this was due to isomerization of the double bond from the C_5 to the C_4 position.

Masui *et al.* (136) have shown that carp liver microsomes contain the necessary 5α-reductase and a 26-hydroxylase to convert 7α,12α-dihydroxycholest-4-en-3-one (XXXVI) to 27-deoxy-5α-cyprinol when fortified with NADPH, and that a mitochondrial enzyme system requiring NADPH converts the latter 5α-sterol to 5α-cyprinol. They suggested that the microsomal system hydroxylated one of the terminal methyl groups in a stereospecific form (25β-form), and the mitochondrial system hydroxylated the 25α-form.

In relation to the pathway of metabolism of cholesterol to the 12-deoxy bile acids as proposed by Mitropoulos and Myant (132), the metabolism of the antiatherogenic cholestane-3β,5α,6β-triol-4-^{14}C in the rat is of particular interest. Most of the fecal or biliary radioactivity was associated with the bile acid fraction. The major biliary metabolite was identified as 3β,5α,6β-trihydroxy-cholanic acid (137).

VIII. METABOLIC TRANSFORMATIONS*

A. Allolithocholic Acid

Like its 5β-isomer this acid undergoes 7α-hydroxylation in the liver to provide allochenodeoxycholic acid (41). Several other metabolites have not been characterized. Following intracecal administration lithocholic, allochenodeoxycholic and its 5β-isomer have been identified in bile.

B. Allochenodeoxycholic Acid

In contrast to its 5β-isomer this dihydroxy acid is not metabolized extensively after intraperitoneal administration (41, 171). From experiments with cholestanol-^{14}C in the hyperthyroid rat (122) proportionately more β-muricholate was found in bile than in the normal rat, but less than 2% of the chromatographed biliary ^{14}C was found in these fractions, suggesting that the 5α-cholestane (or 5α-cholane) nucleus was not metabolized in rat liver as extensively as the 5β ring system (40). Kallner (41) suggested that the minor unidentified metabolites derived from allochenodeoxycholic acid are related to the muricholic acids derived in the rat from chenodeoxycholic and lithocholic acids (138, 139, 140). Less than 5% of the chromatographed ^{14}C is present in these fractions; the major metabolite is allocholic acid and a minor constituent may be allohyocholic acid (171).

C. Allodeoxycholic Acid

In the rat this acid is converted to allocholic acid (55); in the intestine it is regenerated from allocholic acid (41, 123). After intracecal administration 34% of the biliary dihydroxy acid and 47% of the fecal dihydroxy acid is deoxycholic acid (45).

D. Allocholic Acid

Like cholic acid this allo-acid is not metabolized by the rat on passage through the liver; in the intestine it is converted to cholate, allodeoxycholate, and deoxycholate (41, 123).

---

*The transformations of the 3β-hydroxy-5α acids will not be reviewed here, since little information is available. Adequate evidence for an active 3β-hydroxysteroid dehydrogenase of rat liver is at hand (165, 166, 167).

E. 3β,6β-Dihydroxyallocholanoic Acid

After intraperitoneal administration to the rat, 90% of this acid is recovered unchanged in bile (141); 0.6% was present in monohydroxy acids of which 60% was identified as 3β-hydroxy allocholanoic acid. About 9% was converted to unidentified trihydroxy acids. *In vitro* experiments with liver homogenates indicate less extensive metabolism of this dihydroxy allo-acid; in this case allolithocholate appears to be formed instead of the 3β-isomer.

F. 3β-Acetoxychol-5-enoic Acid

Although this derivative is not an allo-acid, recent studies (89) with the radioactive acid have shown that 5β and 5α acids are present in rat bile; thus, allolithocholic and its 3β-isomer and 3ξ,6ξ-dihydroxy-5α-cholanoic acids were obtained.

IX. SUMMARY

Although this class of substances has observed a fiftieth anniversary in man's knowledge of bile acids, the subject has not sparked biochemical interest until recent times, when, aided by modern techniques, the biochemist has been able to study smaller quantities of materials at the subcellular level. It seems likely that this is the level where new knowledge of the biological importance of allo bile acids will be generated. Where there are current explanations for the roles of the 5β bile acids in nature, these answers have not yet appeared for the allo-acids, partially because of the newness of the subject and the paucity of materials. Until the role of the allo-acids in nature is fully detailed, investigations in this area will continue.

ACKNOWLEDGMENTS

The author is pleased to acknowledge the contributions of former and present colleagues, particularly Drs. M.N. Mitra, S.A. Ziller, Jr., H.J. Karavolas, B.W. Noll, E.A. Doisy, Jr., Messrs. W. Sweet, M.A. Thorne, H. Robinson, Rev. L.B. Walsh, and Misses M.M. Mui and P.A. Roberts.

REFERENCES

1. L. F. Fieser and M. Fieser, "Steroids," Reinhold, New York (1959).
2. H. Wieland and F. J. Weil, *Z. Physiol. Chem.* **80,** 287 (1912).
3. A. Windaus and K. Neukirchen, *Ber.* **52,** 1915 (1919).
4. A. Stoll, A. Hofmann, and A. Helfenstein, *Helv. Chim. Acta* **18,** 643 (1935).
5. A. Windaus and A. Bohne, *Ann.* **433,** 278 (1923).
6. H. Wieland, E. Dane, and C. Martius, *Z. Physiol. Chem.* **215,** 15 (1933).
7. L. Ruzicka, *Helv. Chim. Acta* **18,** 68 (1935).
8. A. Windaus, *Ann.* **447,** 233 (1926).
9. M. Mizuguchi, *Proc. Japan Acad.* **30,** 209 (1954).
10. R. G. Curtis and H. Silberman, *J. Chem. Soc.* **1952,** 1187.
11. J. Jacques, H. Kagan, and G. Ourisson, "Selected Constants. Optical Rotatory Power. Ia. Steroids" (S. Allard, ed.), Pergamon Press, Oxford (1965).
12. H. Wieland and E. Dane, *Z. Physiol. Chem.* **212,** 41 (1932).
13. I. Ushizawa, S. Takani, and K. Yamasaki, *Yonago Acta Medica* **2,** 50 (1957); *cited in C. A.* **51,** 14679 (1957).
14. K. Ohta, *Z. Physiol. Chem.* **259,** 53 (1939).
15. H. Isaka, *Z. Physiol. Chem.* **266,** 117 (1940).
16. K. Yamasaki, *J. Biochem. (Tokyo)* **38,** 93 (1951).
17. G. A. D. Haslewood and I. D. P. Wootton, *Biochem. J.* **63,** 3p (1956).
18. I. G. Anderson, G. A. D. Haslewood, and I. D. P. Wootton, *Biochem. J.* **67,** 323 (1957).
19. I. G. Anderson and G. A. D. Haslewood, *Biochem. J.* **85,** 236 (1962).
19a. I. G. Anderson and G. A. D. Haslewood, *Biochem. J.* **81,** 15p (1961).
20. G. A. D. Haslewood, *Biochem. J.* **78,** 352 (1961).
21. G. A. D. Haslewood, *in* "The Biliary System" (W. Taylor, ed.), p. 106, F. A. Davis, Philadelphia (1965).
22. K. Amimoto, *J. Biochem. (Tokyo)* **59,** 340 (1966).
23. G. A. D. Haslewood, *Ann. N. Y. Acad. Sci.* **90,** 877 (1960).
24. T. Sasaki, *J. Biochem. (Tokyo)* **60,** 56 (1966).
25. G. A. D. Haslewood, "Bile Salts," Methuen and Co., London (1967).
26. A. R. Tammer, *Biochem. J.* **98,** 25 (1966).
27. P. Eneroth, B. Gordon, and J. Sjövall, *J. Lipid Res.* **7,** 524 (1966).
28. H. J. Karavolas, W. H. Elliott, S. L. Hsia, E. A. Doisy, Jr., J. T. Matschiner, S. A. Thayer, and E. A. Doisy, *J. Biol. Chem.* **240** (4), 1568 (1965).
29. H. J. Karavolas and W. H. Elliott, *in* "The Biliary System" (W. Taylor, ed.), p. 175, F. A. Davis, Philadelphia (1965).
30. B. W. Noll and W. H. Elliott, *Fed. Proc.* **26,** 851 (1967).
31. T. Hoshita, S. Shefer, and E. H. Mosbach, *J. Lipid Res.* **9,** 237 (1968).
32. S. Kishi, *Z. Physiol. Chem.* **238,** 210 (1936).
33. K. Matumoto, *J. Biochem. Japan* **36,** 183 (1944).
34. H. Danielsson, A. Kallner, and J. Sjövall, *J. Biol. Chem.* **238,** 3846 (1963).
35. M. Yukawa, *Hiroshima J. Med. Sci.* **11,** 167 (1962).
36. A. F. Hofmann and E. H. Mosbach, *J. Biol. Chem.* **239,** 2813 (1964).
37. A. F. Hofmann, V. Bokkenheuser, R. L. Hirsch, and E. H. Mosbach, *J. Lipid Res.* **9,** 244 (1968).
38. V. Bokkenheuser, T. Hoshita, and E. H. Mosbach, *J. Lipid Res.* **10,** 421 (1969).
39. T. Hoshita, K. Amimoto, T. Nakagawa, and T. Kazuno, *J. Biochem.* **61,** 750 (1967).
40. S. A. Ziller, Jr., E. A. Doisy, Jr., and W. H. Elliott, *J. Biol. Chem.* **243**(20), 5280 (1968).

41. A. Kallner, *Arkiv Kem.* **26,** 567 (1967).
42. E. Fernholz, *Z. Physiol. Chem.* **232,** 202 (1935).
43. K. Amimoto, *Hiroshima J. Med. Sci.* **15,** 225 (1966).
44. K. Okuda, M. G. Horning, and E. C. Horning, "Proc. 7th Intern. Congr. Biochem.," Tokyo, Vol. IV, Abstracts, p. 721, Sci. Council Japan, Tokyo (1967).
45. A. Kallner, *Acta Chem. Scand.* **21,** 315 (1967).
46. P. J. Thomas, S. L. Hsia, J. T. Matschiner, S. A. Thayer, W. H. Elliott, E. A. Doisy, Jr., and E. A. Doisy, *J. Biol. Chem.* **240,** 1059 (1965).
47. A. Kallner, *Opuscula Medica,* **Suppl. IV,** 11 (1967).
48. A. Norman and J. Sjövall, *J. Biol. Chem.* **233,** 872 (1958).
49. J. T. Matschiner, T. A. Mahowald, W. H. Elliott, E. A. Doisy, Jr., S. L. Hsia, and E. A. Doisy, *J. Biol. Chem.* **225,** 771 (1957).
50. K. Amimoto, *Hiroshima J. Med. Sci.* **15,** 213 (1966).
51. P. Eneroth, *in* "Lipid Chromatographic Analysis" (G. V. Marinetti, ed.), Vol. 2, p. 149, Marcel Dekker, New York (1969).
52. P. Eneroth, *J. Lipid Res.* **4,** 11 (1963).
53. T. Sasaki, *Hiroshima J. Med. Sci.* **14,** 85 (1965).
54. K. Takeda, T. Komeno, and K. Igarashi, *Pharm. Bull. (Tokyo)* **2,** 352 (1954).
55. A. Kallner, *Acta Chem. Scand.* **21,** 87 (1967).
56. A. Kallner, *Arkiv Kem.* **26,** 553 (1967).
57. T. Masui, K. Amimoto, and T. Kazuno, *J. Biochem. (Tokyo)* **62,** 495 (1967).
58. T. Hoshita, *J. Biochem.* **61,** 440 (1967).
59. F. Hodosan and A. Pop-Gocan, *Rev. Roumaine Chim.* **9**(8–9), 523 (1964); *cited in C. A.* **62,** 14773h (1965).
60. W. H. Elliott, L. B. Walsh, M. M. Mui, M. A. Thorne, and C. M. Siegfried, *J. Chromatog.* **44,** 452 (1969).
61. M. Makita and W. W. Wells, *Anal. Biochem.* **5,** 523 (1963).
62. P. Eneroth and J. Sjövall, *in* "Methods in Enzymology" (S. P. Colowick and N. O. Kaplan, ed.), Vol. 15, p. 237, Academic Press, New York (1969).
63. K. Tsuda, Y. Sato, N. Ikekawa, S. Tanaka, H. Higashikuye, and R. Ohsawa, *Chem. Pharm. Bull.* **13,** 720 (1965).
64. A. F. Hofmann and D. M. Small, *Ann. Rev. Med.* **18,** 333 (1967).
65. T. Hoshita, T. Sasaki, and T. Kazuno, *Steroids* **5,** 241 (1965).
66. A. Kallner, *Acta Chem. Scand.* **21,** 322 (1967).
67. M. N. Mitra and W. H. Elliott, *J. Org. Chem.* **33,** 175 (1968).
68. S. A. Ziller, Jr., M. N. Mitra, and W. H. Elliott, *Chem. Ind.* **1967,** 999.
69. G. A. D. Haslewood and I. G. Anderson, *Biochem. J.* **93,** 34 (1964).
70. T. Masui, *J. Biochem.* **54,** 41 (1963).
71. M. N. Mitra and W. H. Elliott, *J. Org. Chem.* **33,** 2814 (1968).
72. K. Bieman and J. Seibl, *J. Am. Chem. Soc.* **31,** 3149 (1959).
73. V. I. Zaretskii, N. S. Wulfson, V. G. Zalkin, L. M. Kogan, N. E. Voishvillo, and I. V. Torgov, *Tetrahedron* **22,** 1399 (1966).
74. H. Egger and G. Spiteller, *Monatsh. Chem. Verwandte Tiele Anderer Wiss.* **97,** 579 (1966).
75. W. H. Elliott, *in* "Biochemical Applications of Mass Spectrometry," Chapter 11, John Wiley, New York (1971).
76. J. Sjövall, *in* "Gas-Liquid Chromatography of Steroids," Symposium Society of Endocrinol., Glasgow (1966) (J. K. Grant, ed.), p. 243, Cambridge Univ. Press, London (1967).
77. J. G. Allen, G. H. Thomas, C. J. W. Brooks, and B. A. Knights, *Steroids* **13,** 133 (1969).
78. H. Budzikiewicz and C. Djerassi, *J. Am. Chem. Soc.* **84,** 1430 (1962).
79. I. G. Anderson and G. A. D. Haslewood, *Biochem. J.* **74,** 37p (1960).
80. O. Wintersteiner and M. Moore, *J. Am. Chem. Soc.* **65,** 1503 (1943).
81. Y. M. Y. Haddad, H. B. Henbest, J. Husbands, and T. R. B. Mitchell, *Proc. Chem.*

Soc. **1964,** 361.
82. P. A. Browne and D. N. Kirk, *J. Chem. Soc. C* **1969,** 1653.
83. B. Riegel and A. V. McIntosh, Jr., *J. Am. Chem. Soc.* **66,** 1099 (1944).
84. D. Chakravarti, R. N. Chakravarti, and M. N. Mitra, *Nature* **193,** 1071 (1962).
85. M. N. Mitra and W. H. Elliott, *J. Org. Chem.* **34,** 2170 (1969).
86. R. Mozingo, *in* "Organic Syntheses," Coll. Vol. III, p. 181, John Wiley, New York (1955).
87. H. H. Inhoffen, G. Stoeck, G. Kolling, and U. Stoeck, *Ann.* **568,** 52 (1950).
88. T. Goto, *Proc. Japan Acad.* **31,** 466 (1955); *cited in C. A.* **50,** 11357 (1956).
89. F. Nakada, R. Oshio, S. Sasaki, H. Yamasaki, N. Yamaga, and K. Yamasaki, *J. Biochem. (Tokyo)* **64,** 495 (1968).
90. P. Ziegler, *Can. J. Chem.* **34,** 1528 (1956).
91. S. A. Ziller, Jr., "Metabolism of 5α-Cholestan-3β-ol-^{14}C in the Hyperthyroid Rat," Ph.D. dissertation, p. 121, St. Louis University (1967).
92. A. C. Maehly, ed., "Biochemical Preparations," Vol. II, p. 58, John Wiley, New York (1966).
93. G. Lehmann, J. Tepper, and G. Hilgetag, *Biochem. Z.* **340**(1), 75 (1964); *cited in C. A.* **61,** 10916 (1964).
94. J. Kawanami, *Bull. Chem. Soc. Japan* **34,** 671 (1961); *cited in C. A.* **56,** 11662a (1962).
95. A. Corbellini, G. Nathanson, V. Gurdjan, and O. Cerri, *Rend. Ist. Lombardo Sci.* Pt. 1, **91,** 147 (1957); *cited in C. A.* **52,** 11878a (1958).
96. A. Corbellini and G. Nathanson, *Gazz. Chem. Ital.* **86,** 1240 (1956); *cited in C. A.* **53,** 2280f (1959).
97. P. Ziegler, *Can. J. Chem.* **37,** 1004 (1959).
98. M. Ogura and K. Yamasaki, *Steroids* **9,** 607 (1967).
99. I. Ushizawa, *Yonago Igaku Zasshi* **11,** 14 (1960); *cited in C. A.* **55,** 6534i (1961).
100. W. M. Hoehn, J. Linsk, and R. B. Moffett, *J. Am. Chem. Soc.* **68,** 1855 (1946).
101. A. F. Hofmann, *Acta Chem. Scand.* **17,** 173 (1963).
102. A. Norman, *Arkiv Kem.* **8,** 331 (1955).
103. R. Schoenheimer, D. Rittenberg, and M. Graff, *J. Biol. Chem.* **111,** 163 (1935).
104. R. Schoenheimer, D. Rittenberg, and M. Graff, *J. Biol. Chem.* **111,** 183 (1935).
105. R. S. Rosenfeld and T. A. Webster, *Biochem. J.* **37,** 513 (1943).
106. G. G. Baker and D. M. Greenberg, *Cancer Res.* **9,** 701 (1949).
107. H. S. Anker and K. Bloch, *J. Biol. Chem.* **178,** 971 (1949).
108. W. M. Stokes, W. A. Fish, and F. C. Hickey, *J. Biol. Chem.* **213,** 325 (1955).
109. H. Werbin, I. L. Chaikoff, and M. R. Imada, *J. Biol. Chem.* **237,** 2072 (1962).
110. H. Werbin, I. L. Chaikoff, and B. P. Phillips, *Biochem.* **3,** 1558 (1964).
111. R. S. Rosenfeld and I. Paul, *Fed. Proc.* **24,** 661 (1965).
112. R. S. Rosenfeld, B. Zumoff, and L. Hellman, *J. Lipid Res.* **8,** 16 (1967).
113. K. Yamasaki, F. Noda, and K. Shimizu, *J. Biochem.* **46,** 747 (1959).
114. F. M. Harold, S. Abraham, and I. L. Chaikoff, *J. Biol. Chem.* **221,** 435 (1956).
115. D. D. Chapman and I. L. Chaikoff, *J. Biol. Chem.* **234,** 273 (1959).
116. E. G. Tombropoulos, H. Werbin, and I. L. Chaikoff, *Proc. Soc. Exptl. Biol. Med.* **110,** 331 (1962).
117. S. Shefer, S. Milch, and E. H. Mosbach, *J. Biol. Chem.* **239,** 1731 (1964).
118. S. Shefer, S. Milch, and E. H. Mosbach, *J. Lipid Res.* **6,** 33 (1965).
119. S. Shefer, S. Hauser, and E. H. Mosbach, *J. Biol. Chem.* **241,** 946 (1966).
120. S. Shefer, S. Hauser, and E. H. Mosbach, *J. Lipid Res.* **7,** 763 (1966).
121. H. J. Karavolas and W. H. Elliott, *Fed. Proc.* **22,** 591 (1963).
122. S. A. Ziller, Jr., and W. H. Elliott, *Fed. Proc.* **25,** 221 (1966).
123. A. F. Hofmann, E. H. Mosbach, and C. C. Sweeley, *Biochem. Biophys. Acta* **176,** 204 (1969).
124. F. M. Harold, D. D. Chapman, and I. L. Chaikoff, *J. Biol. Chem.* **224,** 609 (1957).
125. W. H. Elliott and S. A. Ziller, Jr., *Fed. Proc.* **27,** 821 (1968).
126. B. W. Noll and W. H. Elliott, *Fed. Proc.* **28,** 884 (1969).

127. S. Shefer, S. Hauser, and E. H. Mosbach, *J. Lipid Res.* **9,** 328 (1968).
128. J. Björkhem, *Eur. J. Biochem.* **8,** 345 (1969).
129. O. Berséus, H. Danielsson, and A. Kallner, *J. Biol. Chem.* **240,** 2396 (1965).
130. K. Einarsson, *Eur. J. Biochem.* **6,** 299 (1968).
131. B. W. Noll and W. H. Elliott, to be published.
132. K. A. Mitropoulos and N. B. Myant, *Biochem. J.* **103,** 472 (1967).
133. T. Hoshita, *Steroids* **3,** 523 (1964).
134. T. Hoshita, *J. Biochem. (Tokyo)* **52,** 125 (1962).
135. T. Hoshita, *J. Biochem. (Tokyo)* **61,** 633 (1967).
136. T. Masui, Y. Hijikato, and T. Kazuno, *J. Biochem. (Tokyo)* **62,** 279 (1967).
137. S. Kikuchi, Y. Imai, Sukuoki-Ziro, T. Matsuo, and S. Noguchi, *J. Pharmacol. Exp. Therap.* **159,** 399 (1968).
138. T. A. Mahowald, J. T. Matschiner, S. L. Hsia, R. Richter, E. A. Doisy, Jr., W. H. Elliott, and E. A. Doisy, *J. Biol. Chem.* **225,** 781 (1957).
139. T. A. Mahowald, M. W. Yin, J. T. Matschiner, S. L. Hsia, E. A. Doisy, Jr., W. H. Elliott, and E. A. Doisy, *J. Biol. Chem.* **230,** 581 (1958).
140. P. J. Thomas, S. L. Hsia, J. T. Matschiner, E. A. Doisy, Jr., W. H. Elliott, S. A. Thayer, and E. A. Doisy, *J. Biol. Chem.* **239,** 102 (1964).
141. M. Shimao, *Yonago Igaku Zasshi* **15,** 35 (1964); *cited in C. A.* **62,** 7014 (1965).
142. A. A. Kandutsch, *J. Lipid Res.* **4,** 179 (1963).
143. A. V. Chobanian, *J. Clin. Invest.* **47,** 595 (1968).
144. O. Dalmer, F. Von Werder, H. Honigmann, and K. Heyns, *Ber.* **68,** 1814 (1935).
145. I. Ushizawa, I. Yamane, and K. Yamasaki, *Proc. Japan Acad.* **33,** 159 (1957); *cited in C. A.* **52,** 10127 (1958).
146. L. Ruzicka, M. Oberlin, H. Wirz, and S. Meyer, *Helv. Chim. Acta* **20,** 1283 (1937).
147. A. Stoll and J. Renz, *Helv. Chim. Acta* **24,** 1380 (1941).
148. E. Fernholz and P. N. Chakravorty, *Ber.* **67,** 2021 (1934).
149. O. Stange, *Z. Physiol. Chem.* **220,** 34 (1933).
150. F. Hodosan, A. Pop-Gocan, and N. Serban, *Rev. Roumaine Chim.* **10** (1), 7 (1965); *cited in C. A.* **63,** 10013e (1965).
151. R. Justoni and R. Pessina, *Il Farmaco (Pavio), Ed. Sci.* **11,** 72 (1956); *cited in C. A.* **50,** 15562a (1956).
152. A. Schubert and C. Damker, *J. Prakt. Chem.* **4,** 260 (1957).
153. J. Kawanami, *Bull. Chem. Soc. Japan* **34,** 509 (1961); *cited in C. A.* **56,** 7389g (1962).
154. G. A. D. Haslewood, *Biochem. J.* **70,** 551 (1958).
155. K. Matumoto, *J. Biochem. (Tokyo)* **36,** 173 (1944).
156. M. Tukamoto, *Z. Physiol. Chem.* **260,** 211 (1939).
157. K. Sasaki and T. Mochizuki, *J. Biochem. (Japan)* **40,** 317 (1953).
158. T. F. Gallagher and J. R. Xenos, *J. Biol. Chem.* **165,** 365 (1946).
159. J. Hattori and N. Yamada, *J. Pharm. Soc. Japan* **61,** 466 (1941).
160. K. Takeda and K. Igarashi, *J. Biochem. Japan* **46,** 1313 (1959).
161. R. L. Ratliff, J. T. Matschiner, E. A. Doisy, Jr., S. L. Hsia, S. A. Thayer, W. H. Elliott, and E. A. Doisy, *J. Biol. Chem.* **236,** 685 (1961).
162. W. Dirscherl, *Z. Physiol. Chem.* **237,** 268 (1935).
163. Z. Hattori, *J. Pharm. Soc. Japan* **59,** 32 (1939); *cited in C. A.* **33,** 3800 (1939); see also *J. Pharm. Soc. Japan* **59,** 416 (1939); *cited in C. A.* **33,** 8622 (1939).
164. W. Klyne, "The Chemistry of the Steroids," p. 55, Methuen and Co., Barnes and Noble, New York (1965).
165. T. Usui, R. Oshio, M. Kawamoto, and K. Yamasaki, *Yonago Acta Medica* **10,** 252 (1966).
166. G. Waller, H. Theorell, and J. Sjövall, *Arch. Biochem. Biophys.* **111,** 671 (1965).
167. H. Theorell, S. Taniguchi, A. A. Keson, and L. Skursky, *Biochim. Biophys. Res. Com.* **24,** 603 (1966).
168. T. Hoshita and T. Kazuno, *in* "Advances in Lipid Research" (R. Paoletti and D.

Kritchevsky, eds.), Vol. 6, p. 207, Academic Press, New York (1968).
169. H. Budzikiewicz, C. Djerassi, and D. H. Williams, "Mass Spectrometry of Organic Compounds," p. 8, Holden-Day, San Francisco (1967).
170. P. A. Roberts and W. H. Elliott, unpublished observations.
171. M. M. Mui and W. H. Elliott, *J. Biol. Chem.* **246,** 302 (1971).

Chapter 4

HYOCHOLIC ACID AND MURICHOLIC ACIDS

S. L. Hsia

University of Miami School of Medicine
Miami, Florida

I. INTRODUCTION

The occurrence of a species-specific bile acid in pig bile [Haslewood (1); Haslewood and Sjövall (2)] and of two such acids in rat bile [Bergström and Sjövall (3); Hsia *et al.* (4); Matschiner *et al.* (5)] was observed almost simultaneously. After their isolation and characterization, these acids were found to be isomeric 3α,6,7-trihydroxy-5β-cholanic acids. The acid from pig bile was named hyocholic acid [Haslewood (6); Ziegler (7)], and the two acids from rat bile were named α- and β-muricholic acids [Hsia *et al.* (8)]. The fourth isomer of these glycols was identified as a metabolite of hyodeoxycholic acid (3α,6α-dihydroxy-5β-cholanic acid) in the rat [Matschiner *et al.* (9, 10)], and was named ω-muricholic acid [Hsia *et al.* (8)]. The vicinal glycol structures in ring B of these acids are unique features, but even more unique are their species-specific characteristics which are particularly demonstrated in the metabolic pathways that lead to their formation.

In a discussion of the biosynthesis of bile acids from cholesterol, Bergström *et al.* (11) pointed out that the modifications of the steroid nucleus, including epimerization of the 3β-hydroxyl group, saturation of the Δ^5-double bond, and hydroxylation at positions 7α and 12α appear to precede degradation of the side chain. Once the side chain is shortened to that of a C_{24} bile acid, the molecule is incapable of undergoing hydroxylation at position 12α. For example, chenodeoxycholic acid (3α,7α-dihydroxy) is not metabolized in the rat to cholic acid (3α,7α,12α-trihydroxy), instead it is hydroxylated at 6β, forming α-muricholic acid (3α,6β,7α-trihydroxy), which is further converted to the 7β-epimer, β-muricholic acid [Bergström and Sjövall (3); Mahowald *et al.* (12); Samuelsson (13); Cherayil *et al.* (14)]. A number of other 5β-cholanic acids are also known to be metabolized similarly in the rat, leading not to 12α-hydroxy derivatives, but to the 6β-hydroxylated α- and β-muricholic acids; this group includes the unsubstituted 5β-cholanic acid [Ray *et al.* (15)], lithocholic acid (3α-hydroxy) [Thomas *et al.* (16)], ursodeoxycholic acid (3α,7β-dihydroxy) [Samuelsson (17)], 7-keto-lithocholic acid [Mahowald *et al.* (18); Samuelsson (19)], and hyodeoxycholic acid (Matschiner *et al.* (18)]. Whereas α- and β-muricholic acids are formed from chenodeoxycholic acid in both the rat and the mouse, ω-muricholic acid (3α,6α,7β-trihydroxy) is formed from chenodeoxycholic acid only in the latter animal [Ziboh *et al.* (20, 21)]. In the pig, chenodeoxycholic acid is hydroxylated neither at 12α nor at 6β, but at 6α, to form hyocholic acid [Bergström *et al.* (22)], which is a characteristic acid of this species. Thus, apart from the formation of cholic acid, which is a common acid in a large number of species, the formation of the 6,7-hydroxylated bile acids appears to constitute a separate pathway; and it is in this pathway that the species-specific characters of bile acids are particularly expressed. In a recent review, Haslewood (23) has designated such species-specific bile acids as "unique" cholanic acids.

This chapter brings together the information concerning the isolation, the elucidation of structures, and the methods of chemical synthesis of the four 3α,6,7-trihydroxy-5β-cholanic acids. Recent findings of taurochenodeoxycholate 6β-hydroxylase are also discussed since the enzyme system directly concerns the formation of α-muricholic acid.

II. ISOLATION OF HYOCHOLIC ACID

Haslewood and Sjövall (2) first observed a substance moving more slowly than glycocholic acid in paper chromatograms during an analysis of pig bile. In subsequent studies, Haslewood (6) isolated from pig bile a trihydroxy acid which he named hyocholic acid. Independently, the same acid was isolated by Ziegler (7). The procedures for isolation in both laboratories required alkaline hydrolysis of the conjugated bile acids. The ethyl esters were prepared and separated by paper chromatography or on alumina columns [Haslewood (6)]. The acetates of the esters were crystallized to afford an effective purification, and subsequently hydrolyzed to yield pure hyocholic acid [Haslewood (24); Ziegler (7)]. Another procedure for the isolation of hyocholic acid was reported by Matschiner *et al.* (9). The crude bile acid mixture obtained from pig bile was separated by chromatography on a Celite partition column* [Matschiner *et al.* (9)], a method frequently employed in studies of bile acids by the team at St. Louis University. Hyocholic acid

---

*See footnote p. 113.

was isolated from the trihydroxy acid fractions (Fractions 80-1, 80-2, 80-3); the small amount of cholic acid present was removed by fractional crystallization. The amount of hyocholic acid isolated by this method was 15–20% of the pig bile acid mixture.

Haslewood (24) pointed out that hyocholic cannot be separated from hyodeoxycholic acid, another characteristic acid of the pig, by the usual procedures of crystallization. A sample of hyodeoxycholic acid with a sharp melting point of 191–192°C was shown to contain more than 20% of hyocholic acid in his laboratory. Caution therefore must be exercised in interpreting prior data obtained on hyodeoxycholic acid which might have been contaminated with hyocholic acid.

III. ISOLATION OF α-, β-, AND ω-MURICHOLIC ACIDS

The α- and β-muricholic acids are minor components of rat bile (approximately 0.25 mg/ml fistula bile), which contains much larger amounts of cholic acid (2.9–3.8 mg/ml) [Matschiner *et al.* (5)]. An important factor that contributed to the success in the isolation of the muricholic acids in the presence of large amounts of cholic acid was the effective separation afforded by the chromatographic method used. The discovery of α- and β-muricholic acids was a result of metabolic studies of chenodeoxycholate-24-^{14}C which was administered to rats with bile fistulas. The bile was collected, and after alkaline hydrolysis of the conjugated bile acids, the free bile acids were separated on the Celite column. In the trihydroxy acid fractions, two radioactive metabolites were detected: one was less hydrophilic (Acid I in Fractions 60–3, 60–4, 80–1) and the other more hydrophilic (Acid II in Fractions 80–3, 80–4, 100–1) than cholic acid (in Fractions 80–1, 80–2). These findings were in accord with an earlier observation by Bergström and Sjövall (3) who reported that the administration of chenodeoxycholic acid-24-^{14}C to the rat led to the formation of two labeled acids chromatographically similar to but not identical with cholic acid. With the use of these radioactive metabolites as markers, the St. Louis group processed several liters of rat fistula bile and finally succeeded in isolating the two acids in crystalline form [Matschiner *et al.* (9)]. These acids were distinguished from cholic acid by their failure to give a color test characteristic for cholic acid [Reinhold and Wilson (25)]. The structures of the two acids were subsequently elucidated and they were named α-muricholic acid (Acid II) and β-muricholic acid (Acid I) [Hsia *et al.* (8)].

ω-Muricholic acid [Hsia *et al.* (8)], the fourth isomer of these glycols, is a metabolite of hyodeoxycholic acid in the rat and was isolated from urine of surgically jaundiced rats after the administration of relatively large amounts of hyodeoxycholic acid (40 mg per rat per day) [Matschiner *et al.* (10)]. The urine was concentrated by lyophilization, and after alkaline hydrolysis, the acidic products were subject to chromatographic separation on the Celite column. In addition to α- and β-muricholic acids, ω-muricholic acid (Acid IV) was identified as another metabolite. This acid forms a di- and tri acetate, which upon chromatography on the Celite column are eluted in Fractions 20–2, 20–3 and in Fractions 0–3, 0–4, respectively, whereas the corresponding acetates of cholic acid are slightly more polar and are respectively eluted in Fraction 40–1 and Fractions 20–1, 20–2. After separation of the acetates, crystalline ω-muricholic acid was obtained by hydrolysis of the purified acetates.

A study by Einarsson (26) has confirmed an earlier observation of Lin *et al.* (27), and has established that hyodeoxycholic acid is formed in the intact rat. It has been suggested that the formation of hyodeoxycholic acid is a result of the interplay of metabolic activities of the liver and intestinal microorganisms [Einarsson (26); Okishio and Nair (28)]. Although isolation of ω-muricholic acid from normal rat bile has never been attempted, it is likely to be present in normal rat bile since it is a metabolite of hyodeoxycholic acid.

The physical properties of the four isomeric acids and derivatives are listed in Table I.

IV. ELUCIDATION OF STRUCTURES

The chromatographic characteristics of these acids suggested that they were trihydroxycholanic acids. The sulfuric acid absorption spectra of these acids determined according to the procedure of Bernstein and Lenhard (29) exhibited three maxima at 309, 368, and 418 mμ, also suggesting the presence of three hydroxyl groups [Hsia *et al.* (30)]. Results of elemental analyses were in agreement of the empirical formula of $C_{24}H_{40}O_5$, that of a trihydroxycholanic acid.

A. Structure of Hyocholic Acid

The structure of hyocholic acid was proposed by Haslewood (24) and by Ziegler (7) to be $3\alpha,6\alpha,7\alpha$-trihydroxy-5β-cholanic acid (I, Fig. 1). Since it was known that pig bile contains hyodeoxycholic acid ($3\alpha,6\alpha$-dihydroxy) and chenodeoxycholic acid ($3\alpha,7\alpha$-dihydroxy) the bile was assumed to contain possibly also an acid with both 6α- and 7α-hydroxyl groups. Chemical evidence for the vicinal glycol structure in hyocholic acid was found after chromic oxidation. The product, 3-keto-6,7-secocholanic acid-6,7-dioic

TABLE I. Physical Properties of 3α,6,7,-Trihydroxy-5β-cholanic Acids and Derivatives

Acid	Melting point, °C	$[\alpha]^{25}_{D}$	Infrared maxima, cm^{-1}	References
hyocholic	183–185	+4.59		(7)
	185–186	+8±1	3597,3460 3111,1704 1075,1047	(30)
	188–189	+5.5±0.5		(24)
ethyl ester	76-78	+27.8±1		
ethyl ester triacetate	185-187			
ethyl ester 3α, 6α-diacetate	121-123			
methyl ester triacetate	184–185	+28.5±1		
	188–190	+20.95	1040,1165 1225,1250 1362,1735–1740 2950	(7)
α-muricholic	199–200	+38±2	3344,3367 1706,1166 1127,1080 1009, 733	(39)
triacetate	117–120			
methyl ester diacetate	91–92			
β-muricholic	226–228	+61.5±2	3509,3425 1695,1658 1047	(43) (30)
methyl ester	75–76		3436,3289 1739,1160 1050	
ω-muricholic	189–191	+36±2	3289,1686 1096,1066 1044,1032 1012	(44)
diacetate	167–168			
triacetate	189–191			

acid (II), previously undescribed, was subjected to the Huang-Minlon's modification of the Wolff–Kishner reduction [Huang-Minlon (31)] and thilobilianic acid of Wieland and Dane (32) (6,7-secocholanic acid-6,7-dioic acid, III) was identified. As supporting evidence for the structure of II, Ziegler reduced it with sodium borohydride to form lactone V, identical with a known product obtained by nitric acid oxidation of 3α-acetoxy-6- (or 7)-keto-5β-cholanic acid [Yamasaki and Chang (33)]. The fact that ring B was ruptured by chromic oxidation forming the 6,7-seco acid (II) was strong evidence for the 6,7-dihydroxy structure in ring B of hyocholic acid.

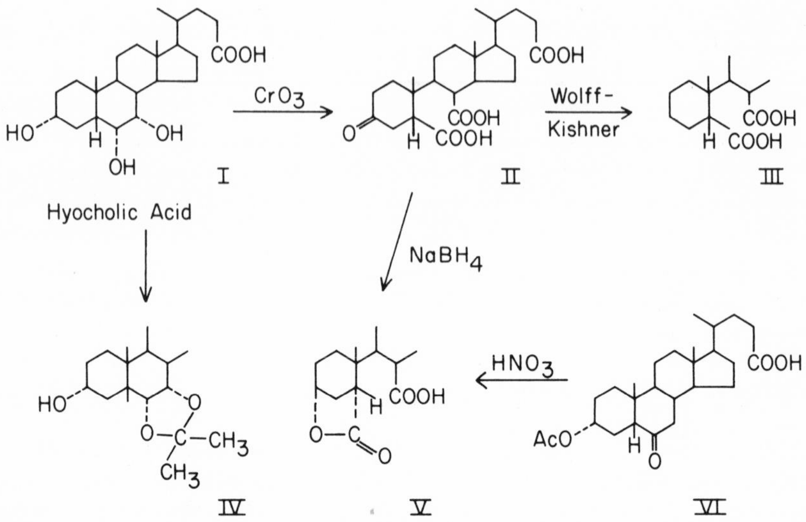

Fig. 1. Structure of hyocholic acid. [Adapted from Haslewood (24) and Ziegler (7).]

Hyocholic acid forms an acetonide (IV). Although it could not be crystallized [Ziegler (7); Haslewood (24)], its formation was substantiated by chromatographic mobility and data of quantitative acetylation. Formation of the acetonide gave evidence for the *cis*-glycol structure in hyocholic acid. The α-orientation of this 6,7-glycol was deduced from data of molecular rotations. Based on values from Barton and Klyne (34), the calculated molecular rotation of $3\alpha,6\alpha,7\alpha$-trihydroxy-5β-cholanic acid would be -13 and that of the $3\alpha,6\alpha,7\beta$-isomer, $+249$. The observed molecular rotation of hyocholic acid was $+19$. It was therefore concluded that hyocholic acid is the $3\alpha,6\alpha,7\alpha$ rather than the $3\alpha,6\beta,7\beta$-isomer [Ziegler (7)].

Confirmative evidence of the proposed structure was obtained from partial synthesis of hyocholic acid [Hsia *et al.* (30)]. An important intermediate in the synthesis was $3\alpha,6\alpha$-dihydroxy-7-keto-5β-cholanic acid (VII, Fig. 2), first prepared by Takeda *et al.* (35). The 3α- and 6α-hydroxyl groups in VII were established by the formation of hyodeoxycholic acid (IX) after hydrogenolysis of the ethylenedithioketal derivative (VIII) with Raney nickel. Hyocholic acid was obtained from VII either by reduction with sodium borohydride or by hydrogenation in the presence of platinum; both methods were known to produce the axially oriented 7α-hydroxy from 7-keto bile acids [Mosbach *et al.* (36); Iwasaki (37)]. More direct evidence for the 7α-hydroxyl group in hyocholic acid was found in a later study [Hsia *et al.* (8)], when hyocholic acid was derived from bromohydrin acetate XII (Fig. 3), which after debromination with Raney nickel and hydrolysis yielded chenodeoxycholic acid (XIV).

The stereochemistry with regard to the formation of hyocholic acid from XII deserves some discussion. Since the bromine at position 6 in XII is of β-orientation, the replacement with a 6α-hydroxyl group apparently involves an inversion. The mechanism of acetolysis of transbromohydrin acetates as proposed by Winstein and Buckles (38) involves the formation of a cyclic orthoacetate intermediate which gives rise to the *cis*-glycol in the presence of water and to the *trans*-glycol under the anhydrous conditions. The *cis*-glycol, hyocholic acid, was obtained by acetolysis of XII in the presence of water; but formation of the *trans*-glycol, α-muricholic acid could not be demonstrated under anhydrous conditions [Hsia *et al.* (8)]. Failure in the formation of the *trans*-glycol was interpreted as due to steric hindrance. The conformation of the 5β-steroids is such that the 4α-hydrogen is in close proximity at positions 6α and 7α and apparently can prohibit participation of the 7α-acetoxy group in XII from forming the cyclic orthoacetate. In forming hyocholic acid, the replacement of the 6β-bromide might have been accomplished not by the mechanism of neighboring group participation, but by a direct attack from the α-side to result in a 6α-configuration with Walden inversion.

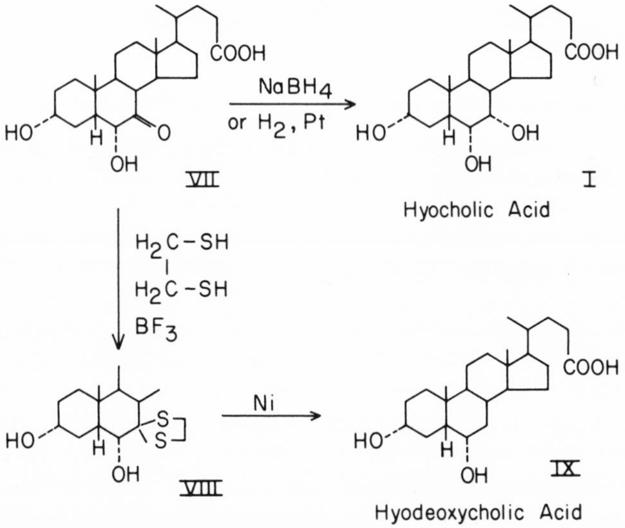

Fig. 2. Evidence for the 6α-hydroxyl group in hyocholic acid. [Adapted from Hsia *et al.* (30).]

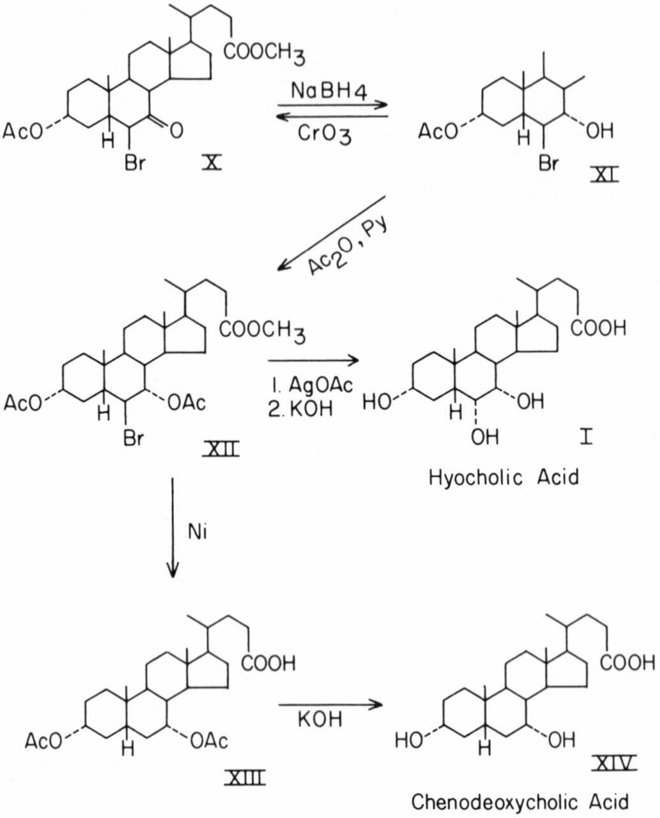

Fig. 3. Evidence for the 7α-hydroxyl group in hyocholic acid. [Adapted from Hsia *et al.* (8).]

B. Structures of the Muricholic Acids

Upon oxidation with chromic anhydride, the three muricholic acids, like hyocholic acid, yielded 3-keto-6,7-secocholanic acid-6,7-dioic acid (II) [Hsia *et al.* (30)]. Independent of the studies of Haslewood (24) and Ziegler (7), the St. Louis group identified this product by deriving it from lactone IV [Yamasaki and Chang (33)] which was hydrolyzed and subjected to chromic oxidation. The 3-keto-seco acid (II) was difficult to crystallize, while its 2,4-dinitrophenyl hydrazone derivative was found to be crystalline and easy to characterize and was suitably used for identification.

Since an identical product, II, was obtained from muricholic acids and hyocholic acid by chromic oxidation, the most obvious explanation was that the four acids are stereoisomers with hydroxyl groups at positions 3, 6, and 7.

Further information concerning the vicinal glycol structure in these acids was obtained from studies with periodic acid. Of the four glycols, ω-muricholic acid was the most prone to periodic oxidation, and β-muricholic acid was oxidized at a rate slower than ω-muricholic acid but faster than hyocholic acid, while α-muricholic acid was practically unattacked by periodic acid under the usual experimental conditions [Hsia *et al.* (39, 8)]. The resistance of α-muricholic acid toward periodic oxidation was interpreted as supporting evidence for the structure of a diaxial *trans*-glycol, i.e., the 6β,7α-glycol. The accepted concept concerning the mechanism of periodic oxidation is the formation of a cyclic intermediate between the glycol and periodic acid [Criegee (40)]. The projected angle of 180° between the 6β- and 7α-hydroxyl groups appears to be too great for the formation of such an intermediate and hence oxidation does not proceed. On the other hand, the projected angle of 60° in the other three isomers presents no such difficulty.

The diaxial *trans*-glycol structure in α-muricholic acid was substantiated by partial synthesis. α-Muricholic acid could be derived from either the 6α,7α-epoxide XVI [Hsia *et al.* (39)] or the 6β,7β-epoxide XV [Hsia *et al.* (41)] (Fig. 4) by scission of the epoxide rings and hydrolysis of the acetates thus formed. In accordance to the Fürst and Plattner rule [Fürst and Plattner (42)], ionic opening of an ethylene oxide results in the diaxial *trans*-glycol. The structure of α-muricholic acid therefore should be 3α,6β,7α-tri-hydroxy-5β-cholanic acid (XVII). The orientation of the epoxide ring in XV and that

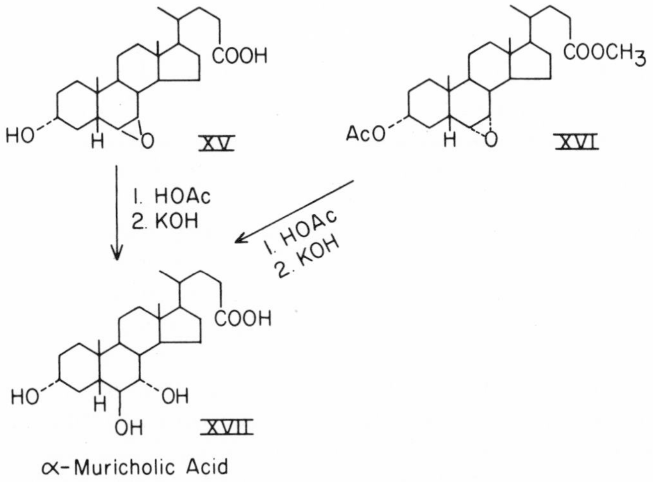

Fig. 4. Partial synthesis of α-muricholic acid. [From Hsia *et al.* (39, 41).]

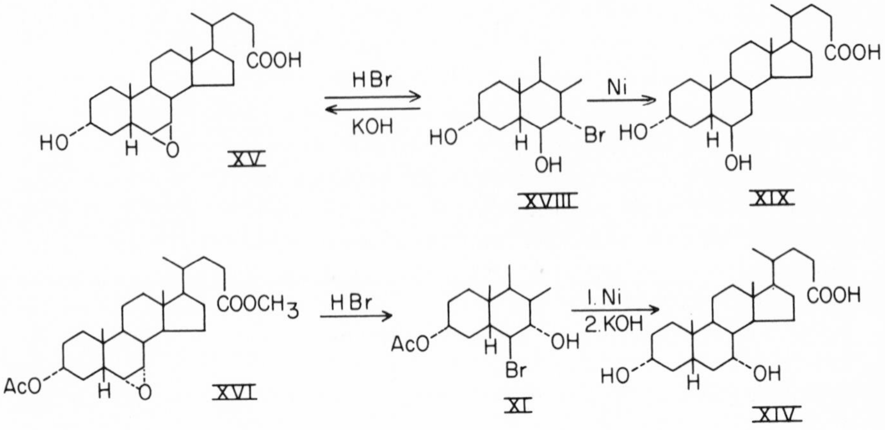

Fig. 5. Configurations of the 6β,7β- and the 6α,7α-epoxides. [From Hsia *et al.* (49, 41).]

in XVI were established by formation of 3α,6β-dihydroxy-5β-cholanic acid (XIX) and chenodeoxycholic acid (XIV), respectively. The dihydroxy acids were identified after the epoxides were treated with HBr to form the corresponding bromohydrins XVIII and XI which were then debrominated with Raney nickel (Fig. 5.). Since none of the reactions involved an inversion of the oxygen functional groups, the structures of the two dihydroxy acids gave credence to the 3α,6β,7α-trihydroxy structure of α-muricholic acid. The formation of hyocholic acid from bromohydrin acetate XII was alluded to previously (p. 100). Thus hyocholic acid and α-muricholic acid are both hydroxylated at 7α, and are isomeric at position 6.

The structure of β-muricholic acid was established by several routes of partial synthesis to be 3α,6β,7β-trihydroxy-5β-cholanic acid (XXI) (Fig. 6). Kagan (43) in studies on the hydroxylation of Δ^6-cholenates with OsO_4, concluded from data of molecular rotation that the products were 6β,7β-*cis*-glycols. A sample of the glycol synthesized in Kagan's laboratory from methyl 3α-acetoxy-Δ^6-cholenate (XX) by hydroxylation with OsO_4 was found to be identical with β-muricholic acid isolated from rat bile. Independently, the synthesis of β-muricholic acid by the same route was reported by the St. Louis group [Hsia *et al.* (44)]. Osmium tetroxide is known to be a reliable reagent for the cis addition of hydroxyl groups to double bonds. The hydroxylation of Δ^6-cholenates by OsO_4 would be expected to proceed by "β" or "frontal" attack, since in the 5β steroids, ring A folds toward the α-side of the molecule giving steric hindrance which presumably can thwart any "α" or "rear" attack by OsO_4.

The orientation of the 6β-hydroxyl group in β-muricholic acid and in α-muricholic acid was confirmed by their synthesis from bromohydrin acetate XVIIIa, which after debromination with Raney nickel and hydrolysis yielded 3α,6β-dihydroxy-5β-cholanic acid (XIX). The formation of β-muricholic acid from XVIIIa was accomplished by refluxing with silver acetate in wet acetic acid, whereas α-muricholic acid was formed as the major product when the reaction was carried out under anhydrous conditions [Hsia *et al.* (45)]. The replacement of the 7α-bromide in VIIIa apparently took place by mechanisms similar to those studied by Winstein and Buckles (38) and by Roberts *et al.* (46), who showed that *trans*-bromohydrin acetates form *trans*-glycols without inversion by the action of silver acetate in anhydrous acetic acid, but inversion occurs during acetolysis in the presence of water. It follows that the 7-hydroxyl group in β-muricholic acid formed in the presence of water should be of the β-orientation, and that in α-muricholic acid, of the α-orientation. These results were in accord with the concept of formation of a cyclic orthoacetate intermediate by participation of the neighboring 6β-acetyl group. Apparently the "β" side of the steroid molecule is open and offers no resistance to the formation of such an intermediate. Similar circumstances exist during the reaction between OsO_4 and methyl 3α-acetoxy-Δ^6-cholenate (XX). The circumstances are, however, quite different in the acetolysis of the 6β-bromo-7α-acetoxy trans-homohydrin (XII). In this instance, the formation of the cyclic orthoacetate intermediate by participation of the 7α-acetyl group is prohibited by steric hindrance of ring A in the "α" side of the molecule.

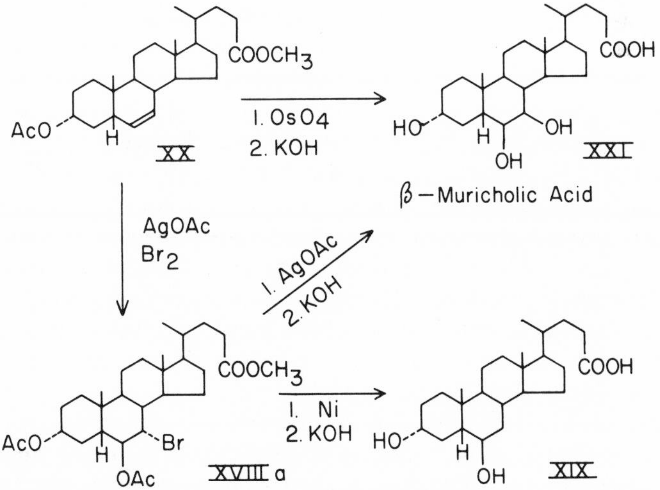

Fig. 6. Partial synthesis of β-muricholic acid. (From Kagan (43) and Hsia *et al.* (41).]

In an earlier paper, Hsia *et al.* (30) reported the formation of β-muricholic acid by Meerwein–Ponndorf reduction of 3α,6α-dihydroxy-7-keto-5β-cholanic acid (VII). Since several 7-keto compounds, including 7-ketolithocholic acid and its 6α- and 6β-bromo derivatives could not be reduced by this reaction under the same conditions, it was doubtful that the formation of β-muricholic acid was a result of direct reduction of the 7-keto group. On the other hand, 6-ketocholanic acids were reduced as expected [Tukamoto (47); Hsia *et al.* (41)]. It was proposed that VII underwent rearrangement under the influence of the aluminum alkoxide to form 3α,7β-dihydroxy-6-keto-5β-cholanic acid and subsequent reduction of the 6-keto group led to the 6β(axial)-hydroxyl group [Hsia *et al.* (41)]. Had it not been for the convincing evidence of the 6β-hydroxyl group in β-muricholic acid as discussed in the preceding paragraphs, the results of Meerwein–Ponndorf reduction could easily lead to an erroneous proposal of 6α-hydroxyl structure in β-muricholic acid. It is of interest to note also that this novel reaction seemed to proceed stereospecifically to yield the 6β,7β-*cis* glycol in excellent (70%) yield, with no evidence of allomerization.

After the structures of hyocholic, α-muricholic, and β-muricholic acids were established, the only possible structure of the remaining fourth isomer, ω-muricholic, was 3α,6α,7β-trihydroxy-5β-cholanic acid (XXIV). In verifying this structure an attractive approach to partial synthesis of ω-muricholic acid appeared to be reduction of the 7-keto group in 3α,6α-dihydroxy-7-keto-5β-cholanic acid (VII) stereospecifically to the equatorially orientated 7β-hydroxyl group. An example of such a reaction among the bile acids could be found in the formation of ursodeoxycholic acid (3α,7β-dihydroxy-5β-cholanic acid) by reduction of 7-ketolithocholic acid with metallic sodium in *n*-propanol [Kanazawa *et al.* (48); Samuelsson (49)]. In an experiment with VII, the product of sodium reduction was, however, not identical with ω-muricholic acid isolated from the rat, but had characteristics of a 5α-acid (m.p. 240–243 °C, infrared absorption band at 1009 cm^{-1}). The unstable ketol structure in ring B of VII apparently underwent rearrangement which led to the thermodynamically more stable 5α-configuration under the influence of alkali. It was likely that allomerization proceeded through a 6-keto intermediate which was formed as a result of tautomerism of the 7-ketone (VII) [see Takeda *et al.* (50)]. Sodium reduction of the product of rearrangement probably gave rise to the stable diequatorial 6α,7β-glycol, i.e. the 5α-isomer of ω-muricholic acid.

A successful synthesis of ω-muricholic acid was achieved by the sequence of reactions shown in Fig. 7 [Hsia *et al.* (44)]. The structure in ring B of VII was stabilized by the formation of the tetrahydropyranyl ether derivative (XXII). The ether linkage at position 6 would prevent the formation of the

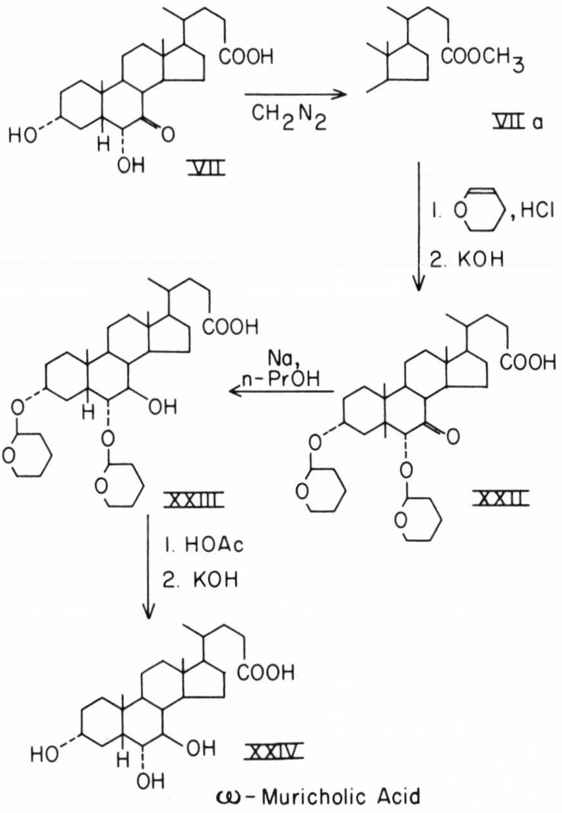

Fig. 7. Partial synthesis of ω-muricholic acid. [From Hsia *et al.* (44).]

6-keto intermediate and hence would also prevent allomerization. This derivative (XXII) was reduced with metallic sodium in boiling *n*-propanol, and ω-muricholic acid was obtained after the tetra-hydropyranyl ether linkages were cleaved by refluxing in acetic acid.

The 6α-hydroxyl group in ω-muricholic and hyocholic acid was verified by the identification of hyodeoxycholic acid derived from these trihydroxy acids. After acetylation in a mixture of acetic anhydride and pyridine at room temperature, hyocholic acid yielded a diacetate whereas ω-muricholic acid yielded both a diacetate and a triacetate. The diacetates were oxidized with chromic anhydride and an ethylenedithioketal was prepared, which after desulfuration with Raney nickel yielded hyodeoxycholic acid [Hsia *et al.* (45)].

V. PARTIAL SYNTHESES

The four isomeric acids can be derived by a series of chemical reactions from cholic acid, which is commercially available (Sigma Chemical). Some of the reactions have already been discussed with regard to structures. This section is intended to provide practical methods by which the four acids can be prepared. Some of the intermediates are by themselves interesting derivatives with regard to the stereochemistry of bile acids. The procedures described here are taken from published works with certain modifications and have been tested in the author's laboratory.

Hyocholic acid is a major acid in pig bile and isolation of hyocholic acid from this natural source is feasible. Partial synthesis of hyocholic acid from cholic acid is therefore of interest in verifying its structure rather than for the purpose of obtaining the material. On the other hand, the muricholic acids are minor constituents of rat bile. Isolation of these acids in large amounts would not be practical, and partial syntheses remain the necessary means to provide pure muricholic acids in useful quantities.

A. Preparation of Chenodeoxycholic Acid

The method (Fig. 8) is based on that described by Fieser and Rajagopalan (51) with modification.

1. Methyl Cholate (XXVa)

Cholic acid (XXV, 30 g) is dissolved in methanol (200 ml) by warming on a hot water bath, and to the warm solution conc. HCl (6 ml) is added. As the mixture cools, crystals of methyl cholate usually begin to deposit on the side of the container. Seeding with a few crystals of methyl cholate is helpful in initiating crystallization, which is enhanced by stirring and scratching with a glass rod. After the mixture is kept at $-4°C$ overnight to allow completion of crystallization, the product is collected on a Büchner funnel and washed with ice cold methanol. Methyl cholate is recrystallized by dissolving in warm methanol (4.5 ml per g of methyl cholate) and chilling in ice; m.p. 155–157°C; yield, over 90%.

2. Methyl 3α,7α-Diacetoxy-12-hydroxy-5β-cholanate (XXVI)

Selective acetylation at positions 3α and 7α of methyl cholate (XXVa, 25 g) is effected in benzene (125 ml) containing acetic anhydride (30 ml) and pyridine (30 ml). The mixture is kept at room temperature for 24 hr and then poured into ice water in a separatory funnel. The benzene layer is washed

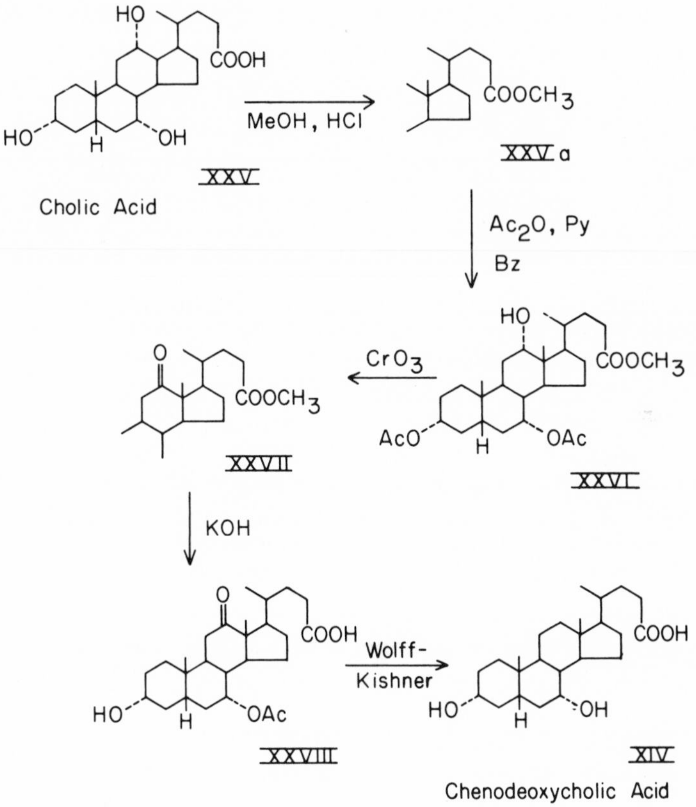

Fig. 8. Partial synthesis of chenodeoxycholic acid. [Adapted from Fieser and Rajagopalan (51).]

repeatedly with water, then dried by filtration through anhydrous sodium sulfate. The semicrystalline residue obtained after evaporation of benzene is washed with hexane to remove the lingering pyridine. After the hexane is decanted, the product (XXVI) is dissolved in warm methanol (100 ml) and crystallized by cooling in an ice bath. Additional crystals are brought about by gradual dilution with water (20 ml). Crystallization is repeated from methanol and water; m.p. 185–189 °C; yield, approximately 70%.

3. Methyl 3α,7α-Diacetoxy-12-keto-5β-cholanate (XXVII)

A solution of CrO_3 (2.7 g) in water (5 ml) is diluted with acetic acid (20 ml) and added slowly to a solution of the above 12α-hydroxy methyl ester (XXVI, 19g) in acetic acid (100 ml) while the mixture is cooled under a water tap. After remaining at room temperature for 4 hr, the mixture is diluted with water, and crystals of methyl 3α,7α-diacetoxy-12-keto-5β-cholanate (XXVII) precipitate copiously. The product can be recrystallized easily from acetic acid and water; m.p. 183–185 °C; yield, over 90%.

4. 3α-Hydroxy-7α-acetoxy-17-keto-5β-cholanic Acid (XXVIII)

In the original procedure described by Fieser and Rajagopalan (51) for the preparation of chenodeoxycholic acid, methyl esters were used throughout the sequence of reactions. Since methyl esters form hydrazides with hydrazine used in the Wolff–Kishner reduction, this side reaction is avoided by hydrolysis of the methyl ester (XXVII) to the free acid (XXVIII) before Wolff–Kishner reduction is attempted.

Methyl 12-keto-3α,7α-diacetoxy-5β-cholanate (XXVII, 20 g) is refluxed in methanol (100 ml) with KOH (6 g/200 ml water) for 45 min till a clear solution is obtained. After the methanol is partially removed by evaporation, the solution is cooled, diluted with water (200 ml) and acidified with acetic acid. The product is 3α-hydroxy-7α-acetoxy-12-keto-5β-cholanic acid [XXVIII, Wieland and Kapital (52)]; m.p. 238–242 °C; yield, 87%.

5. Chenodeoxycholic Acid (XIV)

In a round-bottom flask equipped with a reflux condenser, the above acid (XXVIII, 10 g) is dissolved in triethylene glycol (150 ml) containing hydrazine hydrate (85%, 7.5 ml). A solution of KOH (5 g/6 ml of water) is added and the mixture is heated at 110 °C for 1½ hr. The condenser is then removed to allow evaporation of water and the temperature is gradually raised to 190 °C. The condenser is reinstalled and the heating continued at 190 °C for 4 hr. After cooling, the content of the flask is diluted with an equal volume of water and acidified with acetic acid. Chenodeoxycholic acid settles out as a flocculent precipitate and is collected on a Büchner funnel. This product can be purified through its sodium salt. It is dissolved in a warm solution of 0.5% NaOH, and sodium chenodeoxycholate is salted out by the addition of solid NaCl. The sodium chenodeoxycholate is collected by filtration, washed with a saturated solution of NaCl, and then dissolved in warm water. Chenodeoxycholic acid is precipitated by acidification with acetic acid. The precipitated acid is collected and heated with a small amount of benzene on a water bath so that the oily product separates from the water which is decanted. The remaining small amount of water is removed by azotropic distillation with benzene. The dried residue is recrystallized from ethyl acetate and hexane; m.p. 138–141 °C; yield, 70–75%. For further purification, the acid is chromatographed on the Celite column. Chenodeoxycholic acid is eluted in Fractions 40–1 and 40–2.

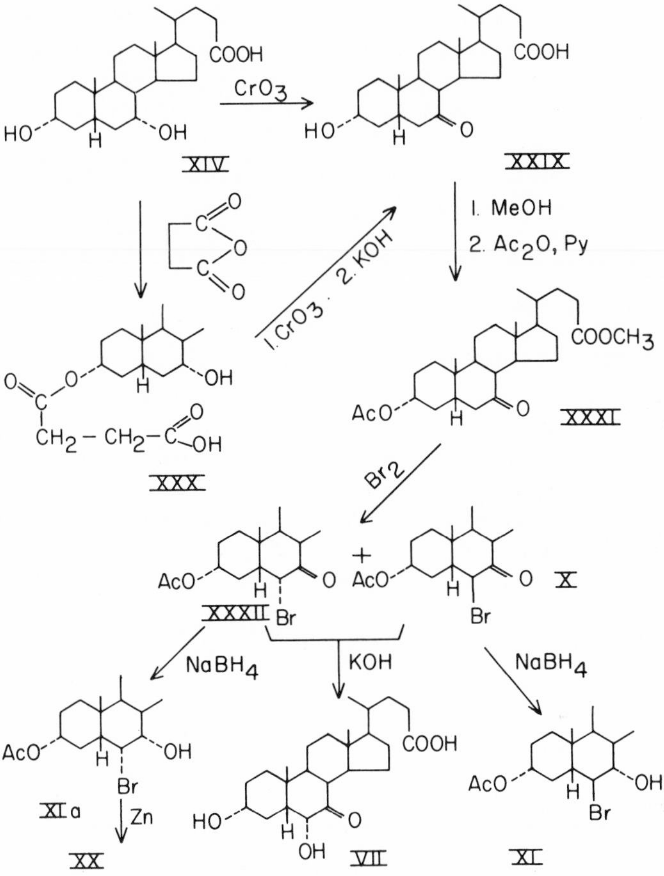

Fig. 9. Partial synthesis of 3α,6α-dihydroxy-7-keto-5β-cholanic acid. [From Takeda *et al.* (35) and Hsia, unpublished data.]

B. Preparation of 3α, 6α-Dihydroxy-7-keto-5β-cholanic Acid (VII. Fig. 9)

1. *7-Ketolithocholic Acid (XXIX)*

This acid may be obtained by partial oxidation of chenodeoxycholic acid [Iwasaki (37); Fieser and Rajagopalan (51); Takeda *et al.* (35)]. Chenodeoxycholic acid (XIV, 3 g) is dissolved in acetic acid (30 ml) and the solution is chilled to slightly above its freezing point in an ice bath. A solution of CrO_3 (523 mg) in 80% acetic acid (5 ml) is added dropwise with stirring.

After all the CrO_3 solution has been added, the mixture is stirred at room temperature for another hour and then diluted with water. The product (XXIX) precipitates in fine platelets and is recrystallized several times from acetic acid and water, and then from acetone and water; m.p. 201–203 °C; yield, 55–60 %.

A somewhat better yield of 7-ketolithocholic acid can be obtained by an alternative procedure, in which the 3α-hydroxyl group of chenodeoxycholic acid is protected from chromic oxidation by the formation of succinic ester. The procedure of Heusser and Wuthier (53) for the preparation of 3α-succinic ester of methyl cholate requires the use of succinic anhydride in pyridine. The use of pyridine is obnoxious in many respects, and it is omitted in the following modified procedure for the preparation of 7-ketolithocholic acid.

Chenodeoxycholic acid (15 g) is refluxed with succinic anhydride (4 g) in benzene (270 ml) overnight, and the solvent is then evaporated. Without isolation of the succinate (XXX), the residue (containing succinic anhydride) is dissolved in acetic acid (200 ml) and oxidized at room temperature with a solution of CrO_3 (3.5 g) in 80 % acetic acid (5 ml). After 3 hr, the mixture is diluted with water, and the green precipitate formed is collected by filtration, and then boiled in 3 % KOH for 15 min to hydrolyze the succinic ester. The green flocculent precipitate of $Cr_2O_3 \cdot H_2O$ is removed by filtration and the filtrate containing the potassium salt of 7-ketolithocholic acid is acidified with acetic acid. 7-Ketolithocholic acid (XXIX) precipitates as crystals, and is recrystallized from acetic acid and water, or from acetone and water; yield, 65–70 %.

2. Methyl 3α-Acetoxy-7-keto-5β-cholanate (XXXI)

7-Ketolithocholic acid (10 g) is methylated by dissolving in methanol (200 ml) containing conc. HCl (12 ml). After being kept at room temperature overnight, the mixture is diluted with water and extracted with ether. The semicrystalline residue from the ether extract is subjected to acetylation with acetic anhydride (10 ml) in pyridine (10 ml) at room temperature overnight. The methyl ester acetate (XXXI) crystallizes upon addition of ice water to the reaction mixture, and is recrystallized from acetone and water; m.p. 140–142 °C; yield, almost quantitative.

3. Methyl 3α-Acetoxy-6α-bromo-7-keto-5β-cholanate (XXXII) and Methyl 3α-Acetoxy-6β-bromo-7-keto-5β-cholanate (X)

Bromination of the above 7-keto derivative (XXXI) has been described by Takeda *et al.* (50). In a glass-stoppered bottle, the methyl ester acetate (XXXI, 3 g) is dissolved in acetic acid (50 ml) and an acetic acid solution (10 ml) of bromine (1.08 g) is introduced. Bromination is initiated by addition of HBr (1 ml 30 % acetic acid solution). The color of bromine fades in about

5 min. Water is added and a crystalline product (2.6 g) is obtained. Although this product has a sharp melting point of 118–120 °C after several crystallizations from acetone and water, it is a mixture of the 6α- and the 6β-bromides. This mixture can be resolved by chromatography on silica gel column, or by washing with a small volume of ether. The α-isomer (XXXII) remains insoluble and forms platelets after recrystallization from acetone and water, m.p. 172–174 °C. The β-isomer is found in the ether solution, and after recrystallizations from acetone and water forms needles, m.p. 166–168 °C. The proportions of the two isomers vary with the length of time that the product is allowed to remain in the reaction mixture after the bromine color has faded. The axial 6β-bromide undergoes inversion in acetic acid in the presence of HBr to the thermodynamically more stable equatorial α-isomer. If the reaction is allowed to proceed for 45 min after bromination has taken place, the product is practically all in the α-form.

4. 3α,6α-Dihydroxy-7-keto-5β-cholanic Acid (VII)

This acid can be obtained by alkaline hydrolysis of either the 6α-bromide (XXXII) or the 6β-bromide (X) [Takeda *et al.* (35)]. The 6β-isomer of this acid has not been isolated. Apparently the 6β(axial)-ketol easily rearranges to assume the more stable 6α-orientation as in VII.

For practical purposes, the racemic mixture of 6α- and 6β-bromides obtained in the previous preparation can be used without resolution into the two isomers. The mixed crystals (1 g) are dissolved in methanol (70 ml), and a KOH solution (2.4 g in 10 ml of water) is added dropwise with stirring. The mixture is kept at room temperature overnight before dilution with water and acidification with acetic acid. The product is extracted with ether, and the residue (800 mg) after evaporation of ether is chromatographed on the Celite partition column.* The desired acid is in Fractions 60–1 and 60–2,

---

*This chromatographic method has proved useful in the isolation and identification of a number of bile acids, and merits recording here. Celite washed with methanol and dried at 100° is used as the supporting medium, and a mixture made of 7 volumes of glacial acetic acid and 3 volumes of water and equilibrated with hexane is used as the stationary phase. In a regular experiment, 10 g of Celite is mixed with 8 ml of the stationary phase and then packed uniformly into a column of 18 mm diameter. The bile acid mixture, up to 100 mg, is mixed with 0.5 g Celite in 0.4 ml of the stationary phase and applied to the top of the column. The chromatogram is developed with solvent mixtures that have been equilibrated with the stationary phase. The elution begins with 100 ml of hexane followed by 100-ml portions of hexane containing increasing amounts of benzene, the increment being 20% in each successive mixture. The eluates are collected in 25-ml fractions. When larger amounts of bile acids are separated, the size of column and the amounts of Celite and solvents used are increased proportionally. Monohydroxy bile acids are usually eluted in hexane, dihydroxy acids in 20 to 40 % benzene, and trihydroxy acids in 60, 80, or 100 % benzene. For convenience, the chromatographic fractions are designated according to the percentage of benzene in hexane. For example, Fraction 20-1 refers to the first fraction of the eluant containing 20% benzene in hexane.

and can be crystallized from acetone and hexane; m.p. 186–187 °C; yield, 50–60 %.

C. Preparation of Hyocholic Acid from 3α,6α-Dihydroxy-7-keto-5β-cholanic Acid (VII)

A solution of VII (200 mg) in methanol (20 ml) is chilled in an ice bath, and sodium borohydride (1 g) is added in small portions. After the evolution of gas has ceased, the mixture is allowed to remain at room temperature for an additional hour. The solution is then diluted with water, acidified with HCl, and extracted with ether. The residue after evaporation of the ether is chromatographed on the Celite partition column, and hyocholic acid is eluted in Fractions 80–1, 80–2, and 80–3 (mainly in 80–1). It is crystallized from acetone and hexane; m.p. 187–188 °C; yield, 75 %.

Hyocholic acid has been prepared from VII in almost quantitative yield by hydrogenation in the presence of platinum in methanol, or acetic acid, or 5 % aqueous KOH solution [Hsia *et al.* (30)]. It has also been obtained in 62 % yield by refluxing methyl 3α,7α-diacetoxy-6β-bromo-5β-cholanate (XII, Fig. 3) with silver acetate followed by alkaline hydrolysis of the product [Hsia *et al.* (8)].

D. Preparation of ω-Muricholic Acid from VII (Fig. 7)

3α,6α-Dihydroxy-7-keto-5β-cholanic acid (VII, 500 mg) is methylated with diazomethane (generated from 1 g of Diazald, Aldrich Chemical Co.). The methyl ester (VIIa) obtained is treated with freshly distilled dihydropyran (4 ml) and the formation of tetrahydropyranyl ether (XXII) is catalyzed by conc. HCl (4 drops). The mixture is allowed to remain at room temperature overnight and the methyl ester is then hydrolyzed by refluxing in methanol (20 ml) with KOH (600 mg) for 30 min. After dilution with water (40 ml) the neutral materials are removed by extraction with ether and discarded. The aqueous solution is acidified with acetic acid and extracted with ether. The residue from the ether extract containing XXII is dissolved in *n*-propanol (25 ml) and sodium (2.5 g) is added in small pieces to the refluxing solution. After the sodium has been consumed the product of reduction (XXIII) is extracted with ether from the acidified solution. The residue from the ether extract is refluxed in acetic acid (25 ml) for 16 hr to cleave the tetrahydropyranyl ether linkages. The product is then hydrolyzed with KOH (4 % solution, 25 ml) on a boiling water bath for 10 min, and the acidic products obtained after acidification are chromatographed on the Celite column. ω-Muricholic acid (XXIV) is found in Fractions 80–2 and 80–3. After crystallization from acetone and hexane, it melts at 150–153 °C, resolidifies at about 160 °C, and remelts at 184–188 °C; yield, 25 %. ω-Muricholic is immediately preceded by another acid in Fractions 60–2 and 60–3. This by-product has a m.p. of 240–243 °C and an infrared absorption band at 1008 cm^{-1}. It is presumably the 5α-isomer of ω-muricholic acid (see p. 106).

E. Preparation of α-Muricholic Acid (XVII)

1. Methyl 3α-Acetoxy-6α-bromo-7α-hydroxy-5β-cholanate (XIa, Fig. 9)

Methyl 3α-acetoxy-6α-bromo-7-keto-5β-cholanate (XXXII) (2.4 g) is dissolved in methanol (200 ml) and the solution is chilled in an ice bath. Sodium borohydride (3 g) is added in small portions. The mixture is kept at ice-bath temperature for 15 min and allowed to remain at room temperature for 45 min, then acidified with acetic acid and evaporated at room temperature to a small volume. The crystalline product, methyl 3α-acetoxy-6α-bromo-7α-hydroxy-5β-cholanate (XIa) is recrystallized from methanol; m.p. 149–152 °C; yield, 85 %.

2. Methyl 3α-Acetoxy-Δ^6-5β-cholenate (XX)

This Δ^6 derivative (XX) is obtained by refluxing the above bromohydrin (XIa, 1 g) in acetic acid (20 ml) with zinc dust (1 g) which is added in small portions through 1.5 hr. The excess of zinc is removed by filtration, and the product (XX) is obtained from the filtrate in platelets by the addition of water and is recrystallized from methanol; m.p. 123–125 °C; yield, 66 %.

3. Methyl 3α-Acetoxy-6α,7α-epoxy-5β-cholanate (XVI)

The above Δ^6 derivative (700 mg) is added to an ether solution containing excess of monoperphthalic acid prepared according to Bohmn (54). The mixture is allowed to remain at room temperature for 3 days, then the excess of reagent and phthalic acid are removed by extraction with 5 % sodium carbonate. The ether solution is evaporated to dryness and a crystalline product (XVI) is obtained which is recrystallized from benzene and hexane; m.p. 172–174 °C; yield, 85 %.

4. α-Muricholic Acid from XVI

The α-epoxide (XVI, 120 mg) is refluxed in acetic acid (20 ml) for 3.5 hr. The solution is diluted with water and extracted with ether. The product (methyl 3α,6β-diacetoxy-7α-hydroxy-5β-cholanate) is purified on a silica gel column and is eluted by 20 % ether in benzene. After hydrolysis by methanolic KOH (2 %, 20 ml) at room temperature overnight, α-muricholic acid is obtained after acidification and extraction with ether. It forms crystals from an acetic acid solution after the addition of water. These crystals contain acetic acid, which is removed by recrystallization from acetone and hexane; m.p. 199–200 °C; yield, 75%.

α-Muricholic acid can be prepared from 3α-hydroxy-6β,7β-epoxy-5β-cholanic acid (XV, Fig. 4) by acetolysis followed by hydrolysis in methanolic KOH in the same manner as described above. The α-muricholic acid obtained by this procedure is purified by chromatography on the Celite column (eluted in Fractions 80–3, 80–4, and 100–1). The yield is comparable to the preparation from the α-epoxide (XVI). The 6β,7β-epoxy acid (XV) can be conveniently prepared by refluxing methyl 3α,6β-diacetoxy-7α-bromo-5β-cholanate (XVIIIa, see below) in 4% methanolic KOH (Fig. 6).

F. Preparation of β-Muricholic Acid

1. Methyl 3α,6β-Diacetoxy-7α-bromo-5β-cholanate (XVIIIa, Fig. 6)

A solution of bromine (384 mg) in acetic acid (1.2 ml) is added slowly with stirring to a suspension of finely pulverized silver acetate (410 mg) in acetic acid (130 ml). The color of the bromine disappears immediately with the precipitation of silver bromide. The supernatant pale yellow solution was obtained by decantation, and to the solution (20 ml), methyl 3α-acetoxy-Δ^6-5β-cholenate (XX, 550 mg) is added. After 30 min at room temperature, the mixture is diluted with water and extracted with ether. The residue from the ether extract (XVIIIa) is crystallized from acetone and water; m.p. 153–154 °C; yield, 88%.

2. β-Muricholic Acid (XXI) from XVIIIa

The bromohydrin acetate (XVIIIa, 200 mg) is refluxed for 1.5 hr with silver acetate (67 mg) in acetic acid (25 ml) containing water (1 ml). After cooling the mixture is diluted with water and extracted with ether. The residue from the extract is hydrolyzed by 4% methanolic KOH. The acidic product obtained from hydrolysis is purified by chromatography on the Celite partition column. β-Muricholic acid is eluted in Fractions 60–4 and 80–1, and crystallized from methanol and water; m.p. 226–228 °C; yield, 60–70%.

β-Muricholic acid has been prepared in 25–30% yield from methyl 3α-acetoxy-Δ^6-5β-cholenate by hydroxylation with osmium tetroxide [Kagan (43); Hsia *et al.* (44)], and by Meerwein–Ponndorf reduction of 3α,6α-dihydroxy-7-keto-5β-cholanic acid (VII) in 70% yield [Hsia *et al.* (30)].

VI. TAUROCHENODEOXYCHOLATE 6β-HYDROXYLASE OF RAT LIVER

As noted in the introduction, a number of bile acids are metabolized in the rat by 6β-hydroxylation, giving rise to α- and β-muricholic acids. The 6β-hydroxylase thus appears to be an important enzyme system concerning bile acid metabolism in this species. Recent studies have demonstrated *in vitro,* 6β-hydroxylase activity in cell-free preparations of rat liver, utilizing taurochenodeoxycholate as the substrate in the formation of α- and β-muricholic acids [Voigt *et al.* (55)]. Like a number of steroid hydroxylases, the 6β-hydroxylase system requires NADPH and O_2 for activity, and is therefore a mixed function oxidase [Mason (56)]. The activity is located in the microsomal fraction after differential centrifugation, and can be augmented considerably by the addition of the 105,000*g* supernatant fluid, which by itself is inactive in hydroxylating taurochenodeoxycholate. It is interesting that the 105,000*g* supernatant fluids obtained from homogenates of chicken liver and rat kidney also augmented the activity of rat liver microsomes [Voigt and Hsia (57)]. This experiment demonstrated that a common factor (or factors) exists in the 105,000 *g* supernatant fluids from all three tissues studied and since neither chicken liver nor rat kidney hydroxylates taurochenodeoxycholate, the factor that imparts the species-specific characteristics in bile acid metabolism, i.e., hydroxylation at 6β, obviously resides in rat liver microsomes.

The rat liver system definitely prefers taurochenodeoxycholate to the unconjugated chenodeoxycholate as a substrate. When the latter was incubated, the small amounts of α- and β-muricholic acids produced were in the conjugated form, indicating that the substrate might have been conjugated before it was hydroxylated. This view finds its support in an experiment in which CoA, ATP, and taurine were added to the incubation medium and the amount of 6β-hydroxylated products formed from chenodeoxycholate was comparable to that formed from taurochenodeoxycholate. These added cofactors were presumably required for the formation of taurochenodeoxycholate which was subsequently utilized by the hydroxylase. In a previous study by Bergström and Gloor (58) of the transformation of deoxycholate to cholate, it was observed similarly that conjugation probably preceded 7α-hydroxylation.

The 6β-hydroxylase activity in the microsomes and the stimulatory effect of the 105,000 *g* supernatant fluid were found to vary from preparation to preparation, the more active the microsomes, the less stimulatory the supernatant. These observations led to further studies which showed that the ionic strength of the homogenizing medium effects the distribution of enzymic activity. The 105,000 *g* pellet prepared in 0.01 *M* phosphate buffer (*p*H 7.6) was found to contain full activity which was not further augmented by the 105,000 *g* supernatant fluid, whereas the 105,000 *g* pellet prepared in 1.0 *M* phosphate had relatively little activity, but the activity was considerably augmented by the supernatant. From these results a new procedure evolved for the preparation of the 6β-hydroxylase. Rat liver was homogenized in 0.01 *M* phosphate (*p*H 7.6) containing 0.01 *M* $MgCl_2$ and 0.03 *M* nicotinamide, and the 105,000 *g* pellet obtained after differential centrifugation was suspended for 20 min at 0° in 1.0 *M* phosphate buffer (*p*H 7.6) containing 0.01 *M* $MgCl_2$ and 0.03 *M* nicotinamide, and then centrifuged at 105,000 *g* for 1 hr. The new pellet was almost depleted of its 6β-hydroxylase activity, while the supernatant extract contained a complete 6β-hydroxylase system when supplemented with NADPH and O_2.

The 1.0 *M* phosphate extract appeared optically clear, but attempts at resolution into components of the enzyme complex had little success. Apparently the components of the enzyme complex were not fully dissociated from each other by the treatment with 1.0 *M* phosphate. Nonetheless, pertinent information concerning the components of the 6β-hydroxylase system has been obtained by studies of the extract. The 6β-hydroxylase activity was inhibited by cytochrome *c* and by dichlorophenolindophenol, suggesting the possibility of participation by NADPH–cytochrome *c* reductase. These inhibitors may compete with the 6β-hydroxylase system for electrons from NADPH, similar to the inhibition of microsomal hydroxylation of acetanilide [Krisch and Staudinger (59)] and of the adrenal steroid 21-hydroxylase [Ryan and Engel (60)]. It has been proposed by Phillips and Langdon (61) that although cytochrome *c* is not a participant in the microsomal transport of electrons, NADPH–cytochrome *c* reductase may be involved in an early step in the transport chain. Supporting this view is the inhibition of 6β-hydroxylase activity by cupric ion, a recognized inhibitor of microsomal NADPH–cytochrome *c* reductase.

Difference spectroscopy showed that the extract contained the hemoproteins P-450 and cytochrome b_5. Of special interest is the inhibition of 6β-hydroxylase activity by CO and the reversal of this inhibition by light, suggesting that P-450 may be involved in the reaction. In a subsequent experiment, the 6β-hydroxylase activity was assayed in the presence and absence of CO, and under light of various wavelengths between 400–500 mμ. Reactivation of the CO inhibited enzyme was maximum at 450 mμ. An action spectrum obtained from these data exhibited a maximum at 450 mμ and minima at 400 and 500 mμ, closely resembling the absorption spectrum of the P-450 CO complex [Voigt (62)]. It thus appears that the 6β-hydroxylase system, like several other steroid hydroxylases, requires the participation of microsomal electron transport with P-450 as the terminal oxidase.

These studies of the 6β-hydroxylase are only preliminary, and much remains to be clarified. Since there are species variations in bile acid metabolism, the enzymes concerned with bile acid metabolism must have species-specific characteristics. Potentially, comparative studies of these enzymes in various species may be an interesting area for future exploration.

REFERENCES

1. G.A.D. Haslewood, *Biochem. J.* **56,** xxxviii (1954).
2. G.A.D. Haslewood and J. Sjövall, *Biochem. J.* **57,** 126 (1954).
3. S. Bergström and J. Sjövall, *Acta Chem. Scand.* **8,** 611 (1954).
4. S.L. Hsia, J.T. Matschiner, and S.A. Thayer, *Fed. Proc.* **15,** 277 (1956).
5. J.T. Matschiner, T.A. Mahowald, W.H. Elliott, E.A. Doisy, Jr., S.L. Hsia, and E.A. Doisy, *J. Biol. Chem.* **225,** 771 (1957).
6. G.A.D. Haslewood, *Biochem. J.* **56,** 581 (1954).
7. P. Ziegler, *Can. J. Chem.* **34,** 523 (1956).
8. S.L. Hsia, W.H. Elliott, J.T. Matschiner, E.A. Doisy, Jr., S.A. Thayer, and E.A. Doisy, *J. Biol. Chem.* **235,** 1963 (1960).
9. J.T. Matschiner, T.A. Mahowald, S.L. Hsia, E.A. Doisy, Jr., W.H. Elliott, and E.A. Doisy, *J. Biol. Chem.* **225,** 803 (1957).
10. J.T. Matschiner, R.L. Ratliff, T.A. Mahowald, E.A. Doisy, Jr., W.H. Elliott, S.L. Hsia, and E.A. Doisy, *J. Biol. Chem.* **230,** 589 (1958).
11. S. Bergström, H. Danielsson, and B. Samuelsson, *in* "Lipide Metabolism" (K. Bloch, ed.), Wiley, New York (1960), p. 291.
12. T.A. Mahowald, J.T. Matschiner, S.L. Hsia, R. Richter, E.A. Doisy, Jr., W.H. Elliott, and E.A. Doisy, *J. Biol. Chem.* **225,** 781 (1957).
13. B. Samuelsson, *Acta Chem. Scand.* **13,** 976 (1959).
14. G.D. Cherayil, S.L. Hsia, J.T. Matschiner, E.A. Doisy, Jr., W.H. Elliott, S.A. Thayer, and E.A. Doisy, *J. Biol. Chem.* **238,** 1943 (1963).
15. P.D. Ray, E.A. Doisy, Jr., J.T. Matschiner, S.L. Hsia, W.H. Elliott, S.A. Thayer, and E.A. Doisy, *J. Biol. Chem.* **236,** 3158 (1961).
16. P.J. Thomas, S.L. Hsia, J.T. Matschiner, E.A. Doisy, Jr., W.H. Elliott, S.A. Thayer, and E.A. Doisy, *J. Biol. Chem.* **239,** 102 (1964).
17. B. Samuelsson, *Acta Chem. Scand.* **13,** 970 (1959).
18. T.A. Mahowald, M.W. Yin, J.T. Matschiner, S.L. Hsia, E.A. Doisy, Jr., W.H. Elliott, and E.A. Doisy, *J. Biol. Chem.* **230,** 581 (1958).
19. B. Samuelsson, *Acta Chem. Scand.* **13,** 236 (1959).
20. V.A. Ziboh, J.T. Matschiner, E.A. Doisy, Jr., S.L. Hsia, W.H. Elliott, S.A. Thayer, and E.A. Doisy, *J. Biol. Chem.* **236,** 387 (1961).
21. V.A. Ziboh, S.L. Hsia, J.T. Matschiner, E.A. Doisy, Jr., W.H. Elliott, S.A. Thayer, and E.A. Doisy, *J. Biol. Chem.* **238,** 3588 (1963).
22. S. Bergström, H. Danielsson, and A. Göransson, *Acta Chem. Scand.* **13,** 776 (1959).
23. G.A.D. Haslewood, *J. Lipid Res.* **8,** 535 (1967).
24. G.A.D. Haslewood, *Biochem. J.* **62,** 637 (1956).
25. J.G. Reinhold and D.W. Wilson, *J. Biol. Chem.* **96,** 637 (1932).
26. K. Einarsson, *J. Biol. Chem.* **241,** 534 (1966).
27. T.H. Lin, R. Rubinstein, and W.L. Holmes, *J. Lipid Res.* **4,** 63 (1963).
28. T. Okishio and P.P. Nair, *Biochemistry* **5,** 3662 (1966).
29. S. Bernstein and R.H. Lenhard, *J. Org. Chem.* **18,** 1146 (1953).
30. S.L. Hsia, J.T. Matschiner, T.A. Mahowald, W.H. Elliott, E.A. Doisy, Jr., S.A. Thayer, and E.A. Doisy, *J. Biol. Chem.* **225,** 811 (1957).
31. Huang-Minlon, *J. Am. Chem. Soc.* **71,** 3301 (1949).
32. H. Wieland and E. Dane, *Hoppe-Seyler's Z. Physiol. Chem.* **210,** 268 (1932).
33. K. Yamasaki and Y.L. Chang, *J. Biochem. (Japan)* **39,** 185 (1952).
34. D.H.R. Barton and W. Klyne, *Chem. Ind.* **1948,** p. 755.
35. K. Takeda, K. Igarashi, and T. Komeno, *Pharm. Bull.* **2,** 348 (1954).
36. E.H. Mosbach, W. Meyer, and F.E. Kendall, *J. Am. Chem. Soc.* **76,** 5799 (1954).

37. T. Iwasaki, *Hoppe-Seyler's Z. Physiol. Chem.* **244,** 186 (1936).
38. S. Winstein and R.E. Buckles, *J. Am. Chem. Soc.* **64,** 2780 (1942).
39. S.L. Hsia, J.T. Matschiner, T.A. Mahowald, W.H. Elliott, E.A. Doisy, Jr., S.A. Thayer, and E.A. Doisy, *J. Biol. Chem.* **226,** 667. (1957).
40. R. Criegee, *Ber.* **68,** 665 (1935).
41. S.L. Hsia, J.T. Matschiner, T.A. Mahowald, W.H. Elliott, E.A. Doisy, Jr., S.A. Thayer, and E.A. Doisy, *J. Biol. Chem.* **230,** 573 (1958).
42. A. Fürst and P.A. Plattner, Abstract, 12th International Congress of Pure and Applied Chemistry, New York (1951).
43. H.B. Kagan, *Compt. Rend. Acad. Sci.* **244,** 1373 (1957).
44. S.L. Hsia, J.T. Matschiner, T.A. Mahowald, W.H. Elliott, E.A. Doisy, Jr., S.A. Thayer, and E.A. Doisy, *J. Biol. Chem.* **230,** 597 (1958).
45. S.L. Hsia, W.H. Elliott, J.T. Matschiner, E.A. Doisy, Jr., S.A. Thayer, and E.A. Doisy, *J. Biol. Chem.* **233,** 1337 (1958).
46. R.M. Roberts, J. Corse, R. Boschan, D. Seymour, and S. Winstein, *J. Am. Chem. Soc.* **80,** 1247 (1958).
47. M. Tukamoto, *J. Biochem. Japan* **32,** 451 (1940).
48. T. Kanazawa, A. Shimazaki, T. Sato, and T. Hoshino, *Proc. Japan. Acad.* **30,** 391 (1954).
49. B. Samuelsson, *Acta Chem. Scand.* **14,** 17 (1960).
50. K. Takeda, T. Komeno, and K. Igarashi, *Pharm. Bull.* **2,** 352 (1954).
51. L.F. Fieser and S. Rajagopalan, *J. Am. Chem. Soc.* **72,** 5530 (1950).
52. H. Wieland and W. Kapitel, *Hoppe-Seyler's Z. Physiol. Chem.* **212,** 269 (1932).
53. H. Heusser and H. Wuthier, *Helv. Chim. Acta* **30,** 2165 (1947).
54. H. Bohmn, *in* "Organic Synthesis," Collective Vol. 3 (1955), p. 619.
55. W. Voigt, P.J. Thomas, and S.L. Hsia, *J. Biol. Chem.* **243,** 3493 (1968).
56. H.S. Mason, *Adv. Enzymol.* **19,** 79 (1957).
57. W. Voigt and S.L. Hsia, *Fed. Proc.* **26,** 341 (1967).
58. S. Bergström and U. Gloor, *Acta Chem. Scand.* **8,** 1373 (1954).
59. K. Krisch and H. Staudinger, *Biochem. Z.* **334,** 312 (1961).
60. K.J. Ryan and L.L. Engel, *J. Biol. Chem.* **225,** 103 (1957).
61. A.H. Phillips and R.G. Langdon, *J. Biol. Chem.* **237,** 2652 (1962).
62. W. Voigt, *Fed. Proc.* **27,** 523 (1968).

Chapter 5

EXTRACTION, PURIFICATION, AND CHROMATOGRAPHIC ANALYSIS OF BILE ACIDS IN BIOLOGICAL MATERIALS*

P. Eneroth and J. Sjövall

Department of Chemistry
Karolinska Institutet
Stockholm, Sweden

I. INTRODUCTION

Several reviews covering different aspects of bile acid analysis have appeared in recent years (1–15). These articles cover a wide range of techniques with particular emphasis on thin-layer and gas chromatography. In this chapter we will try to present established methods for bile acid analysis as they can be combined to form a complete analytical procedure, including extraction, purification, identification, and quantitation. It has not been possible to include all methods and the various modifications of standard procedures that have been published. References to most of these can be found in the reviews (1–15). Early development in the field is described in Sobotka's books on bile acids and steroids (16, 17). Whenever equally efficient or similar methods exist we have generally preferred to describe those of which we have personal experience. Many of the techniques used for bile acids are similar to those used in steroid analysis and advance in this area is highly relevant for workers in the bile acid field.

Bile acids having the most familiar and commonly used trivial names are listed in the footnote.† Except for these compounds it is usually simpler and

*Supported by grants from the Swedish Medical Research Council (13x-219 and 13x-2520).

†The following systematic names are given bile acids referred to by trivial names: cholic acid, 3α,7α,12α-trihydroxy-5β-cholanoic acid; deoxycholic acid, 3α,12α-dihydroxy-5β-cholanoic acid; chenodeoxycholic acid, 3α,7α-dihydroxy-5β-cholanoic acid; ursodeoxycholic acid, 3α,7β-dihydroxy-5β-cholanoic acid; hyodeoxycholic acid, 3α,6α-dihydroxy-5β-cholanoic acid; lithocholic acid, 3α-hydroxy-5β-cholanoic acid.

clearer to use systematic names. The term "free bile acids" will be used to mean a substituted 5α (allo) or 5β-cholan-24-oic acid unless otherwise indicated. Bile acids in peptide linkage with glycine or taurine will be referred to as conjugated bile acids.

II. EXTRACTION

There are many different ways to extract bile acids from different organs and fluids. They may be divided into two types: extraction with solvents and extraction with solids.

A. Extraction with Solvents

Binding to protein, or other polymers, and the salt form in which the bile acid occurs constitute important factors in determining extractability and solubility of bile acids in different solvents. Sodium and potassium salts are more polar than magnesium, calcium, or ammonium salts. The protonated conjugated bile acids are sufficiently nonpolar to be extracted from an aqueous solution by *n*-butanol. Regardless of the salt form, conjugated or free bile acids can always be dissolved in chloroform/methanol; a 1:1 (v/v) mixture is very useful for extraction of bile acids from organ homogenates, feces, or body fluids. Refluxing or Soxhlet extraction is sometimes necessary (18, 19). If quantitative extraction is not obtained with this general method, the residue may be further extracted with 0.2 M ammonium carbonate in 80% aqueous ethanol to ensure as complete a recovery as possible. In general, alkaline conditions favor extraction of bile acids. Some investigators have experienced difficulties in recovering bile acids from tissues and microorganisms. If the above-mentioned procedure should fail, it is most likely due to firm binding of bile acids to tissue constituents such as proteins. In those instances it may be worth trying to homogenize the material in a suitable buffer and to make an enzymatic digestion before proceeding with the extraction as above. Since some bacteria can hydrolyze bile acid conjugates and/or attack the hydroxyl groups, bacterial contamination must be avoided while working in aqueous solution.

To recover bile acids or their conjugates from biological fluids, such as bile or intestinal contents, several different procedures are available. When the bile acid concentration is sufficiently high (above 1 meq/liter), as, e.g., in bile, 5–100 μl of the fluid may be directly applied to the starting line for paper or thin-layer chromatography. When low concentrations of bile acids are present, bile may be added drop by drop into 10–20 volumes of ethanol during agitation in an ultrasonic bath. After brief boiling and subsequent cooling the extract is filtered and concentrated for further purification. In all solvent extractions it is important that the solvent volume is kept at least 10–20 times larger than the water volume. Addition drop by drop to give a fine protein precipitate is also very important.

Many good methods for extracting bile acids from specific tissues or body fluids have been described. It should be remembered that these methods can usually be used only for the specific analytical situation for which they were developed.

Bile

Extraction in 20 vol. of ethanol as described above is likely to give the most complete recovery of all types of bile acids. Alkaline conditions favor extraction. Folch extraction has been used to get a simultaneous purification, and the conjugated bile acids are then found in the upper phase (20). A different type of extraction is that described by Hofmann, who used a liquid anion exchanger, tetraheptylammonium chloride (21). This was added to bile and the salts formed with bile acids could be quantitatively extracted with ethyl acetate. The use of this solvent makes the evaporation of bile extracts simple. The amine can be removed with a cation exchanger, a procedure first described by Anderson (22).

Intestinal Contents

Intestinal contents can be extracted with ethanol or chloroform/methanol (1, 20). Shioda *et al.* (20) used a Folch extraction for duodenal bile and recovered the conjugated bile acids in the top water phase. Free bile acids are often present in intestinal contents (usually distally) and may be partly lost in this procedure.

Feces

Many methods have been published. We have used 48 h Soxhlet extraction with chloroform/methanol, 1:1, for human feces (18) and found quantitative recoveries of endogenously labeled bile acids. Evrard and Janssen (23) used 3 ml glacial acetic acid, 1 h at 120 °C, for 100–300 mg freeze-dried powder of human feces. This is an interesting, simple procedure, but it has to be followed by hydrolysis since acetates may be formed.

Liver

Okishio, Nair, and Gordon (24) extracted freeze-dried whole or fractionated rat liver homogenates by refluxing 30 min in 100 ml of 95% ethanol containing 0.1% ammonium hydroxide. This procedure was repeated once.

After evaporation of the solvents the sample was dissolved in 3–7 ml 0.1 M NaOH and subjected to Amberlyst A-26 extraction/purification (see Section II. B).

Blood Plasma or Serum

Most methods based on solvent extraction employ ethanol, 10–20 volumes, as described above (1). It is advisable to add sodium hydroxide to the sample since protein binding is inhibited at *p*H 11 (25). In the older methods based on colorimetric estimation, barium hydroxide, barium acetate, or zinc sulfate were used to improve yield and to decolorize the extract (see 26, 27).

Urine

If urine is acidified (*p*H 1) and then extracted three times with equal volumes of *n*-butanol, bile acid conjugates can be quantitatively recovered (28). However, due to the time-consuming evaporation of *n*-butanol and the formation of emulsions other ways to extract bile acids have been tried. The method described by Hofmann (21) may be useful.

Bile acids have also been studied in other organs. Lithocholic acid was found in the brain of guinea pigs with experimental allergic encephalomyelitis (29). In this case as well as in studies of fecal bile acids (30), hydrolysis was the first step in the work-up of the samples. It is probable that some bile acids undergo structural changes under the vigorous conditions for hydrolysis, especially when they only represent trace components in these biological materials.

B. Extraction with Solids

These extraction techniques are relatively new in the bile acid field, and improvements and simplifications of such methods are likely to appear. They offer extraction and purification in one step and conditions can often be kept very mild.

Bile acids in plasma or serum (and probably in other biological fluids) can be extracted with the macroreticular strong anion exchanger Amberlyst A-26 (Rohm and Haas Co., Philadelphia) (31). Two to five ml of plasma are diluted with an equal volume of water, the *p*H is adjusted to about 11 with 1 M NaOH, and the sample is applied to a 3-ml column of the ion exchanger in the OH^- form. After washing with water (until neutral), and 20 ml of ethanol, the bile acids are eluted with 150 ml of 0.2 M ammonium carbonate in 80% aqueous ethanol. Kuron and Tennent (32) found this eluent superior to others in work on purification of fecal bile acids on Dowex ion exchangers. The latter exchangers, however, are not as suitable in quantitative work as the macroreticular ones. Other ion exchange techniques are discussed in the chapter by Kuksis (Chapter 6).

An alternative procedure to extract bile acids from biological fluids can be based on the properties of a neutral polymer, Amberlite XAD-2 (Rohm and Haas, Co.), to adsorb detergent-like compounds. The use of this resin to extract steroids from urine was reported by Bradlow (33). Preliminary experiments with rat bile have indicated that 80–100% of the bile acids are recovered if the bile is first diluted with 50 volumes of 1% NaCl in water and this solution is filtered through a 100-ml column of the resin in water at 4 °C. After rinsing with 2 bed volumes of water, bile acids are eluted at room temperature with methanol and chloroform/methanol, 1:1.

Used as described for the analysis of steroids (33, 34) the method may be particularly useful for urine analyses. Norman obtained a quantitative yield of different types of labeled bile acids in urine and the purification was tenfold (35). The method may also be useful for separation of inorganic salts from bile salts. However, several factors influence the results of XAD-2 extractions, e.g., flow rate, temperature, *p*H, protein and electrolyte concentrations. Further studies of the method are needed.

III. STEPS FOLLOWING EXTRACTION

The extract obtained may contain a mixture of conjugated and free bile acids, sometimes in different salt forms, along with numerous other lipids and compounds of similar polarity. The analysis can continue in two different ways: (1) direct purification and separation of the mixture, and (2) hydrolysis of conjugated bile acids and subsequent purification and separation.

A. Purification and Separation Without Previous Hydrolysis of Conjugated Bile Acids

The first step should preferably lead to a group separation of bile acids and to elimination of non-bile acid contaminants. In quantitative work it may be useful to add—at this stage or earlier—labeled compounds, which permit an estimation of recovery of bile acids of different polarities. Taurocholic and 3-ketocholanoic acids constitute suitable extremes in polarity. If specific bile acids are to be analyzed, one may add a suitable internal standard to the biological material. Thus, Roovers *et al.* (36) used nordeoxycholic acid for gas chromatographic determination of bile acids in feces and plasma.

If the weight of the extract is of the order of 20–25 mg or less, preparative thin-layer chromatography or paper chromatography may be tried (see below). Larger amounts of material are preferably subjected to column chromatography.

1. Counter-current Extraction

A counter-current extraction is sometimes useful as a first step. For purification purposes a three-stage counter-current extraction between 70% aqueous ethanol and petroleum ether is sufficient (1). Other systems have been used for the same purpose (11). Extensive separations of bile acids can be obtained with the counter-current distribution systems developed by Ahrens and Craig (37).

2. Ion-Exchange and Gel Chromatography

Ion-exchange chromatography may be used for purification and group separation. As mentioned, Okishio *et al.* (24) purified a liver extract by applying it in 0.1 M NaOH to an Amberlyst A-26 column as described above for solid extraction. An extract can also be dissolved in chloroform–methanol, 1:1, saturated with water, and applied with a slow flow rate to an Amberlyst A-26 column in OH^- form in the same solvent. A 10-ml column should be used for 250 mg of sample. Neutral lipid contaminants are removed by washing with a suitable organic solvent and bile acids are eluted with 0.2 M ammonium carbonate in 80% ethanol as described above (34).

Group separations of glycine- and taurine-conjugated bile acids can also be achieved on ion exchange columns (discussed in Chapter 6 by Kuksis). Preliminary results indicate that these groups can also be separated (with some overlapping) on carboxymethylated Sephadex LH-20 (38) in the ammonium form. When the ammonium salts of bile acids are chromatographed on these gels (20 ml gel for 100 mg of extract) in chloroform–methanol, 4/1 (v/v), the free bile acids are eluted at approximately 80% of the total bed volume, the glycine conjugates at 100% and the taurine conjugates at about 150%, if a gel with at least 2.0 meq of carboxymethyl groups per gram is used (39).

Purification of free and conjugated bile acids can also be obtained on Sephadex LH-20 in chloroform–methanol, 1:1, containing 0.01 M of a suitable electrolyte (40). Depending on the electrolyte used, bile acids are more or less retarded on the column. This illustrates the difference in polarity between different undissociated salts of bile acids.

If there is reason to suspect the presence of several salt forms, one should convert the bile acids into only one form before attempting a chromatography in neutral organic solvents. This can be done with the cation exchanger Amberlyst A-15 [free and glycine-conjugated bile acids can be extracted with ethyl acetate from an acidified water solution but this cannot be done with taurine conjugates (27)]. To obtain the acids, the material is dissolved in chloroform–methanol, 1:4 (v/v), and the solution is applied to an Amberlyst A-15 column in H^+ form packed in the same solvent. Different salts can be obtained by having the ion exchanger in the appropriate form.

3. Reversed Phase Partition Chromatography

Reversed phase partition systems have been described both for conjugated and free bile acids (1) (straight phase systems for free bile acids are described in Section III.B.2). Suitable solvent systems are summarized in Table I. Three types of support have been employed: acid-washed siliconized Hyflo Supercel, Hostalen, and methylated Sephadex (1, 40). The first-mentioned type is the most versatile one for two-phase solvent systems. The Sephadex derivative should be used only with solvents of the C-type; this material is designed for use with miscible solvent systems (see Section III. B.3). The mobility of some bile acids in different two-phase systems are shown in Table II. These values refer to use of Hyflo Super-Cel which takes up 4 ml stationary phase/4.5 g; retention volumes are smaller with Hostalen (3 ml/4.5 g) and larger with methyl Sephadex (6 ml/4.5 g). A 4.5 g Supercel column should not be loaded with more than 25–30 mg of material—flow rate should be about 0.3–0.6 ml/min/cm^2 column cross sectional area. The

TABLE I. Solvent Systems for Reversed-Phase Partition Chromatography of Bile Acids

Solvent system	Mobile phase	Amount (ml)	Stationary phase	Amount (ml)	Application
D	water	300	*n*-butanol	100	Taurine–conjugated bile acids
C1	methanol–water	150:150	chloroform-isooctanol[a]	15:15	Separation of glycine-conjugated bile acids, trihydroxy- and dihydroxymonoketo bile acids
C2	methanol–water	144:156	chloroform-isooctanol	15:15	Same as for C1
F1	methanol–water	165:135	chloroform-heptane	45:5	Separation of monohydroxy-, diketo-, dihydroxy-, and monohydroxymonoketo bile acids
F2	methanol–water	180:120	chloroform-heptane	45:5	Separation of diketo- and monohydroxy bile acids
G	methanol–water	255:45	heptane	50	Nonpolar bile acid derivatives, e.g., bile acid esters

[a] 2-Ethyl-hexanol-1.

TABLE II. Approximate Retention Volumes of Bile Acids in Reversed Phase Liquid–Liquid Partition Chromatography on Hydrophobic Hyflo Supercel[a]

Bile acid[b]	Solvent system				
	D	C1	C2	F1	F2
taurocholic (TC)	front[c]	front	front	front	front
taurodeoxycholic (TD)	30[d]	front	front	front	front
taurolithocholic (TL)	60	22	-	front	front
glycocholic (GC)	>100	37	44	front	front
glycochenodeoxycholic (GCD)	-	120	-	24	front
glycodeoxycholic (GD)	-	132	-	24	front
glycolithocholic (GL)	-	>250	>250	80	40
$3\alpha,6\beta,7\beta$	-	140	-	front	front
$3\alpha,6\beta,7\alpha$	-	80	-	front	front
$3\alpha,7\beta,12\alpha$	-	50	60	front	front
$3\alpha,7\alpha,12\alpha$ (C)	-	100	130	front	front
$3\alpha,12\alpha$,7-keto	-	65	95	front	front
3α,7,12-diketo	-	-	-	35	20
3,7,12-triketo	-	-	-	90	60
$3\alpha,6\alpha$	-	-	-	30	-
$3\alpha,7\beta$ (UD)	-	-	-	30	-
$3\alpha,7\alpha$ (CD)	-	>250	>250	50	35
$3\alpha,12\alpha$ (D)	-	>250	>250	50	35
3α,12-keto	-	-	-	90	60
3,12-diketo	-	-	-	-	120
3α (L)	-	-	-	-	100

[a] The values in the table refer to results obtained with a 4.5 g column.
[b] Compound designation: Substituted 5β-cholanoic acids. Greek letters denote configuration of hydroxyl groups at C-3, C-6, C-7, or C-12. G = glyco, T = tauro. Other capital letters are abbreviations for common bile acids.
[c] Compound eluted with, or close to, the void volume (12–20 ml).
[d] Milliliters of effluent at peak fraction.

effluent from the columns may be monitored by titration, sulfuric acid spectra (cf. Table III), thin-layer chromatography (TLC) or gas–liquid chromatography (GLC).

In work with crude biological bile acid mixtures it has proved advantageous to make an initial separation with a phase system of type F followed by rechromatography of material eluted with the solvent front in phase systems of the C or D type. The only drawback with these latter systems is that isooctanol and *n*-butanol are difficult to evaporate.

Compounds that are retained on the column after chromatography can be eluted with chloroform–methanol, 1:1. The advantage with the reversed phase partition chromatography on silanized supports is that the loss of polar compounds is kept at a minimum. Conditions for quantitative work with this method have been worked out by Mirvish (41).

TABLE III. Solvent Systems for Quantitative Analysis of Conjugated Bile Acids by Paper Chromatography

Solvent system[a]	Conditions for chromatography	Bile acids[b]	Conditions in 65% sulfuric acid	Absorbance maximum of chromogen (nm)	Approximate $E_{1\,cm}^{1\%}$ at absorbance maximum
I/H/F/W 85:15:70:30 upper phase mobile	equilibration 0.5 hr in vapors of mobile phase. Ascending, 20 hr.	TC	60 min, 20°[c]	320	290
		TCD+TD	15 min, 50°[c]	305	170
E/H/A/W 50:50:70:30 lower phase mobile	equilibration 8 hr in vapors of both phases. Descending, 18 hr.	GC	60 min, 20°[c]	320	350
		GUD	15 min, 60°[c]	305	210
		GCD	15 min, 60°[c]	305	210
		GD	10 min, 60°[c]	308	200

[a] I, isoamyl acetate; H, heptane; F, formic acid; W, water; E, ethylene chloride; A, acetic acid. All solvents are redistilled. Ethylene chloride is washed with concentrated sulfuric acid, water, dried over K_2CO_3, and distilled.

[b] For abbreviations see Table II. Compounds are listed in order of increasing R_f value.

[c] 3,6,7-Trihydroxycholanoic acids do not demonstrate UV absorption under these conditions.

4. Paper and Thin-Layer Chromatography

When the concentration of bile acids in the primary or purified extract is sufficiently high, paper or thin-layer chromatography can be used for further analysis. Of the two methods, paper chromatography with aqueous acetic acid as stationary phase permits a better separation of conjugated bile acids but is more tedious to use. For example, the separation factor between deoxycholic and chenodeoxycholic acids was 1.36 after 18 h of descending chromatography using isopropyl ether–heptane, 20:80 (v/v), as mobile phase and 70% aqueous acetic acid as stationary phase. Under the same conditions, using isopropyl ether–heptane, 60:40 (v/v), as mobile phase the separation factor between the glycine conjugates of these acids was 1.40. The fact that the separation factors are the same for the free and conjugated acids indicates ideal partition chromatographic conditions, where functional groups are expressed to the same extent, irrespective of the presence of other groups (42). Paper chromatography is also the only method that has permitted a separation of the taurine conjugates of the two acids mentioned (46 hr descending chromatography using *n*-amyl acetate–heptane, 80:20, as mobile and 70% acetic acid as stationary phase). Table III lists the solvent systems recommended for quantitative paper chromatography of bile and intestinal contents (43, 44). This method has been reviewed in detail (1, 15). The same types of solvent systems are useful both for conjugated and free bile acids and for bile acid methyl esters. In qualitative work isopropyl ether may be used instead of ethylene chloride (45). The mobility of the compounds is controlled by the relative amount of heptane in the mobile phase and it decreases when the proportion of heptane is increased. Equilibration of the papers with stationary phase becomes increasingly important with the higher proportions of heptane. Failure to obtain equilibration results in tailing or elongated spots. This is prevented by pretreatment of the papers with 70% acetic acid. In descending chromatography, too high flow rates also result in tailing.

Glycine-conjugated dihydroxy bile acids can be separated under the conditions given in Table III. These types of systems are also useful for the separation of some 3,7,12- and 3,6,7-trihydroxycholanoic acids. Glycine conjugation lowers the R_f value somewhat more than the addition of a hydroxyl group. The reduction of a bile acid to a bile alcohol also lowers the R_f value. A methyl ester has a higher R_f value than the acid; the difference corresponds roughly to the removal of one hydroxyl group from the acid. In most cases, an axial hydroxyl group gives a higher mobility than an equatorial one in the same position. Ketonic bile acids analyzed in isopropyl ether–heptane solvents usually have a lower R_f value than the corresponding hydroxy acids. Bile acids with very similar mobilities can often be efficiently

TABLE IV. Thin-Layer Chromatography of Bile Acids on Silica Gel G[a]

Compounds[c]	Solvent Systems[b]														
	BAW-1	BAW-2[d]	CMAW[d]	TU-4	S-VIII	S 6	JG-1	JG-1[e]	JG-2[e]	AK II[f]	AK-IV[g]		JHA[h]	JHB[h]	JHC[h]
(C_{27}) TC		0.38	0.35												
TC	0.08	0.21	0.24	0.12	0.09	0.00	0.00	0.06	0.09	1.60	0.27	0.64	0.03	0.19	0.43
TD	0.21	0.40	0.35	0.25	0.20					2.30	0.27	1.14	0.09	0.56	0.75
TCD	0.21	0.40	0.35		0.20	0.00	0.00	0.12	0.16	2.30	0.27	1.14	0.09	0.55	0.76
TL					0.28	0.00	0.03	0.16	0.25	2.36	0.27	1.70	0.24	0.86	0.88
GC	0.35	0.54	0.57	0.53	0.43	0.04	0.10	0.22	0.32	1.00	0.27	0.64	0.43	0.89	0.24
GD	0.50	0.78	0.86	0.80	0.59	0.25	0.30	0.39	0.53	1.60	0.27	1.14	0.73	1.00	0.56
GCD	0.50	0.78	0.86		0.59	0.25	0.26	0.36	0.53	1.60	0.27	1.14	0.68	1.00	0.61
GUD		0.81	0.88												
GL					0.69	0.66	0.53	0.55	0.73	1.60	0.27	1.70	0.94	1.00	0.82
C	0.64			0.83	0.66	0.48	0.42	0.45	0.61	1.00[f]	0.27	1.00[g]	0.98	1.00	0.33
D	0.68			0.93	0.77	0.84	0.74	0.68	0.79	1.60	0.31	2.22	0.98	1.00	0.78
L	0.66			0.98	0.78	0.93	0.90	0.80	0.89	2.35	0.44	2.64	0.98	1.00	0.92

[a] The values given are R_f values unless otherwise indicated.
[b] See Table X.
[c] For abbreviations see Table II. C_{27} is 5β-cholestanoic acid.
[d] The values in this column refer to chromatoplates developed 13 cm from the starting line.
[e] The figures in this column refer to the use of the wedge technique, i.e. the sample is applied at the upper part of the straight strip of adsorbent between two hexagonal areas (1.5 cm-side) from which the adsorbent layer has been removed (see 130).
[f] The values in this column refer to chromatoplates developed 14 cm from the starting line. Mobilities are given relative to that of cholic acid (C). Its absolute mobility was 4 cm.
[g] The values in the two columns refer to chromatoplates developed 14 cm from the starting line. Mobilities are given relative to that of methyl cholate. Absolute mobility of this compound:2.5 cm. The values in the right column are those obtained after treatment of the compounds listed in the left column with BF_3– methanol (see 131).
[h] These solvents are used with silica gel-impregnated glass fiber paper; see (85) and Table XI.

separated if solvent overflow is permitted in solvent systems giving a low mobility rate of the compounds.

The solvent systems for taurine conjugates can be used for bile alcohol sulfates (10). Improved separations were observed with the upper phase of *n*-amyl acetate–heptane–acetic acid–water, 85:15:103:47 (by vol.) (46). Such solvent systems should also be suitable for the bile acid 3-sulfates recently described 47).

Other acidic and some basic solvent systems for separation of bile acids on paper have been described. These have not been used to the same extent as those described above and references are found in (1). Neutral solvent systems [of the type used in steroid analysis (48)] have been employed by Haslewood (49) for separation of bile acid esters.

Phosphomolybdic acid, 10% in ethanol, is the reagent most commonly used to visualize the spots. Other reagents have been described (see 1, 14).

Although separation factors between conjugated bile acids are not as good in thin-layer as in paper chromatography, the thin-layer method has been much more used in recent years because of its greater simplicity and speed. The mobilities of common conjugated bile acids in different solvent systems are given in Table IV. It should be observed that several systems can be used in two-dimensional development.

TLC of conjugated bile acids has been used in many methods for analysis of different biological fluids, mainly bile, intestinal contents, and blood plasma (4, 11, 12). Common to all methods are two problems: interference by other compounds and difficulty in eluting and/or determining the amount of bile acid in the spots. In many descriptions of TLC methods the claims for specificity are not well founded. Gas chromatographic methods exist whereby the specificity can be checked after hydrolysis (see Section IV.C). In our experience these methods sometimes also fail unless they are combined with mass spectrometry.

Several methods have been used to obtain quantitative extraction of conjugated bile acids from silica layers (4, 11, 12). Palmer (50) moistened the silica (removed from the plate with a razor blade) with water and extracted the bile acid by refluxing with chloroform–methanol (redistilled from 2,4-dinitrophenylhydrazine), 1:1, for one hr in a 250-ml round-bottom flask. He then determined the amount of bile acid with a 3α-hydroxysteroid dehydrogenase reaction. The necessity of using large volumes of polar solvent is also evident from the work by Shioda *et al.* (20) who used 50 ml of methanol–acetic acid, 99:1, and achieved at least 96% recovery of all common conjugated bile acids in human bile labeled *in vivo* by cholesterol.

Other workers have not been successful in eluting conjugated bile acids and have therefore made quantitative determinations by forming sulfuric acid chromogens with the gel still carrying the bile acid (51–53). In such procedures the sensitivity of chromogens to interfering material (e.g., heavy metals) should be remembered (1, 52). Quantitation by densitometry has also been described (54).

Radioactivity on paper or thin-layer plates can be determined by gas-flow counting in scanners, by direct counting of paper pieces or silica gel in liquid scintillation counters, or by autoradiography (12).

5. Gas–Liquid Chromatography

In a short note Hanaineh and Brooks describe separation of methyl ester trimethylsilyl ethers of glycine-conjugated bile acids (55). This could perhaps serve as a basis for direct analysis of bile and a similar method has been employed by M. and E. Horning (56). It is not yet possible to separate taurine-conjugated bile acids by gas chromatography.

B. Hydrolysis of Conjugated Bile Acids and Subsequent Purification Steps

In many studies it is not necessary to isolate conjugated bile acids. It is simpler to work with free bile acids and a hydrolysis can be made at different stages of the work-up procedure. One should realize, however, that the nature of conjugated bile acids is not known in all details; ornithine-conjugated bile acids have been described (57), esterified bile acids occur in feces (19), and sulfates were recently found in bile (47).Information about these compounds is lost upon hydrolysis.

1. Hydrolysis

Enzymatic or alkaline hydrolysis can be made (acid hydrolysis to obtain the amino acid). The enzymatic method is described in detail by Nair in another chapter. It is a mild method, and should be preferred to the alkaline hydrolysis in many analytical procedures. All details on the structural requirements are not yet known and some bile acid types are not hydrolyzed. For this reason both enzymatic and alkaline hydrolysis should be used in parallel in identification work.

The alkaline hydrolysis is vigorous, and some loss of bile acid occurs. Ketonic bile acids (particularly 3-ketones) are most likely to give low recoveries. In our experience the losses are larger with biological extracts than when pure bile acids are subjected to these conditions. The possibility of artifact formation should always be kept in mind. Two methods are used more commonly than others: 1.25 M NaOH, 3 hr, 15 psi, and 15% NaOH in 50% aqueous ethanol 10 hr, 110°C (1, 27, 58). Teflon or nickel vessels should be used instead of glass. Some authors have hydrolyzed bile acids with 20% KOH in ethylene glycol at reflux temperature. This method was carefully tested by Evrard and Janssen who found 80–100% recovery after 15–20 min of refluxing (23).

The rate of hydrolysis may vary with the bile acid structure. The hindered nature of the peptide bond in the taurine-conjugated 3α,7α,12α-trihydroxy-5β-cholestanoic acid may explain why this conjugate is more resistant to saponification than C_{24} acids (59).

Norman and Palmer found that metabolites of lithocholic acid in human bile were not hydrolyzed by 1 *N* NaOH, 6 hr, 110 °C, or by 4.5 N NaOH, 6 hr, 130 °C. Complete hydrolysis was achieved with 5 N NaOH, 48 hr, 135 °C. Such conditions resulted in formation of several products from 3-keto-5β-cholanoic acid (60).

Alkaline hydrolysis of bile acid sulfates may lead to artifacts. Sulfates can be cleaved by treatment with 40 % trichloroacetic acid in dioxane for a few days at room temperature (61). Palmer used a modification of solvolytic procedures for steroid sulfates (47, 62).

After alkaline hydrolysis, the solution is acidified to *p*H 1 and the bile acids are extracted with ethyl acetate (if the water phase contains an alcohol it is diluted with water). We have found ethyl acetate to give better and more consistent recoveries than diethyl ether, which is commonly used.

When mixtures of bile acids are hydrolyzed at an early stage of the analytical procedure, the free bile acids formed may have to be further purified. Extraction, counter-current distribution, reversed phase partition, and paper chromatography have been described (Section III.A) and these methods can also be used for free bile acids.

2. Straight-Phase Partition Chromatography

The straight-phase liquid partition solvent systems that have been successfully applied to bile acid separations contain light petroleum mixed with isopropyl ether or benzene as mobile phase, equilibrated with 1/3 of its volume of 70 % acetic acid (1, 63, 64). Celite 545 is used as support after prewashing, and 10 g of Celite 545 can absorb 8 ml of the stationary phase. An advantage with these systems is that they permit the use of nonsilanized support and gradient elution. The sample (less than 100 mg of crude bile acids) is dissolved in 0.5 ml of stationary phase which is then added to 0.5 g of Celite. The resulting mixture is transferred to the column which is eluted with mobile phases containing increasing amounts of benzene or isopropyl ether in petroleum ether. Deoxy- and hyodeoxycholic acids are eluted in this order with 40 % isopropyl ether, and cholic and 3,7,12-triketo-5β-cholanoic acids in that order with 60 % isopropyl ether in petroleum ether. Ketonic acids are usually somewhat more retarded than the corresponding hydroxy compounds (1, 37). This is in contrast to the polarity relations in reversed phase chromatography with neutral solvents. The tendency for keto bile acids to be more retarded in markedly acidic straight phase systems can also be noted in TLC. However, the mobilities of keto and hydroxy acids in acidic straight phase systems on columns are sufficiently similar to permit a separation of hydroxy and/or keto acids into groups of mono-, di-, and trisubstituted bile acids. This is an advantage that counteracts the inconvenience of handling many fractions containing acetic acid. The formation of artifacts due to esterification with acetic acid has to be considered in quantitative work.

3. Liquid–Gel Chromatography

This is the most recent column chromatographic technique. It employs lipophilic or hydrophobic derivatives of Sephadex (40, 65). Miscible solvent systems are used and the separations depend on the formation of a stationary gel phase and a mobile phase, which are either similar or different in polarity. In the former case molecular sieving effects dominate (40, 66), in the latter case a straight or reversed phase system is obtained, probably depending on whether the polar or nonpolar component of the solvent mixture is enriched in the gel phase. Straight phase separations of bile acids can be obtained using Sephadex LH-20 in chloroform with a low percentage of methanol (39, 40). Neutral lipids are eluted within one column volume whereas lithocholic acid leaves the column only after 1.5–2 column volumes when chloroform–ethanol, 99:1, is used as the solvent. One advantage with these gels is that quantitative recoveries can always be obtained (40, 66). If the gels are used for purification of bile acid methyl esters some useful effects are noted with halogenated solvents such as chloroform (67). Ketonic compounds are eluted much faster than the corresponding hydroxy compounds. Methyl cholate, run on Sephadex LH-20 in chloroform, is eluted after about 3–5 bed volumes.

Reversed phase partition conditions are established with alkylated Sephadex LH-20 containing 50% (w/w) of C_{11}–C_{14} alkyl chains, when solvents such as methanol–water–ethylene chloride, 70:30:10, are used. With this system 3α-hydroxy-, 3β-hydroxy-, and 3-keto-5β-cholanoic acids are completely separated in this order (65). The importance of liquid–gel chromatography in bile acid analysis is likely to increase during the next few years.

4. Adsorption Chromatography

Column adsorption chromatography should be applied when one deals with very crude extracts, such as those obtained from feces, and when a group separation between mono-, di-, and trisubstituted hydroxycholanoic acids is desirable. It may also be used for large-scale purification of a specific bile acid, e.g., in synthetic work. Silicic acid should be used for free bile acids since more active adsorbents may give incomplete recoveries. Methyl ester or methyl ester acetates are usually best separated on aluminum oxide. In

TABLE V. Adsorption Chromatography of Bile Acids

Silicic acid chromatograpy of fecal bile acid fractions						Aluminum oxide chromatography of mono- and disubstituted bile acid methyl esters from feces	
Acetone (% in benzene)[a]	Bile acids eluted[b]	Ethyl acetate (% in benzene)[c]	Bile acids eluted	Ethyl acetate (% in benzene)[d]	Bile acids eluted	Heptane or ethyl acetate (% in benzene)[e]	Bile acid methyl ester eluted
0	cholanoic	5	3-keto	50	—	90	3-keto
							(3β)
2	3-keto[f]	10	3β	55	3α,12α,7-keto	80	3β
	(3β)		3α		3α,7α,12-keto		3α
4	3β	15	3β,12-keto	60	3α,12α,7-keto	70	3α
			(3α)		3α,7α,12-keto		
8	3β				(3α,7β,12α)	60	3α
	3α	20	3β,12-keto				
	(12α,3-keto)		3α,12-keto	65	3α,7β,12α	50	3α
	(7α,3-keto)				3β,7β,12α		
		22	3α,12-keto		(3β,7α,12α)	40	3α
12	3α				(3α,7α,12α)		3β,12-keto
	(12α,3-keto)	24	3β,12α				
	7α,3-keto		3α,12α	70	3β,7α,12α	30	3β,12-keto
	3β,12-keto				3α,7α,12α		(3α,12-keto)
	3α,12-keto				(3α,7β,12α)		
					(3β,7β,12α)		
16	3β,12-keto					20	3α,12-keto
	3α,12-keto			75	3α,7α,12α		
					3β,7α,12α	10	3α,12-keto
20	3β,12α						(3β,12α)
	3α,12α			80	3α,7α,12α		
					(3β,7α,12α)	0	3α,12-keto
24	3α,12α			100	(3α,7α,12α)		(3β,12α)

TABLE V. (Continued)

Silicic acid chromatograpy of fecal bile acid fractions						Aluminum oxide chromatography of mono- and disubstituted bile acid methyl esters from feces	
Acetone (% in benzene)[a]	Bile acids eluted[b]	Ethyl acetate (% in benzene)[c]	Bile acids eluted	Ethyl acetate (% in benzene)[d]	Bile acids eluted	Heptane or ethyl acetate (% in benzene)[e]	Bile acid methyl ester eluted
28	$3\alpha,12\alpha$			—	—		
						10[e]	$3\beta,12\alpha$ ($3\alpha,12\alpha$)
						20[e]	$3\alpha,12\alpha$

[a] Silicic acid Mallinckrodt, 20-fold excess. The column was eluted with 20 ml of each solvent per gram silicic acid.
[b] Substituted 5β-cholanoic acids. Greek letters denote configuration of hydroxyl groups on C-3, C-7, or C-12. Figures in parenthesis indicate overlapping.
[c] Silicic acid Mallinckrodt, 25-fold excess. The column was eluted with 40 ml of each solvent per gram silicic acid.
[d] Silicic acid Mallinckrodt, 50-fold excess. The column was eluted with 20 ml of each solvent per gram of silicic acid.
[e] Aluminum oxide, Woelm, activity grade V, 15-fold excess, eluted with 2 ml of each solvent per gram of aluminum oxide. Figures referring to ethyl acetate in italics.

the choice between separation of free bile acids on silicic acid (Mallinckrodt, Silicic acid AR, 100 mesh-powder, activated for 5 days at 120°C), and of methyl esters on aluminum oxide (Woelm), it should be recalled that aluminum oxide can give better separations of stereoisomers of hydroxycholanoates than silicic acid (e.g., the separation of methyl 3α- and 3β-hydroxycholanoates and the methyl esters of deoxycholic and allodeoxycholic acids) (68–70). Examples of the types of separations obtained are shown in Table V.

When a group separation is attempted on silicic acid, a 25-fold excess of adsorbent is used, and four fractions are collected: (1) benzene 10 ml/g silicic acid, (2) benzene–acetic acid, 99:1, 30 ml/g silicic acid, (3) benzene–acetic acid, 3:1, 40 ml/g silicic acid, and (4) chloroform–methanol, 1:1, 10 ml/g silicic acid. Fatty acids and sterols appear in the first two fractions where also unsubstituted cholanoic acid will be eluted. The third fraction contains mainly mono- and disubstituted bile acids and the last one the bulk of trisubstituted cholanoic acids (18). Silicic acid is sometimes very efficient for the separation of protected bile acids. Thus Wootton (71) obtained a partial separation of the methyl ester diacetates of deoxycholic and chenodeoxycholic acids on silica gel (Davison 923, Davison Chemical Corp., Baltimore).

Other adsorbents may be useful for specific purposes. Florisil, 100–200 mesh, purified by hydrochloric acid, water, and methanol washings, and subsequently activated for 2–3 hr at 280°C, has been sucessfully used for the separation of cholic and allocholic acids (72). Charcoal has been successfully used to purify bile acids in extracts from rat feces (73). The original publication does not appear to have tempted other workers to use this adsorbent in bile acid purification, but the results seem to justify further systematic tests, especially since it is a rapid procedure compared to the adsorption methods mentioned above.

IV. CHROMATOGRAPHIC IDENTIFICATION AND QUANTITATION

Thin-layer and gas chromatography are used to study the bile acid content of various fractions during the work-up of a sample. If an extract is found to be too impure, further purification steps must be carried out. It is important to realize that large amounts of many contaminants may come from solvent residues, reagents, adsorbents, glassware, skin surface, and the like. If these sources supply the contaminants, repeated purification steps are of little value.

When the bile acids constitute at least a few weight percent of the purified material, TLC and GLC can usually be applied. If bile acids in complex mixtures are to be identified, it is usually best to start the analysis by TLC of the free acids, followed by derivatization of the isolated compounds and repeated TLC. The compounds thus obtained should be subjected to GLC and, for definite identification, to gas chromatography–mass spectrometry.

A. Thin-Layer and Glass Fiber Paper Chromatography

Silica Gel G and H (Merck AG, Darmstadt, Germany) and Anasil B (Analytical Engineering Laboratories, Inc. Hamden, Connecticut) are the adsorbents most frequently employed for bile acid separations. A few exceptions can be found in Tables IV, VI, VII, VIII, and IX. As seen from Table X, mainly neutral or acidic solvents have been used. Ascending development is normally performed, but the technically more difficult descending development can also be of value for some types of separations (74). For analytical purposes, plates can be stained with a number of different reagents (see 14, 75, 76). As a charring agent we prefer a saturated solution of potassium dichromate in aqueous 80% sulfuric acid. To produce characteristic colors with some common bile acids the procedure of Kritchevsky *et al.* (75) can be used if it is recalled that other types of compounds may interfere with the reactions. The sensitivity obtained by staining with a 15% ethanolic solution of phosphomolybdic acid has been noted by Hofmann (4). Thus, hydroxy bile acids may be detected in amounts as small as 0.1 μg. In preparative work nondestructive detection is preferable, e.g., staining with iodine or spraying with water (51). Colored marker compounds with known relative mobilities can also be run with the bile acids. Once located, the bile acids may be eluted from the gel with solvents such as methanol (77), methanol–acetic acid (20), water–saturated ethyl acetate (78), or ether containing 2% acetic acid (79). To improve the extraction yields of polar bile acids it may be advantageous in preparative TLC to run nonpolar bile acid derivatives such as methyl esters, methyl ester acetates (4), methyl ester trimethylsilyl ethers (12), or methyl ester oximes (80).

Sometimes the methyl esters are better separated than the free bile acids. This is especially true when neutral solvents are used. In acidic solvents the difference is small, particularly with trisubstituted bile acids. Thus, the same types of acidic solvents can be used for free and methylated bile acids. When bile acid methyl esters are acetylated or converted into trimethylsilyl ethers their mobility is considerably increased. The separation of configurational isomers is then often impaired whereas the separation of positional isomers may be improved (4). This is often the case when neutral derivatives are analyzed in neutral solvent systems (12). The influence of the nature and position of nuclear substituents on the mobility of a bile acid can be estimated

TABLE VI. Thin-Layer Chromatography of Tetra-and Trisubstituted Bile Acids

Substituents[b]	Solvent systems[a] for															
	Bile acids											Bile acid methyl esters				
	S1[c]	S4	S6[d]	S7	S7[e]	CMA[f]	K M	S-VI	S-VIII[d]	A-10	B-70	EA-2	S-IV	S-VII	S6[g]	A-III[h]
(C_{27}) 3α7α12α24ξ				0.61												
3α7α12α23ξ	0.04	0.07	0.28	0.06												
3α7α23ξ	0.15	0.25	0.43	0.26												
3α12α16[i]	0.67	2.50	2.54	1.62												
3α6α7α	0.30	1.30	1.49	1.12		1.06		1.00	1.00				1.85	1.49		1.29
3α6β7α		1.17	1.14	0.82	0.82	0.90		1.14	0.95				2.00	1.18		1.88
3α6α7β						1.10										
3α6β7β		1.11	0.86	0.92	0.90	1.19		1.48	1.01				2.38	1.78		2.53
(C_{22}) 3α7α12α							0.67									
(C_{23}) 3α7α12α							0.83									
3α7α12α	0.17	1.00[j]	1.00[j]	1.00[j]	1.00[j]	1.00[j]	1.00[j]	1.00[j]	1.00[j]	1.00[j]	1.00[j]	1.00[j]	1.00[j]	1.00[j]	1.39	1.00[j]
3α7α12α(5α)				0.86											1.00[j]	
(C_{25}) 3α7α12α							1.17									
(C_{26}) 3α7α12α							1.46									
(C_{27}) 3α7α12α[k]	0.13	1.61	1.88	1.38			1.75									
(C_{28}) 3α7α12α[k]							1.92									
(C_{29}) 3α7α12α[k]							2.12									
3α7β12α	0.30	1.47	1.75	1.22						2.31	1.00	1.25				
3α11β12α									1.17				5.15			4.47
3β7α12α		1.44	1.00	0.87								1.08				
3β7β12α			1.53	1.08												
3α7α,12-keto	0.45	1.94	2.32	1.37									4.77			3.65
3α12α,7-keto	0.45	1.92	2.32	1.37				2.62		2.94	1.22	1.50	4.62			3.30
7α12α,3-keto	1.31	2.94	3.02	1.66									6.77			5.00
3α,7,12-diketo	1.07	2.58	2.65	1.52				4.05		3.69	1.09		9.70			4.18
7α,3,12-diketo								6.62					7.92			7.24
3,7,12-triketo	2.13							7.05		5.00	1.48					7.06
Mobility of reference compounds (cm)	5.4[c]	4.0	5.1	9.0				2.0		2.5	4.0	9.0	1.0	2.5		1.5

[a] See Table X.
[b] For abbreviations see Tables II and V. 5α is 5α-cholanoic acid.
[c] In this particular solvent mobilities are given relative to that of deoxycholic acid (cf. Tables VII and VIII).
[d] This solvent is also used for conjugated bile acids (see Table IV).
[e] Descending TLC for 3.5 hr on 500-μ layers of Silica Gel G (74).
[f] Descending TLC for 4.5 hr on 250-μ layers of Silica Gel G (74).
[g] In this solvent mobilities are given relative to that of methyl allocholate.
[h] With this solvent Anasil B is used as adsorbent.
[i] This compound has been analyzed as the lactone.
[j] Mobilities are given relative to this compound.
[k] 24-Methyl (C_{28}) and 24-ethyl (C_{29}) substituents, respectively, of 5β-cholestanoic acid (C_{27}).

TABLE VII. Thin-Layer Chromatography of Disubstituted Bile Acids

Substituents[b]	Solvent systems[a] for										
	Bile acids							Bile acid methyl esters			
	S1[c]	S11	S12	S-V	S-VI[d]	B-70[d]	A[e]	S12	S-IV[f]	A-II[g]	A-III[h]
3α6α	0.63	0.50	0.56	0.50	0.61		0.55		0.51		0.55
3α6β		0.72		0.63	1.03				1.03		1.15
3α7α	1.00	0.88	1.00	0.79	0.98	0.86	0.75	1.00[i]	0.94		0.89
3α7α(5α)								0.79			
3β7α	1.26	0.95	1.33								
3α7β	1.14	0.78	1.00	0.68	1.06				1.14		1.12
(C_{22}) 3α11α					0.87						
(C_{23}) 3α11α					1.13						
3α11α				0.85	1.20				1.22		1.07
3α11β				1.32	2.32				2.21		1.62
(C_{20}) 3α12α					0.67				0.62		
(C_{22}) 3α12α					0.84				0.84		
(C_{23}) 3α12α					1.00				0.89		
3α12α	1.00[i]	1.00[i]	1.00[i]	1.00[i]	1.00[i]	1.00[i]	1.00[i]		1.00[i]		1.00[i]
3α12α(5α)			1.29								1.26
3β12α	1.36	1.06	1.40						1.38		1.39

TABLE VII. (Continued)

Substituents[b]	Solvent systems[a] for										
	Bile acids							Bile acid methyl esters			
	S1[c]	S11	S12	S-V	S-VI[d]	B-70[d]	A[e]	S12	S-IV[f]	A-II[g]	A-III[h]
3β12α(5α)			1.38								
3α12β	1.58	1.09	1.53								
7α12α	2.51	1.62	2.23						2.00		1.36
3α,6-keto		1.04									
3α,7-keto	1.63	0.95	1.44	0.77							
3α,12-keto	1.78	1.17		0.98		0.99					
7α,3-keto	2.13	1.35	2.11	1.11							
12α,3-keto	1.85	1.45	2.36	1.03							
12α,3-keto(5α)			2.39								
12α,3-keto,Δ[4]			1.96								
3,6-diketo										0.66	
3,6-diketo(5α)										0.37	
3,7-diketo	2.65										
3,12-diketo	2.65					1.04				0.79	
3,12-diketo(5α)										0.59	
Mobility of reference compound (cm)	5.4	9.1	5.7	4.5	6.0	14.0	8.5		6.5		8.5

[a] See Table X.
[b] For abbreviations see Table VI; Δ is a double bond.
[c] This solvent is also used for tri- and monosubstituted bile acids (cf. Tables VI and VIII).
[d] This solvent is also used for trisubstituted bile acids (cf. Table VI).
[e] With this solvent, polyamide (Miramid FP) is used as adsorbent, and the plates are developed 12 cm utilizing the wedge technique [see Table IV, Footnote *e* and (132, 133)]. With this system lithocholic, hyocholic, and cholic acids have a mobility relative to that of deoxycholic acid of 0.30, 0.97, and 1.24, respectively.
[f] This solvent is also used for trisubstituted bile acids (cf. Table VI).
[g] With this solvent Anasil B is used as adsorbent. Mobilities are given relative to that of methyl lithocholate (cf. Table VIII).
[h] With this solvent Anasil B is used as adsorbent. The solvent is also used for trisubstituted bile acids (Table VI).
[i] Mobilities are given relative to this compound.

TABLE VIII. Thin-Layer Chromatography of Monosubstituted Bile Acids

Substituents[b]	Solvent systems[a] for						
	Bile acids				Bile acid methyl esters		
	S1[c]	S13	S14	S15	S-I	S-III	A-II[d]
3α	2.54	1.00[e]	1.00[e]	1.00[e]	1.00[e]	1.00[e]	1.00[e]
$3\alpha(5\alpha)$							1.16
3β		1.13	1.08	1.15	1.62	1.04	1.17
7α	3.20	1.35	1.24	1.40	4.00	1.55	1.66
7β		1.19	1.10	1.21			
12α		1.37		1.40	5.00		
12β		1.28		1.32			
3-keto	3.10	1.50	1.21	1.35	4.00	1.73	1.41
7-keto		1.62	1.27	1.50	6.75	1.92	1.69
12-keto		1.62	1.27	1.50	8.25		
unsubstituted		1.73	1.58	1.89			
mobility of reference compound (cm)	5.4[c]	9.4	8.9	7.5	1.5	7	8

[a] See Table X.
[b] For abbreviations see Table VI.
[c] Mobilities in this system are given relative to that of deoxycholic acid. This solvent is also used for tri- and disubstituted bile acids (cf Tables VI and VII.)
[d] This solvent is also used for disubstituted bile acids (cf. Table VII).
[e] Mobilities are given relative to this compound.

TABLE IX. Solvent Systems for the Separation of Some Mono-, Di-, and Trisubstituted Bile Acids that Overlap in Screening System S1

Substituents[b]	Solvent system[a]				
	S1	S9	S10	S11[c]	S13
$3\alpha,7\alpha,12\alpha$	0.17	0.76	0.50	0.22	
$3\alpha,7\alpha$,12-keto	0.45	0.66	0.62	0.48	
$3\alpha,6\alpha$	0.63	0.67	0.71	0.50	0.10
$3\alpha,12\alpha$	1.00[d]	1.00[d]	0.83	1.00[d]	0.21
3α,7,12-diketo	1.07	0.54	0.53	0.52	
$3\beta,7\alpha$	1.26	0.85	0.73	0.95	
$7\alpha,12\alpha$-3-keto	1.31	0.87	0.81	0.63	
$3\alpha 12\beta$	1.58	1.00	0.87	1.09	0.34
3α-7-keto	1.63	0.88	0.70	0.95	0.45
7α-3-keto	2.13		0.93	1.35	0.74
3,7,12-triketo	2.13	0.55	0.46	0.91	0.46
$7\alpha,12\alpha$	2.51		0.98	1.62	0.80
3α	2.54	1.28	1.00[d]	1.60	1.00[d]
3,7-diketo	2.65	1.10	0.85		0.93
3,12-diketo	2.65	1.17	0.91		0.93
Mobility of reference compound (cm)	5.4	9.8	12.0	9.1	9.1

[a] See Table X.
[b] For abbreviations see Table VI.
[c] This solvent is also used for disubstituted bile acids (cf. Table VII).
[d] Mobilities are given relative to this compound.

TABLE X. Designation, Composition and Application of Solvent Systems in Thin-Layer Chromatography of Bile Acids

Abbreviation[a]	Composition[b]	Proportions (v/v)	Reference	Application
BAW-1	*n*-butanol–HAc–H_2O	10:1:1	(51)	Table IV
BAW-2	*n*-butanol–HAc–H_2O	85:10:5	(134)	Table IV
CMAW	$CHCl_3$–methanol–HAc–H_2O	65:20:10:5	(134)	Table IV
TU-4	ethyl acetate–methanol–HAc	70:20:10	(135)	Table IV
S-VIII	isoamyl acetate–propionic acid–*n*-propanol–H_2O	4:3:2:1	(4)	Tables IV and VI
S-6	cyclohexane–ethyl acetate–HAc	7:23:3	(136,137,138)	Tables IV and VI
JG-1	2,2,4-trimethylpentane–isopropyl ether–isopropanol–HAc	2:1:1:1	(139)	Table IV
JG-2	ethylene dichloride–HAc–H_2O	10:10:1	(139)	Table IV
AK-II	$CHCl_3$–methanol–7 N NH_4OH–H_2O	90:45:10:5	(131)	Table IV
AK-IV	$CHCl_3$–methanol–7 N NH_4OH	80:40:4	(131)	Table IV
JHA	2,2,4-trimethylpentane–isopropyl ether–isopropanol–HAc	1:1:1:1	(85)	Table IV
JHB	$CHCl_3$–isopropyl ether–isopropanol HAc	3:2:2:2	(85)	Table IV
JHC	$CHCl_3$–isopropanol–NH_4OH	20:25:1	(85)	Table IV
S-l	C_6H_6–dioxane–HAc	75:20:2	(136)	Tables VI, VII, VIII, and IX
S-4	C_6H_6–dioxane–HAc	55:40:2	(136)	Table VI
S-7	C_6H_6–isopropanol–HAc	30:10:1	(136,140)	Table VI
CMA	$CHCl_3$–methanol–HAc	80:12:3	(74)	Table VI
KM	$CHCl_3$-ethyl acetate-HAc	45:45:10	(141)	Table VI
S-IV	C_6H_6–acetone	70:30	(4,142)	Tables VI and VII
S-VI	CCl_4–C_6H_6–isopropyl ether–isoamyl acetate–*n*-propanol–HAc	2:1:3:4:1:0.5	(4)	Tables VI and VII
S-VII	$CHCl_3$–acetone–methanol	70:25:5	(4)	Table VI
A-III	di-*n*-butyl ether–acetone	70:30	(4)	Tables VI and VII
A-10	isopropyl ether–butanone–HAc	10:4:1	(143)	Tables VI and VII
B-70	2,2,4-trimethylpentane–isopropyl ether–HAc	10:5:7	(143)	Tables VI and VII

TABLE X. (Continued)

Abbreviation[a]	Composition[b]	Proportions (v/v)	Reference	Application
EA-2	ethyl acetate–acetone	70:30	(144)	Table VI
S-11	2,2,4-trimethylpentane–ethyl acetate–HAc	10:10:2	(136,145,146)	Tables VII and IX
S-12	2,2,4-trimethylpentane–ethyl acetate–HAc	5:25:0.2	(87,136,137,138,147)	Table VII
S-V	2,2,4-trimethylpentane–isopropyl ether–HAc	2:1:1	(4,148)	Table VII
A	H_2O–butanone–25% NH_4OH	80:20:5	(132)	Table VII
A-II	di-*n*-butyl ether–acetone	85:15	(4,142)	Tables VII and VIII
S-13	2,2,4-trimethylpentane–ethyl acetate HAc	50:50:0.7	(136)	Tables VIII and IX
S-14	2,2,4-trimethylpentane–ethyl acetate –HAc	10:10:0.25	(136)	Table VIII
S-15	2,2,4-trimethylpentane–ethyl acetate–HAc	10:10:0.1	(87,136)	Table VIII
S-I	heptane–diethyl ether	70:30	(4)	Table VIII
S-III	C_6H_6–acetone	85:15	(4,142)	Table VIII
S-9	2,2,4-trimethylpentane–isopropanol–HAc	30:10:1	(136)	Table IX
S-10	2,2,4-trimethylpentane–isopropanol HAc	60:20:0.5	(136)	Table IX
A-20	2,2,4-trimethylpentane–butanone–HAc	100:58:2	(143,149)	Table XI
B-5	2,2,4-trimethylpentane–isopropyl ether–HAc	100:40:5	(143,149)	Table XI

[a] Whenever possible, the designation in the original paper is used.
[b] HAc = acetic acid.

from data presented in Table VIII. Substituents in a 5β-cholanoic acid decrease the mobility in the following approximate order: 3α (most polar), 3β, 7β, 12β, 7α, 12α, 3-keto, 12-keto (least polar). Equatorial substituents exert a more pronounced retarding effect than axial ones. Substituents at C-3 decrease the mobility proportionally more than those at C-7 or C-12. Whereas ketonic bile acids in most cases move faster than the corresponding hydroxy compounds, they can be strikingly retarded in some solvents, e.g., S9 and S10 (Table IX). Thus the relative mobility of 3α,7α,12α-trihydroxy- (0.76), 3α, 7α,-dihydroxy-12-keto (0.66), 3α-hydroxy-7,12-diketo- (0.54), and 3,7,12-triketo cholanic (0.55) acids in solvent S9 should be noted. The atypical behaviour of steroids in solvents containing alcohols has been discussed by Nienstedt (81) and Bush (82).

A typical solute behavior is not only caused by solvent effects. The chemical and geometrical structure of the gel is also of importance (83). Thus, methyl esters of allo (5α) bile acids have been better separated on Anasil B than on Silica Gel G (4). With the latter adsorbent, the 5α-isomers of chenodeoxycholic and cholic acids move more slowly than the 5β-epimers. Since allolithocholic (3α-hydroxy, axial) and allodeoxycholic acids in similar solvents have higher R_f values than the corresponding 5β compounds, it is tempting to assume a specific adsorbent-solute interaction when a 7α-hydroxyl group is present in an allo bile acid derivative. That steric factors can prevent compounds containing certain substituents from contact with the adsorbent can also be observed for 3α,12β-dihydroxycholanoic acid. This acid is much less adsorbed than the corresponding 12α-isomer in spite of the equatorial position of the 12β-hydroxyl group and irrespective of the solvent used (Table VII). Similarly, 3α,7β,12α-trihydroxy-5β-cholanoic acid (7β-hydroxy, equatorial) moves faster than the 7α-epimer in all solvents tested (Table VI). Furthermore, the presence of an axial 3β-hydroxy group instead of a 3α group does not necessarily lead to an increased mobility (Tables VI and VII). In summary, it is apparent that different functional groups in bile acids often interact in an unpredictable way. To change the nature of functional group interaction, derivatives should be prepared. For instance, the biologically interesting compounds 3α,7α-dihydroxy-12-keto- and 3α,12α-dihydroxy-7-keto-cholanoic acids, which cannot be separated as such or as methyl esters, are separated as the trimethylsilyl ether derivatives (12).

Many misleading interpretations of TLC data may be made if the results are thought of in terms of adsorption chromatography only. Solvent demixing, as evidenced by the appearance of two solvent fronts, occurs with many solvent systems, especially if the individual solvents differ much in polarity. This indicates that partition between a comparatively stationary liquid phase and a continuously changing moving phase may occur in some parts of the

TABLE XI. R_f Values of Free Bile Acids on Adsorbent Impregnated Glass Paper[a]

	Solvent system[a]											
	Benzene–ethanol (v/v)									$CHCl_3$	A-20[b]	B-5[b]
	200:1	200:2	200:3	200:5	200:7		200:10	200:18	200:22			
Substituents[c]	SA[d]	MP[e]	SA	SA	MP	MP	PS[f]	PS	PS	MP	SA	SA
3α6α7β				0.40		0.42		0.00	0.20			
3α6β7α				0.32		0.56		0.08	0.38			
3α6β7β				0.44		0.76		0.12	0.80			
3α7α12α	0.00	0.00	0.00	0.28	0.15	0.26	0.00	0.08	0.16	0.00	0.35	0.00
3α7β12α											0.47	0.02
3α12α,7-keto	0.00	0.00	0.19	0.63	0.15	0.62	0.00	0.24	0.60	1.00	0.64	0.02
3α,7,12-diketo	0.00	0.21	0.57	1.00	0.67	1.00	0.29	0.55	1.00	0.15	0.74	0.06
3,7,12-triketo	0.25	0.46	0.92	1.00	1.00	1.00	0.47	0.74	1.00	0.30	0.88	0.11
3α7α											0.78	0.18
3α12α	0.00	0.09	0.45	0.84	0.41	0.84	0.13	0.45	0.84	0.05	0.83	0.22
3α,12-keto	0.13	0.27	0.78	1.00	0.83	1.00	0.26	0.67	1.00	0.20	0.87	0.36
3,12-diketo	0.35	0.67	1.00	1.00	1.00	1.00	0.34	0.83	1.00	0.47	0.93	0.51
3α	0.68	0.55	1.00	1.00	1.00	1.00	0.36			0.55		

[a] See 15,84,85,143,149; cf. Table IV.
[b] See Table X.
[c] For abbreviations see Table VI.
[d] Glass paper impregnated with silicic acid.
[e] Glass paper treated with monopotassium phosphate.
[f] Glass paper treated with dilute potassium silicate.

layer. In fact, the behaviour of ketonic bile acids and of some configurational isomers resemble that in a liquid–liquid partition system (see above).

To perform a systematic TLC analysis of a mixture of free bile acids, the solvents, techniques, and adsorbents presented in Table VI–X can be used. Table XI lists some systems for chromatography on adsorbent impregnated glass fiber paper (GFP). An analysis of bile acids with GFP may require only 10–30 min; on the other hand, this method is very sensitive to overloading and is used for bile acids in the 0.1–5 μg range (84, 85).

As is evident from Tables VI–X, solvent S1 can be used for the screening of unknown bile acid mixtures. Complex mixtures may require the use of longer plates, e.g., 30 cm (12). If a particular class of bile acids is to be studied, a suitable system can be found for that class in the appropriate table. A combination of S1 and one "class-specific" solvent system, in a two-dimensional chromatography, is usually efficient to resolve complex bile acid mixtures (12).

For special types of separations it may be advantageous to modify the adsorbent by impregnation, e.g, with a glycol complexing agent (86), or with $AgNO_3$. Thus, Gustafsson *et al.* separated 3α-hydroxy-6-cholenoic acid from the saturated acid using solvent S10 and $AgNO_3$ impregnated silica gel plates (87).

If only incomplete separations are obtained, or when inhomogeneity of the spots is suspected, it is necessary to convert the acid(s) into ester derivative(s). These are then analyzed in the same solvent systems as the acids (spot-shift study). Additional derivatives may be prepared before or after preparative TLC. These derivatives should be suitable for subsequnt GLC analysis.

TLC has been used for quantitative bile acid analysis by several authors. This was discussed partly in section III.A.4. The main alternative methods are based on spectrophotometric determination before or after elution (11, 51–53, 88–91), densitometric or fluorimetric recordings (54, 92), and enzymatic determination after elution (50, 93–94). The major problem with use of colorimetric methods in work with biological materials is that of specificity (as evidenced by the number of method modifications). Although it may be difficult to elute bile acids from the adsorbent, satisfactory methods are now available (see III.A.4) and it is likely that the methods based on elution and determination with 3α-hydroxysteroid dehydrogenase give the most reliable results.

B. Derivative Formation

In the choic between different derivatives for TLC and GLC analysis, those should be selected which are simple to prepare in a nearly quantitative yield on a microscale. The reaction should also be specific for a certain type of functional group in the bile acid molecule. Descriptions of such reactions can be found in (7, 14, 48). In the search for possible reactions (95) and (96) may be consulted. The procedures below are intended for use with submilligram amounts of bile acids.

Methyl Esters

The bile acids are dissolved in diethyl ether–methanol, 9:1, a freshly prepared ethereal solution of diazomethane is added, and the reaction is carried out at 4 °C for 15 min.

Trimethylsilyl Ethers

Partial (3, 97) or complete (98) silylation may be obtained. Hydroxyl groups at C-3, 6α, 7β, and 12β are transformed into silyl ethers when treated with hexamethyldisilazane, 0.03 ml, in dry dimethylformamide, 0.06 ml, at 50 °C for 3hr (77). Complete silylation, both of equatorial and axial hydroyxl groups, is obtained in pyridine(dry)–hexamethyldisilazane–trimethylchlorosilane, 10:5:2 (v/v), 15 min, room temperature (98). Prolonged reaction times may give enol silyl ethers with keto bile acids. In an analogous reaction VandenHeuvel and Brady have prepared chloromethyldimethylsilyl ethers (99). Silyl ethers are very readily hydrolyzed.

Trifluoroacetates

Trifluroroacetates (100) are prepared by dissolving the bile acid methyl ester (or the partial silyl ether) in trifluoroacetic anhydride, 15 min, 35 °C (101). The reagent is removed under a stream of nitrogen. Milder conditions yield partial derivatives. Enol esters of 3-keto-Δ^4 bile acids may be formed as side products. Trifluoroacetates should be analyzed within 1–2 days since signs of decomposition may appear on storgae for more than 48 hr at room temperature (7, 102).

Acetates

Acetates have been frequently used in spot-shift analyses on paper (48). They are equally useful in TLC and GLC. Conditions for partial and complete acetylation are reviewed by Van Belle (5).

Dimethylhydrazones

Dimethylhydrazones (103) are prepared by dissolving the bile acid methyl ester in 0.1 ml of 1,1-dimethylhydrazine in screw-capped vials. The vials are flushed with nitrogen, closed, and left at room temperature overnight (77). The reagent is evaporated under a stream of nitrogen, the residue is dissolved in acetone and the sample is analyzed after 1 hr. Under these conditions only unconjugated 3-keto groups react (77, 104).

O-Methyloximes

O-methyloximes (105) are prepared by dissolving the bile acid methyl ester in 0.2 ml of a pyridine solution of methoxyamin hydrochloride (14 mg/ml). Keto groups at C-3 and C-6 are completely converted into methyloximes after 16 hr at room temperature (106). We have found partial derivative formation with 7- and 12-keto groups under these conditions. The formation of *syn* and *anti* isomers which may separate on TLC and GLC should be remembered.

Oxidation and Reduction

Oxidation and reduction give valuable structural information. Oxidation is performed at 0 °C in 1 ml acetone containing 10 μl of oxidizing reagent (26.72 g CrO_3 and 23 ml of sulfuric acid diluted to 100 ml with water) (77, 107). The reaction is terminated after 10 min by addition of 5 ml of water. The product is extracted with ethyl acetate.

Sodium Borohydride Reduction

Sodium borohydride reduction is carried out with 1 mg reagent in 0.5 ml isopropanol. The rates of reduction differ with the position of the keto group (108). Treatment with lithium aluminum hydride in diethyl ether results in formation of C-24-ols from bile acid methyl esters. This may be valuable in chromatographic identification work, especially if combined with mass spectrometry.

C. Gas–Liquid Chromatography

To obtain good results it is necessary to use columns that give minimal tailing of polar methyl hydroxycholanoates. Several factors of importance for the preparation of such columns are still poorly understood but it is clear that the support and its treatment are of great importance. Among the commercial supports we prefer Gas-Chrom-P (Applied Science Laboratories, State College, Pennsylvania and Chromosorb W 100–120 mesh Johns-Manville Co., New York). Procedures for the silanization of the support and subsequent application of a suitable stationary phase have been reviewed (109) and modified (15). With the procedures cited, satisfactory column packings are usually obtained and columns with at least 1500 theoretical plates/m can be made for the analysis of methyl hydroxycholanoates. Improved results may be obtained with devices developed for coating of supports with stationary phase (110, 111). Column packings are available commercially and although the quality of these materials may vary between batches they permit analysis of most bile acid derivatives.

TABLE XII. Relative Retention Times[a] of Different Bile Acid Methyl Ester Derivatives on Methyl (SE-30 and OV-1) and Methyl-Phenyl (SE-52, PhSi-20, and OV-17; 5, 20, and 50% Phenyl Groups, Respectively) Substituted Silicones

Bile acid methyl ester	3% OV-1[b]				0.5-2.3% SE-30[c]				0.75% SE-52[d]		0.5% PhSi-20[e]		3% OV-17[b]			
	5α		5β		5β				5β		5β		5α		5β	
Substituents[f]	OH	TMS[g]	OH	TMS	OH	TMS	OAc	TFA	OH	OAc	OH	OAc	OH	TMS[g]	OH	TMS
None	0.40	-	0.35	-	0.20	-	-	0.71	-	-	0.23	-	0.24	-	0.20	-
3α	0.72	0.72	0.62	0.85	0.60	0.95	0.83	1.13	0.32	0.72	0.54	0.79	0.58	0.92	0.52	0.93
$3\alpha\Delta^5$								1.09								
$3\alpha\Delta^6$								1.01	0.63							
$3\alpha\Delta^7$								1.14		0.76						
$3\alpha\Delta^{7,9}$								0.98		0.66						
3β	0.72	0.82	0.64	0.81				1.08					0.60	1.20	0.51	0.86
$3\beta\Delta^5$											0.64					
6α			0.65	0.65											0.52	0.72
6β			0.62	0.62											0.47	0.63
7α	0.64	0.57	0.60	0.61		0.66							0.51	0.57	0.44	0.63
7β	0.67	0.78	0.59	0.71		0.77							0.52	0.83	0.46	0.75
12α	0.60	0.60	0.55	0.58		0.63					0.42		0.49	0.64	0.39	0.62
12β			0.56	0.54											0.41	0.56
3-keto	0.76		0.71		0.66								0.71		0.61	
3-keto,Δ^6					0.66											
3-keto,$\Delta^{7,9}$					0.66											
6-keto			0.64												0.58	
7-keto	0.66		0.58										0.57		0.46	
12-keto	0.63		0.57								0.45		0.51		0.43	
$3\alpha,6\alpha$			1.20	1.12	1.26	1.25	1.81	1.28	1.24	1.55	1.34	1.92			1.32	1.08
$3\beta,6\alpha(5\beta)$								1.30								
$3\beta,6\alpha(5\alpha)$								1.55								
$3\alpha,6\beta$			1.11	1.10											1.21	0.98
$3\beta,6\beta(5\beta)$								1.18								
$3\beta,6\beta(5\alpha)$	1.20	1.33											1.29	1.31		
$3\alpha 7\alpha$	1.18	0.84	1.08	1.04	1.11	1.22	1.36	1.17	1.12	1.20	1.16	1.35	1.27	0.92	1.14	1.00

TABLE XII. (Continued)

Bile acid methyl ester	3% OV-1[b]				0.5-2.3% SE-30[c]				0.75% SE-52[d]		0.5% PhSi-20[e]		3% OV-17[g]			
	5α		5β		5β				5β		5β		5α		5β	
Substituents[f]	OH	TMS[g]	OH	TMS	OH	TMS	OAc	TFA	OH	OAc	OH	OAc	OH	TMS[g]	OH	TMS
3β7α(5β)	1.18	1.08	1.06	0.96									1.34	1.00	1.10	0.87
3β7α(5α)[h]								1.13								
3α7β			1.06	1.21	1.09	1.35		1.40	1.22	1.58	1.12				1.13	1.14
(C_{22}) 3α12α												0.55				
(C_{23}) 3α12α												0.86				
3α12α	1.12	1.18	1.00	1.00[g]	1.00	1.00[g]	1.12	1.00[g]	1.00	1.00[g]	1.00	1.15	1.19	1.22	1.00	1.00[g]
3α,12αΔ^{7}[h]								0.89								
3α12α$\Delta^{8(14)}$								0.86	1.04							
3β,12α	1.16	1.23	1.00	1.00									1.20	1.19	1.00	0.90
6β,12α			0.94	0.80											0.91	0.69
7α,12α	0.99	0.63	0.91	0.69									0.99	0.53	0.84	0.60
3α,7-keto	1.23	1.41	1.03	1.34	1.08	1.46		2.00			1.14	1.49	1.42	2.30	1.20	1.98
3β,7-keto	1.26												1.41			
3α,12-keto			1.00	1.33	1.05					1.20	1.15	1.46			1.12	1.91
6α,3-keto			1.31	1.32											1.59	1.99
6β,12-keto			0.99	0.95											1.11	1.27
7α,3-keto	1.27	1.08	1.23	1.17									1.60	1.60	1.43	1.84
7α,12-keto	1.03	0.90											1.12	1.18		
12α,3-keto	1.19	1.17	1.07	1.09									1.41	1.85	1.17	1.67
12α,7-keto	1.09	0.97	1.13	0.93									1.13	1.54	0.88	1.35
3,7-diketo	1.17		1.04		1.03						1.16		1.52		1.21	
3,12-diketo	1.15		1.05		1.06						1.27		1.51		1.29	
7,12-diketo	0.96		0.91										1.06		0.85	
3α6α7α			1.85	1.15	2.21						2.41				2.56	1.11
3α6β7α			1.84	1.12											2.65	0.84
3α6α7β			1.69	1.92											2.25	1.71
3α6β7β	1.78	1.44	1.68	1.45							2.20		2.34	1.17	2.30	1.18
(C_{22})3α7α12α											1.03					

TABLE XII. (Continued)

Bile acid methyl ester	3% OV-1[b]				0.5-2.3% SE-30[c]				0.75% SE-52[d]		0.5% PhSi-20[e]		3% OV-17[b]			
	5α		5β		5β				5β		5β		5α		5β	
Substituents[f]	OH	TMS[g]	OH	TMS	OH	TMS	OAc	TFA	OH	OAc	OH	OAc	OH	TMS[g]	OH	TMS
$3\alpha 7\alpha 12\alpha$	1.77	1.05	1.66	1.09	1.89	1.20	1.60	1.09	1.86	1.44	2.20	1.88	2.73	0.87	2.26	0.90
$(C_{27})3\alpha 7\alpha 12\alpha$											4.37					
$3\beta 7\alpha 12\alpha$	1.88	1.01	1.63	0.96									2.71	0.79	2.08	0.79
$3\alpha 7\beta 12\alpha$			1.52							1.98	2.30				2.07	
$3\alpha 6\alpha$,7-keto			1.48												1.59	
$3\alpha 7\alpha$,12-keto			1.73	1.72				1.97		1.93	2.62	2.57			2.57	1.97
$3\alpha 7\beta$,6-keto	1.52												1.88			
$3\alpha 12\alpha$,7-keto	1.95	1.62	1.59	1.59							2.20	2.24	2.89	1.95	2.27	2.03
$3\beta 12\alpha$,7-keto	1.85	1.96											2.87	2.46		
$7\alpha,12\alpha$,3-keto	1.96	1.09	1.88	1.35									3.12	1.45	2.79	1.68
3α,7,12-diketo			1.50	1.88	1.71						2.18	2.52			2.18	3.38
7α,3,12-diketo			1.91												3.30	
12α,3,7-diketo	1.84	1.78	1.54	1.56									3.04	3.76	2.31	3.10
3,7,12-triketo	1.64		1.39		1.61						2.03		2.98		2.25	

[a] Relative to methyl deoxycholate unless otherwise indicated.
[b] Column temperature: 260 °C (113).
[c] Column temperatures: 210 °C (OH from (100) and unpublished, 238 °C (TMS) from (99), 195 °C (OAc) from (102), 210-220 °C (TFA) from (150).
[d] Column temperatures: 250 °C (OH) from (151), and 260 °C (OH) from (150), 235 °C (OAc) from (150).
[e] Column temperature: 215–220 °C from (152).
[f] Greek letters denote configuration of hydroxyl groups. 5α and 5β refer to 5α- and 5β-cholanoates, respectively. Δ is a double bond; OH denotes free hydroxyl groups; TMS denotes trimethylsilyl ethers; OAc denotes acetate; TFA denotes trifluoroacetate. C_{22}, C_{23}, C_{27} denote homologs with 22, 23, and 27 carbon atoms, respectively.
[g] Retention times given relative to this derivative of deoxycholic acid.
[h] From (153, 154).

TABLE XIII. Relative Retention Times[a] of Different Bile Acid Methyl Ester Derivatives on QF-1

Bile acid methyl ester	0.5-1 % QF-1[c]			3 % QF-1[d]				3 % QF-1[e]				QF-1–SE-30–NGS[f]			
	5β		5α	5β				5α		5β		5β			
Substituents[b]	OH	TFA	OH	OH	TFA	TMS	p-TMS	OH	TMS[g]	OH	TMS[g]	OH	TFA	TMS	p-TMS
none	0.16							0.18	-	0.15	-				
3α	0.49	0.42	0.52	0.53	0.47	0.30	-	0.50	0.89	0.49	0.91	0.34	0.22	0.20	-
3αΔ[5]		0.44													
3αΔ[6]		0.42													
3αΔ[7]		0.45													
3αΔ[7,9]		0.38													
3β	0.44	0.41	0.58	0.48	0.46	0.29	-	0.55	1.20	0.44	0.83	-	-	-	-
3βΔ[5]	0.51	0.46													
6α										0.45	0.67				
6β										0.38	0.63				
7α	0.35	0.26		0.38		0.26		0.40	0.62	0.34	0.65				
7β	0.39	0.32		0.44		0.25		0.43	0.81	0.36	0.71				
12α	0.31	0.22		0.34		0.22		0.37	0.66	0.27	0.59	0.22	0.11		
12β	0.34	0.25		0.37		0.21				0.31	0.55				
3-keto	0.95		1.09	1.00				1.06		1.00		0.53			
3-ketoΔ[4]				1.66											
3-ketoΔ[6]	0.87														
3-ketoΔ[7,9]	0.89														
6-keto										0.78					
7-keto	0.57							0.72		0.57					
12-keto	0.49							0.60		0.49					
3α6α	1.47	0.98		1.48	0.93	0.36				1.50	1.20	1.58	0.45	0.26	
3β6α		0.96													
3β6α(5α)		1.18													
3α6β	1.19	0.91								1.20	1.08	1.28	0.43	0.24	0.56
3β6β		0.88						1.27	1.40						
3β6β(5α)		1.09													
3α7α	1.15	0.85	1.17	1.13	0.83		0.64	1.22	1.00	1.18	1.09	1.10	0.40	0.22	0.50
3β7α	0.94		1.24	0.96	0.69		0.58	1.29	1.14	0.94	0.96				

TABLE XIII. (Continued)

Bile acid methyl ester	0.5-1% QF-1[c]			3% QF-1[d]				3% QF-1[e]				QF-1–SE-30–NGS[f]			
	5β		5α	5β				5α		5β		5β			
Substituents[b]	OH	TFA	OH	OH	TFA	TMS	p-TMS	OH	TMS[g]	OH	TMS[g]	OH	TFA	TMS	p-TMS
$3\alpha 7\beta$	1.25	1.00		1.22	0.95	0.37				1.27	1.24	1.23	0.47	0.28	0.54
$3\alpha 12\alpha$	1.00	0.67		1.00	0.68		0.56	1.07	0.93	1.00	1.00	1.00	0.30	0.22	0.45
$3\alpha 12\alpha \Delta^4$				0.70											
$3\alpha 12\alpha \Delta^{8(14)}$		0.58													
$3\beta 12\alpha$	0.85	0.58		0.88	0.59		0.53	1.16	1.35	0.86	1.02				
$3\beta 12\alpha \Delta^4$				0.77											
$3\alpha 12\beta$	1.08	0.78		1.06	0.78	0.30									
$3\beta 12\beta$				0.96	0.72[h]	0.31									
$6\beta 12\alpha$										0.77	0.79				
$7\alpha 12\alpha$				0.73	0.43			0.90	0.63	0.70	0.72				
3α,6-keto	2.61	1.76										1.66	0.90		
3α,7-keto	1.76	1.60		1.73	1.52	1.01		1.97	4.16	1.80	3.65	1.30	0.84		
3β,7-keto								2.12							
3α,12-keto	1.62	1.54		1.57	1.38	0.98				1.61	3.19				
3β,12-keto	1.32	1.28		1.34	1.23	0.85									
6α,3-keto										2.93	5.04				
7α,3-keto	2.17	1.76	2.29	2.22	1.61[h]			2.32	4.12	2.40	3.81				
7α,3-ketoΔ^4				2.98[h]											
7α,12-keto								1.56	2.43	1.10					
12α,3-keto	1.83	1.42	2.14	1.84	1.36			2.23	3.04	1.82	4.10	1.41	0.64		
12α,3-ketoΔ^4				3.28											
12α,7-keto								1.56	3.42	1.10	2.74				
3,7-diketo	2.83			2.69				3.96		3.05					
3,12-diketo	2.86		3.18	2.71				3.52		2.86					
7,12-diketo				1.41				1.97		1.45					
$3\alpha 6\alpha 7\alpha$	2.61	1.24[h]								3.14	1.38	3.86[i]	0.94[j]	0.31	0.47
$3\alpha 6\beta 7\alpha$	2.50	0.87[h]		2.28	0.87[h]		1.24			2.42	1.05	4.13[i]	0.37[h]		
$3\alpha 6\alpha 7\beta$	2.40	1.60[h]								2.67	1.88	3.34[i]	1.19[j]	0.51	
$3\alpha 6\beta 7\beta$	2.31	1.47		2.20	1.32		0.72	2.10	1.41	2.33	1.44	3.36[i]	0.67		

TABLE XIII. (Continued)

Bile acid methyl ester	0.5-1% QF-1[c]			3% QF-1[d]				3% QF-1[e]				QF-1–SE-30–NGS[f]			
	5β		5α	5β				5α		5β		5β			
Substituents[b]	OH	TFA	OH	OH	TFA	TMS	p-TMS	OH	TMS[g]	OH	TMS[g]	OH	TFA	TMS	p-TMS
3α7α12α	2.33	1.39	2.37[k]	2.14	1.29		1.23	2.68	1.00	2.33	1.05	3.31[i]	0.60	0.23	1.34
3α7α12α															0.47
3β7α12α	1.87	1.17	2.50	1.81	0.95		1.13	2.78	1.18	1.90	0.96				
3α7β12α	2.44	1.33		2.17	1.18		0.74			2.20					
3β7β12α				1.93	1.00		0.65								
3α6α,7-keto										1.78					
3α7β,6-keto								2.06							
3α7α,12-keto	4.00	2.80		3.33	2.42		2.09			3.85	4.37	4.27[i]	1.33		
3α12α,7-keto	3.70	2.54		3.07	1.99		1.88	4.62	5.45	3.51	4.95	4.17[i]	1.00		
3β12α,7-keto								4.50	7.00						
7α12α,3-keto	4.79	2.79	4.50	3.94	2.44			6.10	5.07	4.79	5.84				
7α12α,3-ketoΔ^4				3.33[h]											
3α,7,12-diketo	4.47	4.30								4.55	4.55	3.85[i]	2.51		
12α,3,7-diketo								7.14	12.7	5.38	12.5				
3,7,12-triketo	6.33		7.20	5.60				9.24		6.40		4.62[i]			

[a] Relative to methyl deoxycholate unless otherwise indicated.
[b] For abbreviations see Table XII, [f].
[c] Column temperature 210–220 °C from (101,150,155).
[d] Column temperature 230–235 °C from (77,127,147).
[e] Column temperature 230 °C from (113).
[f] Column temperature 210 °C from (24, 155).
[g] Retention times given relative to the bis(trimethylsilyl) ether of methyl deoxycholate.
[h] These compounds often give two peaks. The retention time of the major peak, which may be a degradation product, is given.
[i] Column temperature 220 °C.
[j] The compounds give two peaks.
[k] The TFA has a relative retention time of 1.37.

The separation of bile acids by gas–liquid chromatography is determined by the choice of stationary phase and bile acid derivative. Data that permit the selection of stationary phase and bile acid derivative to suit most separation problems are shown in Tables XII–XIV. As in the original papers relative retention times have been used in the tables instead of the more acceptable retention index (112) which permits better interlaboratory comparisons. Relative retention times are subject to variation, mainly due to temperature differences and, to a lesser extent, to differences in column preparation. In our experience the temperature-dependence is most pronounced with trimethylsilyl ether derivatives on Hi-Eff-8B (cyclohexanedimethanol succinate) columns. Temperature differences do not affect relative retention times on QF-1 (a trifluoropropyl, methyl siloxane) to the same extent.

Of the substituted polysiloxane phases those with alkyl substituents (SE-30 and OV-1) are less specific than those having phenyl groups (SE-52, PhSi-20, and OV-17) and the different methyl cholanoates are eluted in a more narrow range with the two first mentioned phases (see Table XII). Formation of bile acid acetates or trimethylsilyl ethers may increase separation factors slightly on OV-1 and SE-30 columns. A more pronounced effect is noted with trifluoroacetates of isomeric disubstituted bile acid methyl esters on SE-30 columns and the separation of unsaturated compounds is also good. However, the separation between mono-, di-, and trifluoroacetates of methyl cholanoates is remarkably poor. A small difference in chromatographic properties may be found between SE-30 and OV-1 columns. An example is the separation between the methyl esters of $3\alpha,7\alpha$-dihydroxy-12-keto- and $3\alpha 12\alpha$-7-keto-5β-cholanoic acids achieved with OV-1 but not with SE-30. However, this difference may be temperature related (see Table XII).

Improved separations of bile acid methyl esters can be observed with increasing amounts of phenyl substituents in the stationary phase (5% in SE-52, 20% in PhSi-20, and 50% in OV-17). Positional and configurational isomers are better resolved than on SE-30 or OV-1 columns. For instance the valuable separation of the trimethylsilyl ethers of 3,6,7-substituted methyl cholanoates from the corresponding cholic acid derivative on OV-17 should be noted. The pronounced effect of a 7β-hydroxy substituent on retention times on columns of SE-52 and PhSi-20 is noteworthy. The large separation factors between the diacetate derivatives of chenodeoxy- and ursodeoxycholic acids may be most useful in work with bile acids of biological origin.

Bile acid methyl esters with unprotected hydroxy groups are best separated on QF-1 columns (Table XIII). It is advisable to use 3% columns since it is often difficult to prepare 0.5–1% QF-1 columns that do not give tailing of polar compounds. This difficulty has also been experienced by Okishio and Nair (24) who therefore developed a column having a mixture of QF-1, SE-30, and NGS as stationary phase. On QF-1 columns epimeric alchohols

TABLE XIV. Relative Retention Times[a] of Different Bile Acid Methyl Ester Derivatives on Cyanoethyl Silicones (CNSi and XE-60) and Polyester Phases (Hi-Eff-8B and NGS)

Bile acid methyl ester	2–3% XE-60[c] 0.5% CNSi[d]				0.5% Hi-Eff-8B[e]		0.4% NGS[f]
Substituents[b]	OH	TFA	TMS	*p*-TMS	TMS	TMS	TFA
None	0.09	0.35			0.77		0.44
3α	0.37	0.81	1.02		1.36		1.06
3αΔ^{5}	0.33	0.77					0.94
3αΔ^{6}	0.37	0.73					0.90
3αΔ^{7}		0.78					1.09
3α$\Delta^{7,9}$		0.67					0.91
3β	0.34	0.76			1.09		0.89
7α	0.25		0.67				
12α	0.24		0.64				
12β	0.26						
3-keto	0.49				4.26		3.39
3-ketoΔ^{6}	0.43						
12-keto	0.28						
3α6α	1.48	1.62	1.16		1.26		1.93
3β6α	1.35	1.59					
3β6β(5β)	1.26	1.50					
3β6β(5α)	1.64						
3α7α	1.10	1.53	1.09	0.46	1.08		1.50
3α7β	1.18	1.77	1.26		1.57		2.12
3α12α(5α)					0.88[h]		
3α12α	1.00	1.00[g]	1.00[g]	0.41	1.00[g]		1.00[g]
3α12α$\Delta^{8(14)}$		0.82					0.90
3β12α	0.91						
3α12β	1.06						
7α12α	0.68		0.66				
3α,7-keto	1.41	3.80			4.19		
3α,12-keto	1.23				3.97	1.35	
7α,3-keto	1.74						
12α,3-keto	1.44						
3,7-diketo	1.61					4.47	
3,12-diketo	1.56					4.44	
3α6α7α	3.51				1.01		
3α6β7α	3.71	1.15[i]			0.72		
3α6α7β	3.12				2.01		
3α6β7β	2.82	2.56			1.21		
3α6β12α	3.41						
3α7α12α(5α)					0.61[h]		
3α7α12α	3.08	2.27	0.91	1.24	0.75		1.94
3α7β12α	3.20						
3α7α,12-keto	4.19	6.50			3.08	1.00	
3α12α,7-keto	4.07				3.15	1.00[g]	
7α12α,3-keto	4.97						
3α,7,12-diketo	3.98					3.20	
3,7,12-triketo	4.55						

and the corresponding ketone can be separated. Substituents in the 5β-cholanoate nucleus increase the retention time of the parent compound in the following approximate order: $12\alpha < 12\beta \leqq 7\alpha < 7\beta \leqq 6\beta < 3\beta < 6\alpha \leqq 3\alpha \leqq 12$-keto$<$7-keto$<$6-keto$<$3-keto$<$3-keto-$\Delta^4$. The influence of keto groups is more pronounced with this phase than any other so far discussed. QF-1 columns are also very useful for the separation of trifluoroacetates of methyl cholanoates but the retention times are not as well predictable as with free hydroxyl groups. Although individual hydroxyl and keto groups often contribute to the retention time with constant logarithmic factors, interactions between two or more substituents in a molecule sometimes makes it difficult to calculate the retention time of a bile acid methyl ester (3). Similar conclusions were arrived at by Elliot *et al.* (113) who also compared, calculated and found retention times on OV-1 and OV-17 columns. These authors also point out that the agreement is generally better for the 5β- than the 5α-derivatives of methyl cholanoates.

With the cyanoethyl substituted silicone phases (Table XIV), separations similar to those obtained with QF-1 columns can be obtained. However smaller separation factors between epimeric alcohols and the corresponding ketones are noted. This is an advantage, for instance, when diketo-substituted bile acid methyl esters and trihydroxycholanoates are present in the same extract, since such mixtures cannot be separated on QF-1 columns without prior derivatization. With the even more polar phases Hi-Eff-8B and NGS (neopentyl glycol succinate) only bile acid methyl esters with protected hydroxyl groups can be eluted from the columns and ketonic bile acid derivatives are much delayed. These phases have special merits though. Trimethylsilyl ethers of epimeric hydroxycholanoates appear to be nicely separated on Hi-Eff-8B columns and saturated and unsaturated trifluoroacetates of bile acid methyl esters separate on NGS columns. However, since NGS decomposes at temperatures usually needed for bile acid derivatives, short columns with a low percentage of NGS have to be used. The difficulty of preparing such columns without leaving "active sites" in the support that destroy and adsorb the thermolabile trifluoroacetate derivatives is easily realized.

---

[a] Relative to methyl deoxycholate unless otherwise indicated.
[b] For abbreviations see Table XII.
[c] Column temperature 240 °C [OH, TMS and *p*-TMS (97)] and 200 °C [TFA (102)].
[d] Column temperature 225 °C [OH (31)], 210 °C TFA (150)].
[e] Column temperatures 245 °C left column, 280 °C right column (98).
[f] Column temperature 210 °C (150).
[g] Retention times given relative to this derivative.
[h] Column temperature 230 °C. The relative retention times have been recalculated from (115).
[i] These compounds often give two peaks. The retention time of the major peak, which may be a degradation product, is given.

From the discussion above it is apparent that the choice of stationary phase and bile acid derivative must be based upon desired objectives. The usefulness of a preliminary separation of bile acids according to the number of substituent groups prior to gas chromatography seems evident when complex mixtures of bile acids are to be analyzed.

Before attempting a quantitative gas–liquid chromatographic analysis of a bile acid mixture it is for obvious reasons necessary to establish the identity and homogeniety of the compounds (i.e., peaks) eluted from the gas chromatograph. Since identification with classical methods may be tedious and require large amounts of bile acids, the structure of a bile acid often has to be tentatively elucidated by chromatographic methods including derivatization procedures.

The following general procedure, based on the combination of thin-layer chromatography and gas chromatography, may be used for a complex bile acid mixture present in a low concentration in a biological extract (77). The crude mixture is applied to thin-layer plates for preparative subdivision—into mono-, di-, and trisubstituted compounds—before a more detailed analysis is made. To improve the yields, this may be done after methylation. In work with very impure and complicated mixtures, such as fecal lipids, purification may have to be made repeatedly by column chromatography (Section III). Each group of bile acids can then be further subdivided by preparative thin-layer chromatography. The fractions obtained are analyzed by gas–liquid and/or thin-layer chromatography. The results obtained usually permit a preliminary tentative identification of bile acids in mixtures of up to five compounds. However, other compounds may have the same chromatographic properties as bile acids. Therefore derivatization and subsequent spot-shift (TLC) and /or peak-shift (change in relative retention time after a quantitative reaction) studies have to be carried out. When this is made with too many compounds it is difficult to establish correspondence between the peaks (GLC) and /or spots (TLC) of the different derivatives. The same difficulty exists when a complex mixture is analyzed on several gas–liquid chromatography columns with different stationary phases. If however, the fraction in question is oxidized (Section IV. B) the position of original alcohol substituents and the stereochemistry at C_5 can usually be established by gas–liquid chromatography on a QF-1 column since this column retards ketonic bile acid methyl esters in a very specific way. Some methyl cholanoates, e.g., the 3,6,7-trisubstituted ones and most unsaturated bile acid methyl esters, form several oxidation products and have to be studied by some other means (such as impregnation TLC or GLC of TMS derivatives). When there are reasons to believe that the unidentified compound(s) is a methyl keto-cholanoate without hydroxyl groups, oxidation should still be carried out to ensure that no spot-shift or peak-shift occurs.

The number of hydroxyl groups can often be deduced by studying peak-shifts on QF-1 columns. This procedure involves retention time studies of methyl cholanoates, their partial trimethyl silyl ethers, and trifluoroacetates in that order (see Table XIII). An analysis of pertrimethylsilylated methyl cholanoates (Tables XII and XIV) should also be included especially if oxidation studies have revealed the possible presence of 3,6,7-substituted bile acid derivatives. This will also aid in the interpretation of the stereochemistry of the alchohol substituents.

When simple mixtures are analyzed, the peak- or spot-shift techniques may often be directly used without preliminary purifications; the results will permit the selection of suitable reference compounds and the appropriate comparisons can be made with the bile acids in the sample. When minor components are to be identified, degradation products from major bile acids may be mistaken for separate compounds. The predominant compounds should therefore be removed prior to derivatization of minor bile acids. It should also be ascertained (e.g., by analysis of blanks) that interfering compounds are not introduced during the purification procedure—e.g., from insufficiently prepurified silica gel.

The combination of gas chromatography and mass spectrometry aids in differentiation as to whether a peak is due to a bile acid derivative or to a degradation product from another compound appearing later in the chromatogram. By GC–MS analysis a molecular weight can usually be obtained for the unprotected methyl cholanoate or its trimethylsilyl ether (for details see Chapter 7). Di- and trifluoroacetates usually do not give a molecular ion peak and when these derivatives are used the retention time will indicate the probable number of trifluoroacetoxy groups in the molecule. The presence of keto groups is easily established since a characteristic nuclear fragment containing the ketonic oxygen is always found. Positional isomers are distinguished by their mass spectra whereas stereoisomers may not give mass spectral differences. In such cases the complementary use of gas–liquid chromatography is of value since stereoisomers can usually be separated on QF-1 columns.

Finally it should be mentioned that the bile acid nature of a peak can be made likely from isotope studies *in vivo* using simultaneous mass and isotope (^{14}C or less preferably ^{3}H) determination with radio–gas chromatography. The radioactivity detector, however, requires fairly high specific activities and one may have to collect the compound for conventional isotope determination.

Once the identity and homogeneity of gas–liquid chromatography peaks have been established a quantitation in the submicrogram range can be made by peak area measurement. The areas are compared with those obtained with the appropriate standards, preferably in a linear range, and the results of different analyses are made comparable by peak area measurement of a suitable internal standard added before derivatization. The use of a dilution series of a standard mixture is advisable since possible day-to-day variations of the relative response of different bile acid methyl esters are then controlled. When trifluoroacetates or trimethylsilyl ethers of bile acid methyl esters are determined, the standard sample and the unknown sample should be prepared and analyzed on the same day.

The quantification of methyl cholanoates requires the resolving power of a QF-1 column for complex mixtures. Only little tailing with di- and trihydroxy bile acid methyl esters should be permitted. Under acceptable conditions the response relative to methyl deoxycholate is of the order 0.90–1.25 for mono- and disubstituted methyl cholanoates but may be as low as 0.50 for the trisubstituted compounds (18). Poor responses are probably due to column imperfections and not so much caused by the detection unit in the linear range.

However, when the trifluoroacetate of methyl 3β,7α,12α-trihydroxycholanoate is analyzed the argon ionization detector may respond with a negative peak (101). With several other bile acids studied, the responses relative to that of methyl deoxycholate trifluoroacetate were found to be between 0.75 and 1.20 (13, 18). Columns that have been in use for long periods of time for the analysis of biological extracts, and which give tailing and/or loss of compounds may be restored to previous condition by a silanization *in situ*. The detector is disconnected, the column temperature is set at 150 °C and the carrier gas flow is reduced to a minimun. Hexamethyldisilazane, 100–200 μl, is injected in 10-μl portions and the carrier gas supply is turned off. The column is left overnight without gas flow and is then conditioned with carrier gas at 220–240 °C for about 8 hr (15).

Although factors to correct for the differing responses of various bile acid derivatives have been devised (13, 102, 114) we have not been able to obtain constant factors valid for different columns (3, 7). In only one instance has the same response been found with derivatives of different bile acids. Thus Grundy, *et al.* (30), using a hydrogen flame detector for the analysis of trimethylsilyl ethers, obtained the same peak area per unit mass of the parent underivatized methyl cholanoate. These findings have been confirmed by Hofmann *et al.* (115). Using a Hi-Eff-8B column and an argon ionization detector Makita and Wells (98) found variable molar responses with different methylated bile acid trimethylsilyl ethers. It is clear that the conditions for quantitative gas–liquid chromatography must be individually tested for each column and detector system. When this is done, the accuracy of the determinations varies with the nature and concentrations of the components and is of the order ± 5–$\pm 10\%$.

When only the total bile acid concentration is wanted, the simplest method is to convert all bile acids to one compound. This eliminates problems with diverging detector responses. Thus Evrard and Janssen (23) oxidized methyl esters of bile acids in a fecal extract to the corrseponding ketones and then eliminated the keto groups in a Huang-Minlon modification of the Wolff–Kishner reduction. The resulting methyl 5β- (and possibly 5α-) cholanoate was quantitated on a JXR-column. If the reduction step is omitted (as it is in the Evrard–Janssen routine method) bile acids can be determined as the methyl 3-keto, 3,7- and 3,12-diketo- and 3,7,12-triketocholanoates after correction for response differences. Losses are compensated for with the internal standard 23-nordeoxycholic acid which is added in the initial extraction step. The limitations of the use of chromic acid oxidations must, however, be recalled. Thus, the Evrard–Janssen method cannot be used with extracts containing 3,6,7-substituted or unsaturated bile acids unless these acids are first removed.

In the procedure of Grundy *et al.* (30) the bile acids are quantitated as a group after analysis of their trimethylsilyl ether methyl esters on SE-30 columns. As can be seen from Table XII only small separation factors are obtained with this phase. Fortunately, these authors found a unit response for all bile acids analyzed in this way and were able to measure bile acids as the entire base line deflection obtained after 5α-cholestane had been eluted. With such a procedure it has to be established that compounds other than bile acid derivatives are not determined. Whether this is the case under different experimental conditions, such as drug treatment or change of diet, should always be fully explored.

It is apparent from the discussion above that an entirely satisfactory method for bile acid determination is still lacking. Quantitations of bile acids as group(s) after chemical transformation results in loss of information. With a differentiated quantitation minor bile acids may escape detection and such a method may give too low values for the total bile acid content. The choice of analytical method should be determined by the nature of the tissue or fluid to be analyzed and by the aim of the study.

V. APPLICATION OF CHROMATOGRAPHIC METHODS TO SPECIFIC ANALYTICAL PROBLEMS

Under this heading are listed a few examples of the use of methods described in preceding sections. This is done mainly to illustrate in which ways the individual methods can be combined. The aim is to give a description of principles and not to make a review of the literature regarding the presence of different bile acids in the tissues or body fluids.

A. Bile and Intestinal Contents

In a recent study Hofmann *et al.* (116) studied the bile acid composition of bile from germ-free rabbits. Deproteinized bile was first analyzed by thin-layer chromatography for a tentative identification of the conjugated bile acids. After alkaline hydrolysis and methylation, bile acid methyl esters were analyzed by thin-layer and gas chromatography. Trifluoroacetate esters and trimethylsilyl ethers of methyl cholanoates were run on QF-1 or Hi-Eff-8B columns, respectively. Gas chromatography–mass spectrometry was used for the final identifications.

For the identification of bile acids in bile from man and different animals, Kuksis (13) and co-workers have used ion-exchange chromatography for the separation of glycine- and taurine-conjugated bile acids. After alkaline hydrolysis, extraction, and methylation, the bile acids were analyzed on SE-30 and QF-1 columns. Tentative identifications were supported by additional gas–liquid chromatographic analysis of methyl ester acetates and methyl ester trifluoroacetates. The latter derivatives were also analyzed on OV-17 columns (117).

In quantitative studies, Stiehl *et al.* (91) extracted bile acids from duodenal bile with methanol–acetone and the conjugates were hydrolyzed according to the method of Nair *et al.* (118). The free bile acids were then separated with thin-layer chromatography and determined spectrophotometrically after reaction with a sulfuric acid reagent.

For analysis of the small amounts of bile acids present in gallstones the stones are pulverized and refluxed in chloroform–methanol, 1:1. After evaporation, three-stage countercurrent distribution, and alkaline hydrolysis, the bile acids are determined by gas–liquid chromatography (119).

After deproteinization of bile, Turnberg and Anthony-Mote (94) were able to quantitate bile acids directly with a NAD-dependent steroid oxidoreductase from a *Pseudomonas* strain. The suitable concentration range was 10–80 mg of bile acids per 100 ml. The amounts of individual bile acids could be determined after preparative thin-layer chromatography, either before or after hydrolysis of bile acid conjugates.

B. Plasma and Urine

After purification of human serum bile acids on an Amberlyst A-26 anion exchanger, Sjövall and co-workers (31, 120, 121) hydrolyzed the concentrated bile acids in alkali and extracted the free acids with ethyl acetate. After methylation, the methyl cholanoates were purified on an aluminum oxide column. The bile acid fraction then obtained was subjected to gas–liquid chromatography on CNSi columns after conversion to trifluoroacetate derivatives. Gas chromatography–mass spectrometry supported in the identification work necessary for accurate quantification.

Nair and Garcia (122) extract bile acids from serum (adjusted to *p*H 11) with ethanol. The dried extract is partitioned between an alkaline aqueous phase and diethyl ether and the remaining lower phase is subjected to clostridial cholyl glycine hydrolase at *p*H 5.6. After further acidification free bile acids are extracted with ether. The subsequent purification procedure is similar to that above (31), and a triple component (QF-1/SE-30/NGS) column (see Table XIII) is used for quantitative analysis. With the omission of the diethyl ether extraction of the alkaline aqueous phase the same procedure was used with rat bile.

With a purified enzyme preparation from *Clostridium perfringens*, Roovers *et al.* (36) cleave conjugated bile acids directly in plasma. Proteins are then precipitated with $Ba(OH)_2$-saturated ethanol (123) and the supernatant is taken to near dryness. The residue is dissolved in a toluene–isopropanol–methanol–30% aqueous NaOH (10:20:20:6, v/v) mixture, water is added, and neutral lipids removed by light petroleum extraction. Bile acids are then obtained by acidification and diethyl ether extraction. The bile acids are then methylated and analyzed as above except that the bile acid methyl esters are acetylated before chromatography on 1% XE-60 columns at 250°C. With this column the following retention times relative to that of the diacetoxy derivative of methyl deoxycholate were found for the following acetate methyl ester derivatives: lithocholic, 0.60; 23-nor-deoxycholic acid (internal standard), 0.77; chenodeoxycholic acid, 1.24; and cholic acid, 1.88. The deoxycholic acid derivative was eluted after 9.0 min and methyl 5β-cholanoate after 0.32 min.

Heparinized whole blood could also be subjected to digestion with the bile acid conjugate hydrolase. Except for a slightly modified extraction procedure, the same method as utilized for plasma was applied to the workup of whole blood.

C. Tissues

With the procedure of Okishio *et al.* (24, 124) freeze-dried rat liver homogenates are exhaustively extracted with 95% ethanol containing 0.1% ammonium hydroxide, and the extract is taken to dryness. The residue is dissolved in aqueous NaOH, *p*H 11, and applied to an Amberlyst A-26 anion exchanger. After alkaline or enzymatic hydrolysis the free bile acids are extracted with diethyl ether after acidification. The bile acids are methylated and then purified on aluminum oxide. The bile acid methyl ester fraction eluted from this column is taken to dryness and the residue is trifluoroacetylated and analyzed on a triple-component column (QF-1–SE-30–NGS) for quantitative determination.

In a study on experimentally induced allergic encephalomyelitis in guinea pigs, gray and white matter from the animal brains was analyzed for bile acids (29). The tissue was saponified by autoclaving and the hydrolyzate was acidified and extracted with diethyl ether. After another saponification (at room temperauure) the bile acids were recovered in diethyl ether. By preparative thin-layer chromatography of free and methylated bile acids, a fraction was obtained which could be analyzed by gas–liquid chromatography on a 3% QF-1 column. Subsequent combined gas chromatography–mass spectrometry analysis permitted the identification of methyl lithocholate.

By swabbing the skin of the the back with cotton wool moistened with acetone, Schoenfield *et al.* (125) obtained an extract that was analyzed for bile acids with the methods described above for human plasma bile acids (31).

D. Feces

Quantitative recoveries of endogenously labeled bile acids in homogenized human feces can be obtained by continuous extraction for 48 hr with hot chloroform–methanol, 1:1 (18). After saponification, acidification, and continuous diethyl ether extraction, the bile acids are purified on silicic acid (Section IIIB:4 and Ref. 18) to give one mono- and disubstituted, and one trisubstituted bile acid fraction. For identification purposes further subfractionation can be made [see Table V; (69, 77, 126, 127)]. The subfractions are subsequently subjected to small-scale preparative thin-layer chromatography of methylated bile acids. The fractions eluted from the thin-layer plates are next subjected to peak-shift analyses followed by final identification by gas chromatography–mass spectrometry. When the fecal bile acid composition has been elucidated in this way the mono-, di- and trisubstituted bile acids from the first silicic acid column may be quantitated after methylation and by analysis on QF-1. These results are then compared with those obtained after trifluoroacetylation of the bile acid methyl esters. (18).

In the method of Grundy *et al.* (30) an aqueous homogenate of human feces is made alkaline with 1 M NaOH in 90% aqueous ethanol after the addition of ^{14}C-labeled deoxycholic acid. The mixture is refluxed and neutral lipids are removed by petroleum ether extraction. The acidic compounds remaining in the lower phase are next exposed to strong alkaline conditions to hydrolyze conjugated bile acids. After acidification, free bile acids are extracted from the hydrolyzate with chloroform-methanol, 2:1. After solvent evaporation the residue is purified on a Florisil column. After removal of fatty acids, bile acids are eluted with diethyl ether–acetic acid 9:1, v/v. Ethyl acetate and water are added to the bile acid eluate and the lower phase removed. After methylation, the bile acids are separated by preparative thin-layer chromatography. Mono-, di-, and trisubstituted methyl cholanoates are eluted together from the thin-layer plate and then quantitated on SE-30 columns as the trimethylsilyl ether derivatives.

Ali *et al.* (128 and 129) suspended freeze-dried human feces in alkaline–50% ethanol and extracted with petroleum ether–diethyl ether, 1:1, to remove neutral lipids. The bile acids in the aqueous phase were next subjected to conditions for hydrolysis of conjugates. After acidification, the free bile acids were extracted with diethyl ether, methylated, and purified by preparative thin-layer chromatography. Mono-, di-, and trisubstituted methyl cholanoates are eluted together from the thin-layer plate and then quantitated on QF-1 columns before and after conversion into trifluoroacetates.

In the procedure of Evrard and Janssen (23), freeze-dried human feces is extracted with hot acetic acid containing nordeoxycholic acid (internal standard). Toluene is added and the supernatant removed and taken to dryness. The residue is dissolved in alkaline ethylene glycol and refluxed for 20 min at 220 °C. Unsaponifiable compounds are next removed by petroleum ether extraction, the acidic compounds in the remaining aqueous phase are extracted with diethyl ether after acidification. The free bile acids thus obtained are methylated, oxidized with CrO_3 in 90% aqueous acetic acid, and recovered by diethyl ether extraction. After evaporation of the ether the ketonic bile acid methyl esters are determined by temperature-programmed gas–liquid chromatography on a 1% JXR-column.

VI. CONCLUSION

Significant progress in the analysis of bile acids has been made during the last decade. In spite of this it is often difficult for an investigator to find simple and rapid methods that are directly applicable to this problem. This is particularly true in clinical and physiological studies and may lead to the use of oversimplified modifications or combinations of existing methods. Although reasonable reproducibility and recoveries may be obtained, specificity is often sacrificed in these modifications. Admittedly, methods that give the desired specificity are often tedious and not suitable for serial analysis. However, it is difficult to defend the use of unspecific methods without checks on the nature of the compounds determined in a given type of biological material. At present the combined gas chromatography–mass spectrometry instruments are by far the most powerful tools in identification work. These instruments will, no doubt, become available to many workers in all larger research centers.

Bile acid analysis consists of sample workup and final determination. Developments are necessary in both parts of the analysis. Initial progress has already been made—the enzymatic hydrolysis of conjugated bile acids and the solid extraction techniques in the workup of biological materials. A method that may prove of great value in serial quantitative analysis of common bile acids (after appropriate purification) is the determination with 3-hydroxysteroid oxidoreductases.

Future development is likely to include improvements in column liquid–liquid chromatography—new column packing materials and new detection systems. Stationary phases consisting of specific groups covalently bound to inert matrices are being developed in many laboratories for use in liquid–liquid as well as gas–liquid chromatography. Such phases might be made for use in bile acid analysis.

Gas chromatographic analyses are likely to become automated, with automatic injection and computer evaluation of the chromatograms. Computer treatment is of even greater importance in combined gas chromatographic–mass spectrometric analysis. These improvements will probably be available in many laboratories within the next few years.

REFERENCES

1. J. Sjövall, *in* "Methods of Biochemical Analysis" (D. Glick, ed.), Vol. 12, p. 97, Wiley (Interscience), New York (1964).
2. J. Sjövall, *in* "New Biochemical Separations" (A. T. James and L. J. Morris, eds.), p. 68, Van Nostrand, Princeton (1964).
3. J. Sjövall, *in* "Biomedical Applications of Gas Chromatography" (H. A. Szymanski, ed.), p. 151, Plenum Press, New York (1964).
4. A. F. Hofmann, *in* "New Biochemical Separations" (L. J. Morris and A. T. James, eds.), p. 362, Van Nostrand, Princeton (1964).
5. H. Van Belle, "Cholesterol, Bile Acids and Atherosclerosis," North-Holland, Amsterdam (1965).
6. A. Kuksis, *J. Am. Oil Chemists Soc.* **42,** 276 (1965).
7. A. Kuksis, *in* "Methods of Biochemical Analysis" (D. Glick, ed.), Vol. 14, p. 325, Wiley (Interscience), New York (1966).
8. G. A. D. Haslewood, "Bile Salts," Methuen, London (1967).
9. J. Sjövall, *in* "The Gas–Liquid Chromatography of Steroids" (J. K. Grant, ed.), *Mem. Soc. Endocrinol.* No. 16, p. 243, Cambridge University Press, London (1967).
10. T. Hoshita and T. Kazuno, *in* "Advances Lipid Research," (R. Paoletti and D. Kritchevsky, eds.), Vol. 6, p. 207, Academic Press, New York (1968).
11. B. Frosch and H. Wagener, *Klin. Wschr.* **46,** 913 (1968).
12. P. Eneroth, *in* "Lipid Chromatographic Analysis" (G. V. Marinetti, ed.), Vol. 2, p. 149, Marcel Dekker, New York (1969).
13. A. Kuksis, *in* "Lipid Chromatographic Analysis" (G. V. Marinetti, ed.), Vol. 2, p. 215, Marcel Dekker, New York (1969).
14. B. P. Lisboa, *in* "Methods in Enzymology" (R. B. Clayton, ed.), Vol. 15, p. 3, Academic Press, New York (1969).
15. P. Eneroth and J. Sjövall, *in* "Methods in Enzymology" (R. B. Clayton, ed.), Vol. 15, p. 237, Academic Press, New York (1969).
16. H. Sobotka, "Physiological Chemistry of the Bile," Williams and Wilkins, Baltimore (1937).

17. H. Sobotka, "Chemistry of the Sterids," Williams and Wilkins, Baltimore (1938).
18. P. Eneroth, K. Hellström, and J. Sjövall, *Acta Chem. Scand.* **22,** 1729 (1968).
19. A. Norman, *Brit. J. Nutr.* **18,** 173 (1964).
20. R. Shioda, P. D. S. Wood, and L. W. Kinsell, *J. Lipid Res.* **10,** 546 (1969).
21. A. Hofmann, *J. Lipid Res.* **8,** 55 (1967).
22. I. G. Anderson, *Nature* **193,** 60 (1962).
23. E. Evrard and G. Janssen, *J. Lipid Res.* **9,** 226 (1968).
24. T. Okishio, P. P. Nair, and M. Gordon, *Biochem. J.* **102,** 654 (1967).
25. D. Rudman and F. E. Kendall, *J. Clin. Invest.* **36,** 530 (1957).
26. J. B. Carey, Jr., *J. Clin. Invest.* **37,** 1494 (1958).
27. S. J. Levin, C. G. Johnston, and A. J. Boyle, *Anal. Chem.* **33,** 1407 (1961).
28. T. A. Mahowald, J. T. Matschiner, S. L. Hsia, E. A. Doisy, Jr., W. H. Elliott, and E. A. Doisy, *J. Biol. Chem.* **225,** 795 (1957).
29. S. H. M. Nagvi, B. L. Herndon, M. T. Kelley, V. Bleisch, R. T. Aexel, and H. J. Nicholas, *J. Lipid Res.* **10,** 115 (1969).
30. S. M. Grundy, E. H. Ahrens, and T. Miettinen, *J. Lipid Res.* **6,** 397 (1965).
31. D. H. Sandberg, J. Sjövall, K. Sjövall, and D. A. Turner, *J. Lipid Res.* **6,** 182 (1965).
32. G. W. Kuron and D. M. Tennent, *Fed. Proc.* **20,** 268 (1961).
33. H. Bradlow, *Steroids* **11,** 265 (1968).
34. C. S. Shackleton, J. Sjövall, and O. Wisén, *Clin. Chim. Acta* **27,** 354 (1970).
35. A. Norman, personal communication, 1969.
36. J. Roovers, E. Evrard, and H. Vanderhaege, *Clin. Chim. Acta* **19,** 449 (1968).
37. E. H. Ahrens, Jr., and L. C. Craig, *J. Biol. Chem.* **195,** 763 (1952).
38. E. Nyström, *Arkiv Kemi* **29,** 99 (1968).
39. P. Eneroth and E. Nyström, unpublished results.
40. J. Sjövall, E. Nyström, and E. Haahti, *in* "Advances in Chromatography" (J. C. Giddings and R.A. Keller, eds.), Vol. 6, p. 119, Marcel Dekker, New York (1968).
41. S. Mirvish, *S. Afr. J. Med. Sci.* **23,** 33 (1958).
42. A. J. P. Martin, *Ann. Rev. Biochem.* **19,** 517 (1950).
43. J. Sjövall, *Arkiv Kemi* **8,** 317 (1955).
44. J. Sjövall, *Clin. Chim. Acta* **4,** 652 (1959).
45. J. Sjövall, *Acta Chem. Scand.* **8,** 339 (1954).
46. R. J. Bridgewater, G. A. D. Haslewood, and A. R. Tammar, *Biochem. J.* **85,** 413 (1962).
47. R. H. Palmer, *Proc. Nat. Acad. Sci.* **58,** 1047 (1967).
48. I. E. Bush, "The Chromatography of Steroids," Pergamon Press, Oxford (1961).
49. G. A. D. Haslewood, *Biochem. J.* **56,** 581 (1954).
50. R. Palmer, *in* "Methods in Enzymology" (R. B. Clayton, ed.), Vol. 15, p. 280, Academic Press, New York (1969).
51. H. Gänshirt, F. W. Koss, and K. Morianz, *Arzneimittel-Forsch.* **10,** 943 (1960).
52. B.A. Kottke, J. Wollenweber, and C.A. Owen, *J. Chromatog.* **21,** 439 (1966).
53. J. S. Tung and R. Ostwald, *Lipids* **4,** 216 (1969).
54. G. Semenuk and W. T. Beher, *J. Chromatog.* **21,** 27 (1966).
55. L. Hanaineh and C.J.W. Brooks, *Biochem. J.* **92,** 9 P (1964).
56. M.G. Horning and E.C. Horning, personal communication, 1968.
57. L. Peric-Golia and R.S. Jones, *Proc. Soc. Exp. Biol. Med.* **110,** 327 (1962).
58. L. Irvin, C.G. Johnston, and J. Kopala, *J. Biol. Chem.* **153,** 439 (1944).
59. G. A. D. Haslewood, *Biochem. J.* **52,** 583 (1952).
60. A. Norman and R. H. Palmer, *J. Lab. Clin. Med.* **63,** 986 (1964).
61. R.J. Bridgewater, T. Briggs, and G.A.D. Haslewood, *Biochem. J.* **82,** 285 (1962).
62. S. Burstein and S. Lieberman, *J. Am. Chem. Soc.* **80,** 5235 (1958).
63. E.H. Mosbach, C. Zomzely, and F.E. Kendall, *Arch. Biochem. Biophys.* **48,** 95 (1954).
64. J.T. Matschiner, T.A. Mahowald, W.H. Elliott, E.A. Doisy, Jr., S.L. Hsia, and E.A. Doisy, *J. Biol. Chem.* **225,** 771 (1957).

65. J. Ellingboe, E. Nyström, and J. Sjövall, *J. Lipid Res.* **11,** 266 (1970).
66. R.A. Keates and C.J.W. Brooks, *J. Chromatog.* **44,** 509 (1969).
67. P. Eneroth and E. Nyström, *Biochim. Biophys. Acta* **144,** 149 (1967).
68. E. Heftmann, E. Weiss, H.K. Miller, and E. Mosettig, *Arch. Biochem. Biophys.* **84,** 324 (1959).
69. H. Danielsson, P. Eneroth, K. Hellström, S. Lindstedt, and J. Sjövall, *J. Biol. Chem.* **238,** 2299 (1963).
70. H. Danielsson, A. Kallner, and J. Sjövall, *J. Biol. Chem.* **238,** 3846 (1963).
71. I.D.P. Wootton, *Biochem. J.* **53,** 85 (1953).
72. H.J. Karavolas, W.H. Elliott, S.L. Hsia, E.A. Doisy, Jr., J.T. Matschiner, S.A. Thayer, and E.A. Doisy, *J. Biol. Chem.* **240,** 1568 (1965).
73. H.G. Roscoe and M. J. Fahrenbach, *Anal. Biochem.* **6,** 520 (1963).
74. C.M. Siegfried and W.H. Elliott, *J. Lipid Res.* **9,** 394 (1968).
75. D. Kritchevsky, D.S. Martak, and G.H. Rothblat, *Anal. Biochem.* **5,** 388 (1963).
76. S.K. Goswami and C.F. Frey, *J. Chromatog.* **47,** 126 (1970).
77. P. Eneroth, B.A. Gordon, J. Sjövall, and R. Ryhage, *J. Lipid Res.* **7,** 511 (1966).
78. N. Spritz, E.H. Ahrens, Jr., and S.M. Grundy, *J. Clin. Invest.* **44,** 1482 (1965).
79. A. Kallner, *Acta Chem. Scand.* **21,** 315 (1967).
80. B. Matkovics and Z. Tegyey, *Microchem. J.* **13,** 174 (1968).
81. W. Nienstedt, *Acta Endocrinol.* Suppl. 114 (1967).
82. I.E. Bush, *in* "Methods in Biochemical Analysis" (D. Glick, ed.), Vol. 13, p. 357, Wiley (Interscience), New York (1965).
83. L.R. Snyder, *J. Chromatog.* **16,** 55 (1964).
84. J.G. Hamilton and J.W. Dieckert, *Arch. Biochem. Biophys.* **82,** 203 (1959).
85. H.E. Gallo-Torres and J.G. Hamilton, *J. Chromatog. Sci.* **7,** 513 (1969).
86. L.J. Morris, *in* "New Biochemical Separations" (L.J. Morris and A.T. James, eds.), p. 342, Van Nostrand, Princeton (1964).
87. B.E. Gustafsson, T. Midtvedt, and A. Norman, *J. Exptl. Med.* **123,** 413 (1966).
88. B. Frosch, *Klin. Wochschr.* **43,** 262 (1966).
89. B. Frosch, *Arzneimittel-Forsch.* **15,** 178 (1965).
90. J. Wollenweber, B.A. Kottke, and C.A. Owen, *J. Chromatog.* **24,** 99 (1966).
91. A. Stiehl, J. Wollenweber, and H. Wagener, *J. Chromatog.* **43,** 278 (1969).
92. D.T. Forman, C. Phillips, W. Eiseman, and C.B. Taylor, *Clin. Chem. (New York)* **14,** 348 (1968).
93. T. Iwata and K. Yamasaki, *J. Biochemistry (Tokyo)* **56,** 424 (1964).
94. L. A. Turnberg and A. Anthony-Mote, *Clin. Chim. Acta* **24,** 253 (1969).
95. L.F. Fieser and M. Fieser, "Reagents for Organic Synthesis," Wiley, New York (1967).
96. C. Djerassi, ed. "Steroid Reactions. An outline for organic chemists," Holden-Day, San Francisco (1963).
97. T. Briggs and S.R. Lipsky, *Biochim. Biophys. Acta* **97,** 579 (1965).
98. M. Makita and W.W. Wells, *Anal. Biochem.* **5,** 523 (1963).
99. W.J.A. VandenHeuvel and K.L.K. Braly, *J. Chromatog.* **31,** 9 (1967).
100. W.J.A. VandenHeuvel, J. Sjövall, and E.C. Horning, *Biochim. Biophys. Acta* **48,** 596 (1961).
101. J. Sjövall, *Acta Chem. Scand.* **16,** 1761 (1962).
102. A. Kuksis and B.A. Gordon, *Can. J. Biochem. Physiol.* **41,** 1355 (1963).
103. W.J.A. VandenHeuvel and E.C. Horning, *Biochim. Biophys. Acta* **74,** 560 (1963).
104. E.C. Horning, W.J.A. VandenHeuvel, and B.G. Creech, *in* "Methods of Biochemical Analysis (D. Glick, ed.), Vol. 11, p. 112, Wiley, New York (1963).
105. H.M. Fales and T. Luukkainen, *Anal. Chem.* **37,** 955 (1965).
106. J.G. Allen, G.H. Thomas, C.J.W. Brooks, and B.A. Knights, *Steroids* **13,** 133 (1969).
107. C. Djerassi, R.R. Engle, and A. Bowers, *J. Org. Chem.* **21,** 1547 (1956).
108. O.H. Wheeler and J.L. Mateos, *Can. J. Chem.* **36,** 1049 (1958).
109. E.C. Horning, W.J.A. VandenHeuvel, and B.G. Creech *in* "Methods of Biochemical

Analysis (D. Glick, ed.), Vol. 11, p. 80, Wiley (Interscience) New York, 1963.
110. J.F. Parcher and P. Urone, *J. Gas Chromatog.* **2,** 184 (1964).
111. R.F. Kruppa, R.S. Henly, and D.C. Smead, *Anal. Chem.* **39,** 851 (1967).
112. No author, *J. Gas Chromatog.* **6,** 1 (1968).
113. W.H. Elliott, L.B. Walsh, M.M. Mui, M.A. Thorne, and C.M. Siegfried, *J. Chromatog.* **44,** 452 (1969).
114. D.K. Bloomfield, *Anal. Chem.* **34,** 737 (1962).
115. A.F. Hofmann, V. Bokkenheuser, R.L. Hirsch, and E.H. Mosbach, *J. Lipid Res.* **9,** 244 (1968).
116. A.F. Hofmann, E.H. Mosbach, and C.C. Sweeley, *Biochim. Biophys. Acta* **176,** 204 (1969).
117. M.T. Subbiah, A. Kuksis, and S. Mookerjea, *Proc. Can. Fed. Biol. Soc.* **10,** 5 (1967).
118. P.P. Nair, M. Gordon, S. Gordon, J. Reback, and A.I. Mendeloff, *Life Sci.* **4,** 1887 (1965).
119. L.J. Schoenfield, J. Sjövall, and K. Sjövall, *J. Lab. Clin. Med.* **68,** 186 (1966).
120. K. Sjövall and J. Sjövall, *Clin. Chim. Acta* **13,** 207 (1966).
121. T. Cronholm and J. Sjövall, *Europ. J. Biochem.* **2,** 375 (1967).
122. P.P. Nair and C. Garcia, *Anal. Biochem.* **29,** 164 (1969).
123. B. Josephson, *Biochem. J.* **29,** 1519 (1935).
124. T. Okishio and P.P. Nair, *Biochemistry* **5,** 3662 (1966).
125. L.J. Schoenfield, J. Sjövall, and E. Perman, *Nature* **213,** 93 (1967).
126. H. Danielsson, P. Eneroth, K. Hellström, and J. Sjövall, *J. Biol. Chem.* **237,** 3657 (1962).
127. P. Eneroth, B. Gordon, and J. Sjövall, *J. Lipid Res.* **7,** 524 (1966).
128. S.S. Ali, A. Kuksis, and J.M.R. Beveridge, *Can. Biochem.* **44,** 957 (1966).
129. S.S. Ali, A. Kuksis, and J.M.R. Beveridge, *Can. J. Biochem.* **44,** 1377 (1966).
130. V. Prey, H. Berbalk, and M. Kausz, *Microchim. Acta,* p. 968 (1961).
131. M.T. Subbiah and A. Kuksis, *J. Lipid Res.* **9,** 288 (1968).
132. U. Freimuth, B. Zawta, and M. Büchner, *J. Chromatog.* **30,** 607 (1967).
133. W. Matthias, *Naturwiss.* **41,** 18 (1954).
134. T. Sasaki, *Hiroshima J. Med. Sci.* **14,** 85 (1965).
135. T. Usui, *J. Biochem. (Tokyo)* **54,** 283 (1963).
136. P. Eneroth, *J. Lipid Res.* **4,** 11 (1963).
137. A. Kallner, *Acta Chem. Scand.* **18,** 1502 (1964).
138. A. Kallner, *Arkiv Kemi* **21,** 567 (1967).
139. J.A. Gregg, *J.Lipid Res.* **7,** 579 (1966).
140. T. Sasaki, *J. Biochem. (Tokyo)* **60,** 56 (1966).
141. K. Morimoto, *J. Biochem. (Tokyo)* **55,** 410 (1964).
142. A.F. Hofmann and E.H. Mosbach, *J. Biol. Chem.* **239,** 2813 (1964).
143. J.G. Hamilton, *Arch. Biochem. Biophys.* **101,** 7 (1963).
144. T. Kazuno and T. Hoshita, *Steroids* **3,** 55 (1964).
145. K. Einarsson, *J. Biol. Chem.* **241,** 534 (1966).
146. G. Johansson, *Acta Chem. Scand.* **20,** 240 (1966).
147. A. Kallner, *Arkiv Kemi* **26,** 553 (1967).
148. K.A. Mitropoulos and N.B. Myant, *Biochem. J.* **103,** 472 (1967).
149. J.G. Hamilton, J.R. Swartout, O.N. Miller, and J.E. Muldrey, *Biochem. Biophys. Res. Commun.* **5,** 226 (1961).
150. K. Tsuda, V. Sato, N. Ikegawa, S. Tanaka, H. Higashikuze, and R. Osawa, *Chem. Pharm. Bull. (Tokyo)* **13,** 720 (1965).
151. W.L. Holmes and E. Stack, *Biochim. Biophys. Acta* **56,** 163 (1962).
152. J. Sjövall, C.R. Meloni, and D.A. Turner, *J. Lipid Res.* **2,** 317 (1961).
153. A. Kallner, *Acta Chem. Scand.* **22,** 2361 (1968).
154. A. Kallner, *Acta Chem. Scand.* **22,** 2353 (1968).
155. T. Okishio and P.P. Nair, *Anal. Biochem.* **15,** 360 (1966).

Chapter 6

ION-EXCHANGE CHROMATOGRAPHY OF BILE ACIDS*

A. Kuksis

Banting and Best Department of Medical Research
University of Toronto
Toronto, Canada

I. INTRODUCTION

The common bile acids and their glycine and taurine conjugates are relatively freely soluble in aqueous solutions and possess dissociation constants that range from near neutrality to values exhibited by the strongest of organic acids. Ion-exchange chromatography should therefore be theoretically suitable for an efficient separation of such molecules. In practice, however, the adaptation of ion-exchange techniques to bile acid work has been limited and progress extremely slow. This has been partly due to the great success of the thin-layer and gas–liquid chromatographic techniques as means of analysis of bile acids, as well as to the discovery of extremely selective solvent pairs for the separate extraction of the glycine and taurine conjugates from natural sources. The main difficulty has been the complex behavior of these natural detergents in aqueous solutions. The realized separations apparently have involved certain sorbent properties of the ion exchangers as well as chromatography of ions by ion exchange.

For purposes of the present discussion, ion-exchange chromatography of bile acids will include all chromatographic separations based on ionic and nonionic interactions carried out with ion-exchange material as the stationary phase. The chapter reviews the progress to date and provides critical

*Unless otherwise specified, the term "bile acid" is used in this chapter as a generic term for this class of compounds and includes both free and conjugated bile acids as well as their salts. The term "bile salt" includes the salts of both free and conjugated bile acids.

descriptions of the practical accomplishments. Where possible, an effort has been made to account for both success and failure of the current analytical systems in terms of known principles previously elaborated in work with small molecules.

II. THEORETICAL CONSIDERATIONS

For optimum resolution, the chromatographic conditions must be chosen to allow a maximum exploitation of the differences in the physical properties of the components of the mixture. To facilitate this choice in the present application, a brief account is given of the pertinent chemical and physical properties of the cholanic acids and their conjugates and of the ion-exchange media. The theoretical section concludes with a consideration of the nature of binding of bile acids to ion-exchange resins and a discussion of the principles of ion-exchange chromatography.

A. Some Physical Properties of Bile Acids and Their Amino Acid Conjugates

Chemically, the bile acids are hydroxylated derivatives of cholanic acid, a tetracyclic steroid acid of 24 carbon atoms. The acids occur in nature largely as the water-soluble sodium salts of peptide conjugates of glycine and taurine. The free acids are liberated by saponification or specific enzyme hydrolysis. The chemistry of the bile acids has been reviewed in Chapter 1 of this volume (1). In view of their highly polar nature, special attention is called to the recent discovery of the cholic acid conjugates of ornithine (2, 3) and the 3α-sulfate esters of glycolithocholic and taurolithocholic acids (4).

Physically, the bile acid molecule is a rigid, disc-shaped body kinked at the *cis*-juncture of the A and B rings, with a small side chain that terminates in a highly polar ionic group. One side of the body is hydrophobic, the other is hydrophilic, containing 2 or 3 hydroxyl groups. At *p*H 6.5, sodium cholate, sodium chenodeoxycholate, and sodium deoxycholate precipitate from solution in acid form, while the glycine conjugates of these acids remain in solution to about *p*H 4.5 (5). The taurine conjugates are fully soluble as the ionized or un-ionized sulfonic acid, as are the 3α-sulfate esters of glyco- and taurolithocholic acids (4).

At low concentrations, the bile acids are freely soluble in water and yield molecular solutions. The concentration at which molecular solubility is reached is termed the critical micelle concentration. Above this concentration, aggregation of the molecules occurs and micelles are formed. Hofmann and Small (6) have summarized the effects of temperature, electrolyte concentration, impurities, and the *p*H of the solution upon the critical micelle concentration of bile acids. An increase in temperature increases the solubility of the monomer and therefore increases the critical micelle concentration. The latter is decreased by an increase in concentration of the electrolyte, which decreases the electrostatic repulsion between the charged tails of the bile acid molecules. The addition of polar substituents to the hydrocarbon skeleton (hydroxylation, sulfate ester formation) increases considerably the solubility of the monomer and therefore the critical micelle concentration. Short-chain alcohols, such as ethanol, increase the solubility of the monomer to such an extent that micelle formation may be abolished. On the other hand, long-chain alcohols, such as cholesterol, may decrease the critical micelle concentration.

Table I lists the available critical micelle concentrations of bile acids in water and in a salt solution. Accordingly, during chromatography the concentration of the bile acids should be kept below 0.01 *M* for the trihydroxy salts and below about 0.05 *M* for the dihydroxy derivatives. Further decreases in the concentration of the external solution at least may be required when working in the presence of salt or buffer and swelling amphipaths (monoglycerides, sterols). When more concentrated solutions are desired, the addition of alcohol to the aqueous solution of bile acids should be considered. The presence of simple and/or mixed micelles in the external solution during chromatography, however, may not necessarily be detrimental to the resolution of the acids, as the micelles would be expected to be in rapid equilibrium with the bile acids in the molecular solution. The actual rates of exchange of various bile acids between micelles and molecular solutions have not been determined.

The influence of *p*H on the bile acid solubility and ionization is complex and of particular interest to their ion-exchange behavior. No quantitative studies have been published on the effect of *p*H, concentration, and electrolyte concentration on the solubility of any pure bile acid. Ideally, a series of acids would be expected to be displaced from a basic exchange resin in the order of their dissociation constants. Although the ion-exchange chromatographic properties of bile acids are largely determined by the nature of their ionic groups, both electrostatic and short-range or Van der Waals' forces may be involved in a practical resolution.

TABLE I. Critical Micelle Concentrations of Some Bile Salts (Moles/liter)[a]

Bile salt	Medium	
	Water	0.15 *M* Na+
trihydroxy	0.12	0.03–0.08
dihydroxy	0.04–0.06	0.02–0.04

[a] Modified from Hofmann and Small (6).

TABLE II. True Dissociation Constants and pK_a Values of Bile Acids[a]

Bile acid	Dissociation constants	pK_a values	
	Josephson (7)	Josephson (7)	Small (5)
cholic	64.6×10^{-7}	5.13	5.29
chenodeoxycholic			6.18
deoxycholic	3.8×10^{-7}	6.29	6.29
glycocholic	355×10^{-7}	4.41	3.75
glycodeoxycholic	$1{,}047 \times 10^{-7}$	3.81	4.77
taurocholic	2.75×10^{-7}	1.83	1.85
taurodeoxycholic	1.23×10^{-7}	1.74	1.93

[a] Kumler and Halverstadt (8) report a pK_a of 6.73 for lithocholic acid in 66.6% ethanol.

Table II gives the dissociation constants along with the calculated pK_a values for the common free and conjugated bile acids as determined by Josephson (7) and by Small (5). Both authors agree that as a result of conjugation the average pK_a of bile acids of about 6 is lowered to about 3.8–4.8 by glycine conjugation and to about 1.9 by taurine conjugation. It should be noted, however, that the values for the glycine and taurine conjugates of cholic and deoxycholic acids given by Small are in the opposite order to those values given by Josephson. This discrepancy may possibly be due to the use of supersaturated colloidal solutions in Josephson's electrometric determinations. The dissociation constants reported by Josephson for the free bile acids have been revised upwards by Kumler and Halverstadt (8) who showed that the difference between the strongest and the weakest of the common free bile acids was only 0.09 pK_a units. There was a precise agreement between the experimentally determined strengths of these acids and their relative strengths as predicted from the known acid-strengthening character of the substituents on the steroid nucleus. The following order of acid strength was observed: dehydrocholic > dehydrodeoxycholic > cholic > apocholic > deoxycholic > lithocholic. According to Ekwall (9), the pK_a of a molecular solution of cholic (4.98) and deoxycholic (5.17) is increased to 5.48 and 6.35, respectively, for solutions above the critical micelle concentration. Lithocholic acid was reported (8) to have a pK_a of 6.73 in 66.6% alcohol. Since dissociation constants are somewhat smaller in alcohol–water than in pure water, this value must be a little high in comparison to those given for the other free bile acids in Table II.

In one of the earliest studies on the dissociation constants of bile acids, Henriques (10) observed that the pK_a values of the different bile acids determined in alcohol ranged between 2.46 and 3.23. No recent record of pK_a values of bile acids in alcohol solutions exists although aqueous alcohol solutions have since been used in the titration of the bile acids in the pK_a

determinations, from which extrapolations to pure water solutions were made. Such values would have been of help in the appraisal of the ion-exchange behavior of bile acids when chromatographed in alcoholic solutions. It is possible that the effect of alcohol upon the dissociation of different acids may not be constant.

The pK_a values of the recently discovered ornithine conjugates of bile acids and of the sulfate esters of glyco- and taurolithocholates have not yet been determined.

B. Some Chemical and Physical Characteristics of Synthetic Anion Exchangers

Pertinent accounts of the chemical and physical properties of the ion exchangers have been summarized in texts on ion-exchange chromatography (11–13). The most important ion exchangers for bile acid work are the synthetic resins, containing polar groups of basic nature, which are introduced before or after the polymerization stage. An important property of a synthetic ion-exchange resin is insolubility. This can be obtained only if the polymer forms a three-dimensional network with the long chains of linear polymer connected at intervals by cross-linkages, so that the whole resin particle becomes a single molecule. It has been demonstrated for several chemically different types that the linear polymer formed by end-to-end union of monomer molecules and devoid of cross-linkages is soluble in organic solvents even at high degrees of polymerization. Ion-exchange resins, like other synthetic resin polymers, must therefore have cross-linkages between the linear polymer chains. It will be seen later that the extent of cross-linkage considerably modifies the ion-exchange behavior of the resin toward the relatively large molecules of bile acids.

The strong-base anion-exchange resins, such as Dowex-1 and Dowex-2 (Dow Chemical Company) and Amberlite IRA-400 and Amberlite 410 (Rohm & Haas Company) are prepared by the method of addition polymerization, with the desired basic groups being introduced into a styrene–divinylbenzene copolymer after the polymerization. The styrene and divinylbenzene are usually copolymerized in aqueous suspension in the presence of a catalyst such as benzoyl peroxide. The organic reactants, being almost water insoluble, form small droplets, the size of which can be regulated within wide limits (diameters 1 μ to 2 mm) by the addition of suspension stabilizers of various kinds and by control of the temperature, viscosity of the medium, and degree of mechanical agitation. The final copolymer retains the approximately spherical shape of the original droplets so that it can be obtained in beads of any required size. The beads are fairly uniform, thus eliminating the considerable losses encountered in grinding and screening coarse resin in order to obtain a fraction within a required particle size range. The extent of cross-linking in the copolymer can be varied by alteration of the amounts of divinylbenzene used in the copolymerization. Resins prepared with 1–16% divinylbenzene content are available commercially.

The physical properties of the resin vary in a regular way with the amount of divinylbenzene used in its preparation. There has therefore been a tendency to equate the percentage of divinylbenzene used in the preparation with the amount of cross-linking in the copolymer. This may not be a valid assumption in all cases, and the cross-linking numbers afford only an indication of the extent of cross-linking.

In order to prepare the strong- or weak-base anion-exchange resins, the styrene–divinylbenzene copolymers are reacted with chloromethyl methyl ether, which converts the phenyl residues into benzyl chloride groups that are subsequently allowed to react with either secondary or tertiary amines. The chloromethyl groups supposedly become attached to the 4-position in the phenyl residues. Trimethylamine (Dowex-1, Amberlite IRA 400) and dimethylethanolamine (Dowex-2, Amberlite 410) are typical of the tertiary amines used in the preparation of commercial resins. The quaternary ammonium ion-exchange resins are highly ionized and can be used over the entire *p*H range (14).

The exchange capacities of the commercially available strong-base anion-exchange resins are in the range of 2–3 meq/g of dry resin. Amberlyst XN-1006 or Amberlyst A-26 (Rohm & Haas Company) is an experimental strong-base anion-exchange resin, which is characterized by small changes in volume on change of solvent.

The search for means to prepare ion-exchange papers which would combine the advantages of ion-exchange methods with the simplicity of paper chromatography has led to the introduction of ionic residues into cellulose. The anion exchanger, DEAE-cellulose, is prepared by the condensation of 2-chlorotriethylamine with alkaline cellulose. These materials serve as insoluble stationary phases with a very low effective degree of cross-linking suitable for the ion-exchange chromatography of even macromolecular solutes. A more strongly basic derivative can be prepared by reacting the DEAE-cellulose with ethyl bromide to yield the triethylaminoethyl (TEAE) cellulose. With due care aminoethyl celluloses can be prepared with exchange capacities from 0.45–0.75 meq/g of dry material (12). Such preparations have pK_a values of about 9.3 in 0.5 *M* sodium chloride. The modified cellulose can be made into paper, or into powder for column packing and spreading of thin-layer plates.

For work with large molecules of biochemical interest, Sephadex (Pharmacia, Uppsala, Sweden) has been the most popular of the carbohydrate gels. Sephadex is prepared from polysaccharide dextrans that have been synthesized by the action of a bacterium on sucrose. The dextran chains are cross-linked by a reaction with epichlorohydrin to give pores of desired size. Sephadex gels are insoluble in water and are stable in bases and weak acids. In comparison to DEAE-cellulose, the cross-linked dextran, DEAE-Sephadex, has a higher exchange capacity and provides an ion-exchange material with superior properties (15). There appears to be no interaction between DEAE-Sephadex and molecules containing carbonyl groups such as esters and ketones, which would manifest itself in tailing of these materials during elution.

Since the anion-exchange resins are frequently used in the hydroxy form, certain precautions should be observed when employing them with non-aqueous solvents. If acetone is used, excessive diacetone alcohol may be formed via base-catalyzed aldol condensation. A small amount of water should be present in all organic solvents containing alcohol to repress trans-esterification of esters (15) and esterification of acids (16) with alcohol as solvent. Without water, up to 5% of transesterification in 15 to 20 min has been observed with DEAE-Sephadex. After an overnight contact the trans-esterification approached 35%. With 1% water in the solvent, however, less than 0.3% transesterification and 0.3% saponification occurred when 70 mg of butyl stearate was passed through a DEAE-Sephadex column (15).

Thin-layer applications have called for the selection of the finest resin or modified cellulose particles in order to increase the surface area and insure maximum contact between particles in the thin layer. The addition of inert binders to promote the adherence of the coarse resin particles to the glass plate has met with less success. The best thin-layer plates of anion-exchange resins have been obtained by impregnating fine cellulose powders with liquid anion-exchange resins such as the polyethylenimine (17). The latter material is commercially available with a particle size of approximately 10 μ and an ion-exchange capacity of about 1.0 meq/g.

C. Binding of Bile Acids to Anion-Exchange Resins

It has been demonstrated (18, 19) that with monofunctional strong-base anion-exchange resins, the binding of cholate and glycocholate is due to ion exchange only. Since other free bile acids and their glycine conjugates possess comparable pK_a values, whereas the taurine conjugates are more acidic and therefore more likely to be involved in ion exchange, all bile acids may be bound to ion-exchange resins solely by ion exchange. In a number of experiments, however, the ion-exchange isotherms leveled off at an equivalent ionic fraction of less than one. This indicated an approach to a limiting exchange capacity of the resins for the bile salt anion which was less than the exchange capacity for chloride anion. This type of behavior was attributed

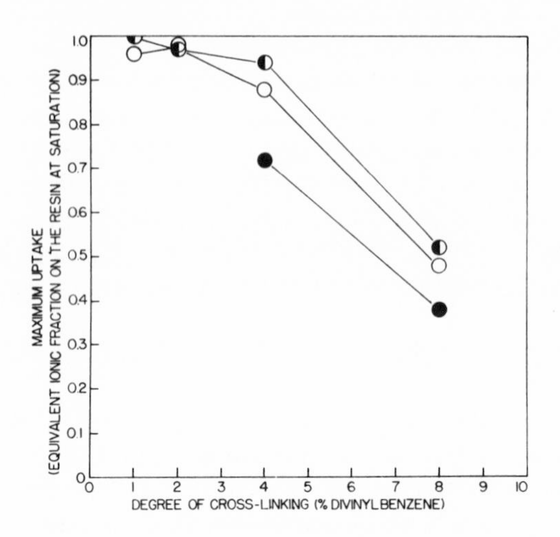

Fig. 1. Effect of the degree of cross-linking on the maximum uptake by the resin. ◐, sodium cholate (50–100 mesh Dowex-1 resin); ●, sodium cholate (20–50 mesh Dowex-1 resin); ○, sodium glycocholate (50–100 mesh Dowex-1 resin). Redrawn with the permission of authors and publishers of *J. Phys. Chem.* **72,** 1204, 1968.

to the possibility that some regions of the resin are inaccessible to large organic ions and that small ions at these sites cannot be replaced by the large organic ions. Such an interpretation was supported by the fact that the fraction of total possible uptake of the bile salt anions by the resins increased with decreasing cross-linking and was essentially unity for resins with lower cross-linkages (18).

Furthermore, the uptake of bile salt anions by ion-exchange resins was increased when the particle size was reduced (19), presumably due to the increase in the number of exchange sites available to the large organic molecules which cannot penetrate the interior of the particle. Figure 1 illustrates the effect of the degree of cross-linking and particle size upon the uptake by the resin. Obviously, the more efficient chromatography of bile acids will be anticipated on the more open resins, with smaller particle size and lower cross-linking. In practice, a Dowex-1 resin with 2% cross-linking and 50–100 mesh particle size has proved satisfactory (2, 16).

There is evidence (18, 19) that the fraction of resin sites entering into mass-action equilibrium with the bile salt anions may not be constant but increases gradually as the concentration of the bile salt anions in the resin increases. The increase in binding could also indicate some form of a cooperative process. A possible explanation is that sodium cholate and sodium glycocholate have a strong tendency to associate in aqueous solution. Therefore, as the concentration of bile salt anion in the resin phase reaches a critical value, possibly the critical micelle concentration, aggregates form, which would enhance binding.

Blanchard and Nairn (19) also found that the uptake of glycocholate was slightly lower than that of the cholate anion by the same resin. This discrepancy was attributed to a difference in the hydrophilicity and the size of these anions. Due to the more hydrophilic nature of the glycine conjugates, it was assumed that the glycocholate anion would be more strongly solvated and hence less free water would be left in the interior of the resin phase than if cholate anion were taken up by the resin. The resin should therefore prefer the less strongly solvated cholate anion as was found in their studies. A similar explanation could be advanced to account for the somewhat longer retention of the dihydroxy when compared to trihydroxy bile acids as observed by Gordon *et al.* (2) during ion-exchange chromatography. The latter workers, however, postulated a sorption effect to account for the differences in the retention. The dihydroxy bile acids being more hydrophobic would be expected to interact with the polystyrene matrix more strongly than the trihydroxy derivatives. The resin would appear to act as a hydrophobic adsorbent and the elution sequence would be in the order of decreasing polarity as observed experimentally.

The binding of the bile acids to the anion-exchange resins appears to have been completely reversible, although actual experimental recoveries of the added bile acids have only occasionally been tested. The most complete data are due to Sandberg *et al.* (20) who added C^{14}-labeled bile acids to an anion-exchange resin along with small amounts of dilute serum and recovered the acids after exhaustive washing of the column by elution with ammonium carbonate in aqueous ethanol. Table III gives the amounts added and the

TABLE III. Recovery of 24-C^{14}-labeled Bile Acids from Amberlyst XN-1006 (20)[a]

Bile acid	Amount added (cpm)	Amount recovered (cpm)	Percentage recovery
	19,600	18,700	95.4
cholic	19,600	18,300	93.4
	31,400	30,200	96.2
deoxycholic	31,400	30,000	95.5
	18,670		
chenodeoxycholic	18,670	18,100	96.9
	29,700	28,400	95.6
taurocholic	29,700	28,200	95.0

[a] Acids added at *p*H 11 to a 1 × 5 cm column of resin in the OH^- form. After washing with water and aqueous alcohol the bile acids were eluted with 150 ml of 0.2 *M* ammonium carbonate in 80% ethanol. The specific activity of the cholic, chenodeoxycholic and deoxycholic acids was approximately 10, 8, and 3 μc/mg, respectively.

percent recoveries obtained. The concentrations of the additions ranged from 0.5 μg to 50 μg per 5 to 10 ml of serum. The radioactivity recoveries in all cases were 95% or better. An assessment of the recovery by gas chromatography of the bile acids usually resulted in considerably lower values due to a partial destruction of the recovered bile acids during saponification, which must precede the gas chromatographic step. Quantitative recoveries of selected bile acids have been obtained also from other anion-exchange resins using acidic eluants and larger columns (21). Binding of bile salt anions by ion-exchange resins for medicinal purposes has been extensively investigated (22–24) where it has been shown that the therapeutic effectiveness of a resin depends on the selectivity between the bile salt anions and the chloride anion and also the capacity of the resin for the organic ion.

In view of the data obtained regarding the binding and release of bile salts by anion-exchange resins, there is good reason to believe that the criteria of reversible equilibrium and reasonably rapid exchange rates are satisfied. Under these conditions, the exchange isotherm determining the retention of the bile acids in the resin phase ought to be linear and should yield sharp zones with Gaussian concentration profiles.

D. Principles of Ion-Exchange Chromatography

The general problem in all chromatographic separations is the selection of conditions such that the distribution ratios of the solutes to be separated differ as widely as possible. With chromatographic columns of practical dimensions, having up to 500 equivalent theoretical plates, differences in distribution ratio of at least 10% are desirable for adequate separations (12). In addition, the maximum amount of solute retained by the resin stationary phase should not exceed about 2% of the maximum exchange capacity, in order to obtain linear exchange isotherms. This precaution also ensures the maximum rate of solute exchange for a particular solute-resin system.

In processes depending wholly on ion exchange, the distribution ratio of a solute between mobile and stationary (resin) phases may be varied in principle in three ways (11, 12, 13). First, the distribution can be changed by varying the concentration of a competing ion present in the mobile phase. An increase in the concentration of the chloride ion in the mobile phase will decrease the affinity of the organic solute for the resin phase, and will accelerate its passage through the column. Secondly the ratio can be changed by variation in the degree of ionization of the active fixed groups in the resin. Thus the basic groups of weak anion exchangers will be nonionized at high *p*H values, so that bound solutes may be eluted from such resins by strongly basic solutions. This method is practical in work with weakly basic exchange resins. Finally, the distribution ratio can also be changed by variation in the degree of ionization of the various solutes, by variation of the *p*H and ionic strength of the mobile phase. The three methods are all valuable for particular purposes and are to some extent interdependent and mutually exclusive.

The first method is most widely used. In cases where solubility data exist for the relative affinities of the displacing ion and the solute in question, it is possible to formulate exact relationships between the solute distribution ratio and the concentration of the eluting ion in the mobile phase.

The second method of varying the distribution ratios by changing the degree of ionization of the fixed groups in the resin is a special case of the first method. The method is not very suitable for strongly basic anion-exchange resins, such as those customarily employed in the work with bile acids.

The third method of varying the distribution ratios of solutes is by selecting the conditions of *p*H and ionic strength so as to produce variation in the degree of ionization of the various solutes in the mixture. The method is best applicable to weak electrolytes, such as the organic acids, and in practice very high selectivity can be obtained with it. In principle, it should be possible to select the operating conditions from a consideration of the dissociation or net charge-*p*H curves of the various bile acids, as shown by Cohn (25) for the nucleotides. Nonionic interactions or retention mechanisms, however, may invalidate the conclusions. In the case of a zonal separation of two weak acids with similar adsorption affinities for the resin, the separation will depend wholly on an ion-exchange mechanism. In this case the optimum *p*H for the separation will be that at which the difference in the degree of dissociation of the two acids is at a maximum. For the acids, HA and HB, the optimum *p*H of resolution is

$$pH = \frac{pK_A + pK_B}{2}$$

The *p*H value, however, is that of the surface of the resin. This will differ from that in the bulk of the external solution on account of the surface charge on the resin (26). In case of the bile acids, the *p*K values should also be those of the surface of these molecules.

With the availability of ion-exchange resins with excellent stability toward organic solvents, it has been possible to extend ion-exchange chromatography to nonaqueous systems. Owing to the effect of the solvent on the ionization of the fixed polar groups in the resin, strongly basic resins are best suited for work with nonaqueous solvents (27). It was also observed that the effect of degree of cross-linking on the capacity for organic solutes was more marked in nonaqueous solvents, so that resins with low degrees of cross-linking are particularly suitable in this case. This effect is probably due to the lower swelling of the resins studied in the organic solvents. In such systems, the ionization of any electrolyte in the aqueous resin phase will be favored relatively to that occurring in the mobile organic phase. Furthermore, separation by a partition mechanism will supplement separation by ion-exchange. Finally, mixed solvents such as water–dioxan have been shown (28) to be very effective in diminishing the effects of molecular adsorption of, for example, aromatic solutes, on polystyrene ion-exchange resins, so that the order of displacement of a series of organic acids from Dowex-2 was in agreement with predictions based on their dissociation constants. Davies and Owen (29) point out that this effect may be due to the fact that the removal of a solute molecule from the adsorption site will usually involve a much smaller energy increment in an organic or mixed solvent than in water.

III. METHODOLOGY

Of the techniques of exploiting differences in pK_a values for the chromatographic resolution of acids, the ion-exchange resins have been best suited for the separation of molecules of relatively low molecular weight. As already noted, the bile acid molecules may be interfered with significantly in their interaction with the active sites of the resin by the presence of high amounts of cross-linking. No such interference ought to take place during the migration of these molecules in a thin film of continuous solution under the influence of an electrical gradient. In both cases, methodology has presented problems which have been further compounded by the difficulty of detecting small amounts of bile acids and their conjugates in dilute aqueous salt solutions. As a result theoretical resolutions have either not been achieved or not clearly recognized.

A. Utilization of Ion-Exchange Resins

The use of ion-exchange resins in bile acid work with columns, paper, or thin layers has been rather limited and about equally disappointing. With a better understanding of the processes involved in the binding and displacement of bile acids from ion-exchange materials, improvements are being currently made in the chromatographic technology. Therefore more extensive use of resins may be anticipated at least in specific resolutions. In this section an attempt has been made to generalize the experience gained in limited applications of ion-exchange processes to bile acid isolation and separation.

1. Columns

The conventional methods of ion-exchange chromatography employ columns of resin of a height:diameter ratio of 10 to 20. The actual size of the column depends on the amount of material to be isolated or separated and the desired degree of resolution. Columns of 0.5 to 1.0 cm in diameter and 5 to 10 cm in height have been successfully employed both for isolation and resolution of the taurine and glycine conjugates of bile acids. The acids are readily taken up by the resin in either chloride, acetate, bicarbonate, or hydroxide form. The uptake of the bile acids by the column, particularly in the presence of such competing surfaces as proteins, should theoretically be further enhanced by using an alkaline solution (*p*H 11), which would favor the ionization of the weakly acidic substances.

a. Preparation of Columns. A suitable anion-exchange resin in the chloride form (Dowex-1, 2% cross-linked, 50–100 mesh, J.T. Baker Chemical Co.) is washed on a sintered glass funnel with 5–10 volumes of the following series of solvents: water, ethanol, hexane, ethanol, and water. To obtain the resin in the hydroxide form, it is recommended (20) that the ion exchanger first be converted into the bicarbonate form (by washing with 1 *N* ammonium bicarbonate and distilled water) and then into the hydroxide form. The resin is stored in the bicarbonate form in water and the columns are treated with sodium hydroxide and washed to neutrality with water just prior to application of the sample.

The preparation of the column is illustrated by a routine established for the isolation of small amounts (up to 1 mg) of bile acids and conjugates from biological fluids such as blood (20). About 5 ml of the anion-exchange resin is pipetted into a chromatographic column 1.0 cm in diameter. The resin is washed with 50 ml of 1 *N* NaOH and 50 ml of 1 *N* NaOH in 80% ethanol, followed by distilled water until the effluent is neutral. The column is now ready for use.

b. Binding of Bile Acids. An aqueous solution of the bile acids (5–10 ml) is adjusted to *p*H 11 with a few drops of 1*N* NaOH and quantitatively transferred into the column and allowed to pass through the resin at a rate of about one drop every 2 sec. The column is then washed to neutrality with distilled water. This is followed by successive washings with 20 ml of 95% ethanol, 20 ml of ethanol-ethylene chloride (1:1, v/v), and 20 ml of 80% ethanol. The bile acids and their conjugates resist elution with either organic solvents alone or with aqueous electrolytes alone, but are readily removed with alcoholic electrolytes. The washing of the column with organic solvents serves to remove any neutral lipids present in the bile acid mixture. If too much neutral lipid is present, the initial adsorption may be accomplished from a 50% alcohol solution, in which case care should be taken to dilute the solution sufficiently to reduce the electrolyte concentration below a level at which bile acids are eluted.

c. Recovery of Bile Acids and Conjugates. Free and conjugated bile acids

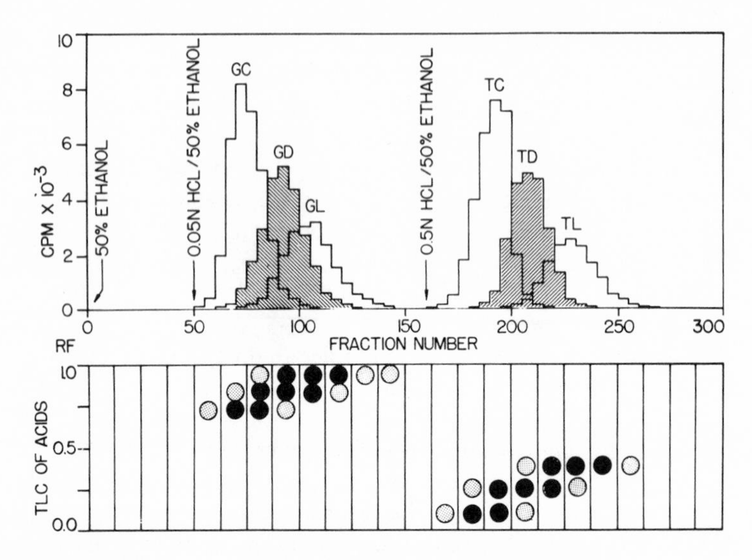

Fig. 2. Ion-exchange chromatography of standard bile acid conjugates. Column: 1 × 10 cm Dowex-1 (50–100 mesh) in OH-form.
A. *Upper Half*—trace of the radioactivity in ^{14}C-labeled glycine and taurine parts of the conjugates. GC, GD, and GL: glycine conjugates of cholic, deoxycholic and lithocholic acids; TC, TD, and TL: taurine conjugates of cholic, deoxycholic, and lithocholic acids, respectively.
B. *Lower Half*—conventional thin-layer chromatography of the conjugates appearing in the eluate (silica gel G; water–*n*-propanol–propionic acid–isoamyl acetate, 5–10–15–20, v/v/v/v).

may be recovered from the above-described column in a minimum volume of eluting solvent by washing with 1 *N* ammonium hydroxide in 80% ethanol. The eluate (a maximum of 25 ml) can be readily concentrated and the mixed solutes recovered in the form of their ammonium salts. Quantitative recovery of all bile acids from the above column may also be obtained by washing it with 150 ml of 0.2 *M* $(NH_4)_2CO_3$ in 80% ethanol (20, 30). The ammonium carbonate sublimes with decomposition on evaporation *in vacuo* at 40°C.

d. Separation of Glycine and Taurine Conjugates. Free bile acids and glycine conjugates may be eluted from anion-exchange columns with dilute (1 *N*) acetic or formic acid in aqueous ethanol, but they are not completely resolved because of the closeness of their pK_a values. Either or both are resolved from the taurine conjugates, which may be recovered with dilute ammonia or hydrochloric acid in aqueous ethanol. Figure 2 shows a typical resolution obtained for a mixture of standard bile acids under the general conditions of chromatography described above (31). The progress of the elution of the bile acids was followed by measuring the radioactivity of the ^{14}C-labeled acids added as markers. The glycine and taurine conjugates

TABLE IV. Recovery of ^{14}C-labeled Bile Acid Conjugates from Dowex 1 (31)[a]

Bile acid conjugates	Amount added (cpm)	Amount recovered (cpm)	Percentage recovery
glycocholic	33,000	32,900	100
	33,000	32,100	97
glycodeoxycholic	26,500	25,800	97
	26,500	25,200	95
glycolithocholic	20,000	19,300	96
	20,000	19,000	95
taurocholic	31,000	30,600	99
	31,000	30,000	97
taurodeoxycholic	22,500	22,000	98
	22,500	21,600	96
taurolithocholic	15,500	14,900	96
	15,500	14,500	94

[a] Chromatography conditions as given in Fig. 2.

in the effluent were independently identified by thin-layer chromatography. Figure 2 also shows the degree of cross-contamination of the eluted fractions. This contamination can be reduced by rechromatography or by a more judicious choice in the eluting solvents and conditions. Table IV gives the recoveries of the radioactivity of the labeled solutes as obtained after correction for cross-contamination and as determined by quantitative gas chromatography of the free bile acids resolved as the conjugates by thin-layer chromatography.

Bile acid conjugates in quantities ranging from a few milligrams to several grams have been resolved (2, 32) on much larger Dowex-1 anion-exchange columns (2 cm in diameter and 30 cm in height). Prior to column packing, the resin was extracted (Soxhlet) continuously (24 hr) with glacial acetic acid and sodium acetate. This treatment removed pigments and other impurities from the resin and converted it into the acetate form. After a transfer of the resin to the column, it was washed with 500 ml of distilled water and 500 ml of 25% ethanol. The bile acids were adsorbed from a 25% ethanol solution, and the column washed with 300–400 ml of 25% ethanol. The glycine conjugates were eluted with 0.05 *N* HCl in 25% ethanol and the taurine conjugates with 0.5 *N* HCl in 25% ethanol. A removal of the glycine conjugates by 1 *N* acetic or formic acid in 80% ethanol required nearly double the volume of eluent compared to the dilute HCl. The conjugates are recovered from the HCl solutions after neutralization with NaOH, evaporation to dryness, and extraction with *n*-butanol.

In addition to the anion exchangers, cation-exchange resins can also be usefully employed in bile acid work. Thus columns of a strong cation-exchange resin (Dowex-50, 2% cross-linked, 50–100 mesh, J.T. Baker Chemical Co.) have been used in the hydrogen form to purify extracts of bile acids prepared from biological materials (2, 30). Prior to use the resin is extracted with glacial acetic acid in order to remove pigments and other impurities. The size of the column used depends on the amount of the foreign material in the sample. When applied in 50% ethanol, the bile acid preparations are effectively freed of most of the pigment and all of the amino acids except taurine.

The ion-exchange celluloses and ion-exchange cross-linked dextrans (Sephadex derivatives) have received only limited attention in bile acid work (33). DEAE-cellulose in the acetate form has been used by Rouser *et al.* (34) for the fractionation of lipids. The DEAE-Sephadex has a higher exchange capacity, shows less bleeding in the presence of organic solvents, and provides a mechanically better ion-exchange material than DEAE-cellulose. The separation of acidic and neutral lipids, such as required in bile acid isolation, can be best accomplished in an organic solvent system with the DEAE-Sephadex in the free base form. The ion exchanger is prepared (15) by washing one-gram portions of it in a sintered-disc glass funnel with small portions of 1 *N* HCl, water, 1 *N* KOH, and water, until the washings are neutral. Finally it is washed with methanol and ether–methanol–water (89:10:1, v/v/v) and transferred to a column 0.5 cm × 8.5 cm where it is allowed to equilibrate overnight with the ether–methanol–water mixture. The material to be fractionated, which should not contain more than 0.5 meq of free acidic material, is dissolved in the minimum amount of the ether–methanol–water mixture and is added to the column. Neutral lipids are washed completely through the column with five 5-ml portions of the same solvent. Fatty acids are eluted (15) from this column with ether–methanol (90:10) saturated with CO_2 by placing the solvent in a round-bottom flask, flushing with CO_2, then shaking it under 8 psi for a few minutes. About 150 ml of the CO_2-saturated solvent are needed to remove quantitatively the common fatty acids. Free bile acids and their glycine conjugates can be removed from such columns with solvents containing 4% formic acid (31). The taurine conjugates are removed with solvent containing ammonia. When a formic acid elution is desired after an elution with CO_2, the column is first washed with the ether–methanol to remove CO_2 and prevent any disruption of the column caused by formation of gaseous CO_2.

2. *Liquids*

Liquid extractants with ion-exchange properties have recently found increasing use in the atomic energy field, where long-chain amines have been extensively used for separation as well as purification of inorganic ions. Several hundred organonitrogen compounds have been investigated for potential analytical and industrial application (35). The anion extraction or ion exchange may be represented as follows:

$$B_{org} + H^{+}_{aq} + A^{+}_{aq} = BHA_{org}$$

where HA is a strong monobasic acid and B is an amine insoluble in the aqueous phase. The solubility in the aqueous phase can be further reduced by employing the liquid anion exchanger as a solution in an organic solvent.

An interesting application of this observation in bile acid work is provided by the use of an oil-soluble quaternary amine for the quantitative extraction of bile acid conjugates from aqueous solution (36). Thus in a test experiment it was shown that when a chloroform solution of tetraheptylammonium chloride (10 m*M*) was used to extract an aqueous solution of ^{14}C-labeled glycine and taurine conjugates (10 m*M* total) over 99% of the radioactivity entered the chloroform phase in two extractions. Chloroform without added amine extracted less than 3% of the radioactivity under similar conditions. Comparable results were obtained with ethyl acetate as the solvent. Extraction with tetraheptylammonium chloride solutions could not be performed with such solvents as petroleum ether or benzene, since the resulting salts formed insoluble oils at the liquid–liquid interface. The tetraheptylammonium chloride was superior to another liquid ion exchanger of the secondary amine type, Amberlite XLA-3 (Rohm and Haas Co.). Extraction of an aqueous solution containing 30 μmoles of mixed taurine-conjugated bile acids with an equal volume of chloroform solution containing 5 g of XDL-3 per 100 ml removed 20% of radioactivity in two extractions. Since tetraheptylammonium chloride is a quaternary amine, it may be used for extraction of aqueous solutions of *p*H 2–11 without *p*H adjustment.

According to Hofmann (36) two schemes have proved consistently satisfactory for the recovery of bile acids from biological systems by extraction with liquid ion exchangers. Dilute aqueous solutions are extracted 2–3 times with an equal volume of ethyl acetate or chloroform, containing 5 g of tetraheptylammonium chloride per 100 ml. Serum or biological fluids containing protein or micellar aggregates are first extracted with a conventional chloroform–methanol mixture, with the modification that the chloroform contains the amine, 5 g/100 ml, and then extracted again with the lower phase containing the amine. The binding of other acidic lipids to the tetraheptylammonium chloride is nonspecific and any anionic lipid is extracted to some extent by the solvents containing the amine.

The extracted anionic lipid is removed from the amine by a second ion-exchange procedure. The solvent is evaporated from the pooled extracts and the residue is dissolved in ethanol–water 1:1. The solution is passed over a cation-exchange column that is in the hydrogen form and that has been packed in the same solvent mixture. The amine is quantitatively bound, the extracted lipid passes through the column and may be quantitatively recovered from the eluent, which, however, also contains hydrochloric acid because of the liberation of hydrogen ions from the resin by the amine cations. The cation-exchange resin Biorad AG-1 (California Corporation for Biochemical

Research, Los Angeles, California), 1 ml (wet volume), was sufficient to bind 1.3 meq, or the cation present in 580 mg of the amine chloride. A sixfold excess of the resin was routinely used. The amine may be recovered from the cation exchange resin by elution with 1 *N* hydrochloric acid in 50% ethanol.

According to Helfferich (13) liquid–liquid extraction with liquid ion exchangers is more akin to nonionic liquid extraction than to ion exchange.

3. Paper

Reeve Angel chromatography papers (Reeve Angel and Co., Clifton, New Jersey) of the ion-exchange resin-loaded type, contain approximately 45–50% of finely divided Amberlite ion-exchange resin supported on a cellulose matrix. The resins used are high-molecular-weight copolymers completely insoluble in the pH range 0–14. Their exchange composition may be modified by irrigation with appropriate acid, base, salt, or buffer solution. Of interest to the resolution of the bile acid conjugates is the strong-base Amberlite IRA-400 or XE-19 type, which has an approximate exchange capacity of 1.5–1.6 meq/g. It is available in the chloride form, but can be changed to any desired other counter-ion form. It has been successfully utilized in the formate form for the separation of ribonucleotides (37).

The strong-base anion-exchange paper has been employed (38) in the hydroxide form for the separation of free and taurine-conjugated bile acids with diisobutyl ketone–acetic acid–water (4:5:1, v/v/v) or chloroform–methanol–ammonia (4:1:0.1, v/v/v) as developing solvents. It would be anticipated that these systems would also be capable of resolving glycine- and taurine-conjugated bile acids.

During chromatography the ion-exchange papers are suspended in suitable glass containers so that the free ends just dip into the developing solution. Depending on the solvent, up to 2 hr may be required for the solvent front to ascend 20 cm. The paper is air-dried and the bile acids are located by spraying with phosphomolybdic acid.

4. Thin Layers

Thin-layer chromatography is superior to all other analytical techniques used for fractionating highly polar substances of relatively low molecular weight. Most remarkable is the speed of this method. Ion-exchange thin-layer chromatography should therefore permit more efficient resolution of bile acids in a shorter time than the ion-exchange columns. Because of greater capacity this technique should be well-suited for the preparation of small quantities of high-purity samples, which is not readily accomplished by ion-exchange paper chromatography.

The fractionations of low-molecular-weight organic acids on thin layers of DEAE-cellulose or on cellulose impregnated with polyethylenimine resin were first described by Randerath (17). He found that thin-layer chromatography on such anion exchangers gave indeed more efficient and more rapid separations of mononucleotides than was possible by column chromatography. Moreover, the detection methods were shown to be more sensitive in thin-layer chromatography than in paper chromatography. According to Randerath (17) an essential advantage of this type of thin-layer chromatography was the finding that the chromatographic behavior of acids in a given solvent could be predicted with a greater degree of certainty than was possible in column or paper chromatography.

The polyethylenimine-impregnated cellulose is also suitable for the anion-exchange thin-layer chromatography of bile acids and their conjugates (39). The layers are prepared by thoroughly homogenizing 20 g of cellulose powder MN 300 (Macherey & Nagel, Duren, Germany) without plaster, with 130–135 ml of dialyzed polyethylenimine solution (Polymin P, Badische Anilin and Sodafabrik, Ludwigshafen, Germany) in an electric mixer for 30 sec, and is spread onto glass plates by one of the standard methods of thin-layer application. To avoid edge effects, the plates are removed from the aligning tray immediately after coating and are laid horizontally to dry at room temperature. Plates spread in this way have a capacity of about 1.5 meq/g. Their mechanical strength is very good. The capacity of the polyethylenimine-cellulose layers can be varied within wide limits by using solutions of different concentration (0.5–1.5%) to prepare the coating suspension. It is not necessary to dialyze the commercial imine if a concentration below 0.7–0.8 is used. In order to obtain a straight solvent front, a dividing line is scratched in the layer with a spatula, 4 to 5 mm in from each edge. It is important to clean the glass plates thoroughly before applying the thin layer. It has been found advantageous to wash the plates in a concentrated soda solution followed by rinsing with cold distilled water. Polyethylenimine layers can be kept for only a limited period of time at room temperature. In the dark and at 0 °C they can be kept several months. Polyethylenimine-cellulose powders for thin-layer chromatography are commercially available (MN-Cellulose Pulver 300PEI, Macherey & Nagel, Duren, Germany). The average particle size is about 10 μ and the exchange capacity about 1.0 meq/g. The properties of the commercially prepared anion exchangers are variable and occasionally differ from the layers prepared in the laboratory.

Figure 3 shows the resolution of standard bile acids and their glycine and taurine conjugates on a thin layer of polyethylenimine-cellulose (39). About 5–10 μg of each acid was applied as a spot and the chromatoplate was developed with 1 *N* acetic acid in 25% ethanol. Under these conditions the bile acids were resolved roughly on the basis of their pK_a values. The

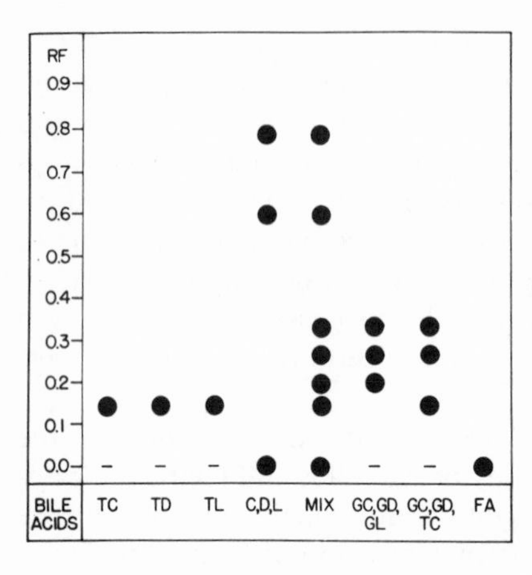

Fig. 3. Schematic chromatogram of the ion-exchange thin-layer chromatography of free and conjugated bile acids. Ion exchanger: polyethylenimine-cellulose (250-μ-thick layer, 20 × 20 mm plate); solvent: 100 ml 1 *N* acetic acid, 50 ml absolute ethanol; development time: 120–150 min. Amount applied: 5–10 μg of each acid; detection: 0.05% 2,7-dichlorofluorescein in 50% methanol, under UV light. Bile acid conjugates identified as in Fig. 2; C, D, and L: free cholic, deoxycholic, and lithocholic acids, respectively; FA: common fatty acids.

free bile acids moved the fastest, with the cholic acid preceding the deoxycholic acid. The free acids were followed by the glycine conjugates, which were resolved on the basis of the number of hydroxy functions in the molecule. The trihydroxy derivatives migrated faster than the dihydroxy, with the monohydroxy acid lagging behind. In this solvent system there was no resolution of the taurine conjugates, all of which, however, were displaced from the origin. When applied as a mixture, the free bile acids were only incompletely resolved, as were the corresponding glycine conjugates. The separation between free, glycine-conjugated, and taurine-conjugated acids was always complete. A more effective resolution of the individual free bile acids and the individual glycine and taurine conjugates by the use of buffer solutions is unlikely because the minor differences in the acidity of these compounds are effectively obscured by differences in the polarity of the steroid nuclei. Free fatty and lithocholic acids remained at the origin (Table V).

The relatively easy resolution of the free and glycine conjugated bile acids on the polyethylenimine–cellulose layers contrasts sharply with the difficulties experienced in the chromatography on ion-exchange columns and on papers loaded with ion-exchange resins. This must be due to the rather ineffective sorption or partition effects operating in the relatively inaccessible hydrophobic interior of the resin particles. The more open cellulose structures and the free polyethylenimine chains apparently permit a more complete equilibration of the bile acid with the active sites (both ionic and hydrophobic) of the ion exchanger and allow a more effective exploitation of the minor differences in the pK_a values between the free and glycine-conjugated bile acids, as well as the differences in the hydrophobic character of the individual molecules. The failure to resolve the taurine conjugates may be due to the higher acidity of these molecules, which overrules the differences in the hydrophobicity. A resolution of the individual taurine conjugates might there-

TABLE V. Relative Mobilities of Bile Acids and Conjugates in Different Solvents (39)[a]

Compounds	System I	System II	System III
bile acids			
cholic	0.78	0.42	0.16
deoxycholic	0.59		
lithocholic	0.0		
conjugates			
glycocholic	0.32	0.42	0.14
glycodeoxycholic	0.25		
glycolithocholic	0.20		
taurocholic	0.14	0.42	0.0
taurodeoxycholic	0.14		
taurolithocholic	0.14		
others			
palmitic acid	0.0		0.14
cholesterol	1.0		

[a] System I: 100 ml 1 *N* acetic acid + 50 ml absolute ethanol; System II: 100 ml 40% ethanol + 1 ml of 1 *N* NH_4OH; System III: 100 ml 1 *N* acetic acid. Ion-exchange thin-layer plates prepared as given in Fig. 3.

fore be possible by a finer adjustment in the *p*H of the developing solvent. In such a case the conjugates would be expected to be eluted in the order of decreasing pK_a values, e.g., with the taurocholate emerging last.

B. Electrochromatography

The differences in the acid dissociation constants of bile acids have been shown (40, 41) to be sufficient for their resolution by electrophoresis. The application of an effective *p*H gradient during the ionophoresis appears to have given the best results. Using a standard paper electrophoresis apparatus and a moving acid boundary, Briggs *et al.* (40) could separate a mixture of bile salts into three groups: unconjugated bile acids, glycine conjugates, and taurine conjugates. For this purpose the cathode compartment was filled with 0.05 *M* acetate buffer, *p*H 4.8, and 0.03 *M* sodium chloride. The anode compartment contained 0.05 *M* monochloroacetate buffer, *p*H 2.7, and 0.03 *M* sodium chloride. The bile salt mixture (up to 500 μg) was applied to a strip of Whatman No. 3 paper (6.5 cm × 46 cm), 15 cm from one end. Prior to the application of the *p*H gradient, the paper was impregnated with acetate buffer and placed in the electrophoresis apparatus with the point of application nearest the cathode chamber. The separation was carried out for 16 hr under a potential gradient of 5 V/cm. Initially at *p*H 4.8, the conjugated bile acids migrated together toward the anode, the weaker unconjugated acids remaining at the origin. As the separation progresses, the acid front advancing from the anode compartment represses the ionization of the glycine conjugates and renders them immobile. The taurine conjugates continue to move through the acid front and are found near the anode at the end of the run. The glycine conjugates are located near the middle of the paper, while the free bile acids are found near the origin. The locations of the bile acids on the paper strip were found by spraying it with 3% (w/v) vanillin and 15% (v/v) phosphoric acid in *sec*-butanol.

Electrophoresis on paper using a single buffer was applied to bile acid separation by Biserte *et al.* (41). The buffer system was made up of 30 ml of pyridine and 100 ml of glacial acetic acid in 5 liters of water. The *p*H was 3.9. In this system, cholic, glycocholic, taurocholic, and taurochenodeoxycholic acids were attracted toward the anode at rates increasing in the order named. Curiously, free deoxycholic acid remained at the origin. The locations of the bile acids on the paper were found by spraying with phosphomolybdic acid. Results with various animal biles were reported.

The electrophoretic mobilities of ^{14}C-labeled cholic, deoxycholic, and chenodeoxycholic acid and their corresponding taurine and glycine conjugates were determined by Norman (42). The paper electrophoresis was performed in barbiturate buffer of ionic strength 0.1, *p*H 8.6, in an electric field of 7.5 V/cm for 3 hr. When 1 μg of each acid, as the sodium salt dissolved in 25 μl of water, was applied to the paper strips, the isotope determination after electrophoresis showed broad peaks all with a mean mobility similar to that of albumin or slightly lower. The electrophoretic mobilities of all of the bile acids were influenced by the concentration in the solution applied and presented difficulties in identifying bile acids in natural extracts. The migration of bile salt–lecithin micelles on paper electrophoresis has been reported by Shimura (43). The micelles were prepared by addition of lecithin to mixed bile salts, which may have also contained cholesterol.

Although both the gradient and the constant-*p*H approaches to the electrophoretic separation of bile acids could be further improved, at the present time these techniques do not seem to offer any advantages over the simpler thin-layer chromatographic methods which handle comparable amounts of bile acid.

C. Detection of Bile Acids

The difficulty of detecting small amounts of sample components in the presence of a large concentration of eluting ion is one of the major disadvantages of ion-exchange column chromatography. Continuous recording is not common, although in specific applications spectrophotometric or radioactivity measurements could have been made on such a basis. The most common practice is to collect numerous small, equal-volume fractions and analyze each fraction for the species sought. Sjövall (33) has discussed a large number of color reactions suitable for quantitative determination of bile acids in biological materials and column effluents. Colorimetric methods, alone, however, have limited applicability and require appropriate standards or mixtures of standards.

Gordon *et al.* (2) used a modification of the Pettenkofer reaction to assess quantitatively the bile acid content of each fraction of the ion-exchange column effluent. This method was adopted by Whiteside *et al.* (24) for the assay of the *in vitro* binding of cholate by cholestyramine. In this reaction conjugated and free cholic acid give approximately the same molar absorption, but other bile acids do not yield sufficient chromogen to be measured reliably. Standard methods of monitoring radioactivity are by far the most sensitive and specific means of assessing the chromatographic behavior of labeled bile acids on ion-exchange columns. These methods have been used for assessing the uptake and recovery of both free and conjugated standard bile acids by anion exchangers (20, 31).

Because of its speed and simplicity thin-layer chromatography on silica gel is often used to follow the course of elution and separation of solutes from ion-exchange columns. It provides an excellent substitute for the paper-chromatographic monitoring of bile acids described by Gordon *et al.* (2).

The detection of components on thin layers of anion exchanger is relatively simple. In case of the polyethylenimine–cellulose layers, the still-wet plate can be examined in the daylight for bright spots of hydrophobic material. By this method 5–10 $\mu g/cm^2$ of bile acid can be detected. Spraying with 0.05% 2,7-dichlorofluorescein in 50% methanol allows the detection of 1–2 $\mu g/cm^2$ of bile acid, provided care is taken to remove the water or solvents prior to the examination under the ultraviolet light (39).

The bile acids on the paper strips from electrophoresis can be identified by spraying with phosphomolybdic acid (41) or with 3% (w/v) vanillin and 15% (v/v) phosphoric acid in secondary butanol (40). The background provided by the resin-loaded paper is frequently colored and unsuitable for the detection of bile acids by spraying with nondestructive agents. Heating with strong acids, however, may attack the carbohydrate matrix of the paper. In any event the color tests are much less sensitive on resin-loaded paper than those obtained on a thin-layer plate of silica gel, for example. The best method under these circumstances appears to be cutting the paper in strips at regular intervals, eluting the sections, and determining the acids in the eluates by conventional methods. Following electrophoresis the ^{14}C-labeled bile acids on paper strips have been detected by counting, either stepwise or continuously, in a paper scanner using a Frieseke–Hoepfner windowless counter with a 1-cm slit (42).

The conjugated bile acids can be identified and estimated quantitatively by determining their amino acid moieties. With the discovery of the ornithocholanic acids in the bile of man and experimental animals, this method of determination of bile acid conjugates may become more important. All three amino acid conjugates give purple color with ninhydrin and can be measured quantitatively after chromatographic resolution and hydrolysis (2).

IV. NATURE OF RESOLUTION

The ultimate goal of theories of chromatography is to provide means of predicting from independently measurable fundamental properties, such as the pK_a values of acids, the optimum conditions for specific separations. In the case of ion-exchange chromatography of bile acids this goal is still remote. The basic difficulties of predicting equilibria and rates with ion exchangers in general are compounded by the detergent nature of the molecules and the use of aqueous organic solvents in the elution systems.

The general elution sequence as observed on both ion-exchange columns and thin layers is that based on the pK_a values. The free bile acids (pK_a 5.29–6.29) are eluted ahead of the glycine conjugates (pK_a 3.75–4.77) which are followed by the taurine conjugates (pK_a 1.85–1.93). Complete resolutions between the free and glycine- and taurine-conjugated bile acids are obtained only on thin layers of ion exchangers. On columns the free and glycine-conjugated bile acids tend to overlap. The pK_a values of ornithocholanic acids and of the sulfates of glyco- and taurocholanic acids have not been determined. Because of the free amino and carboxyl group in the former, they would be expected to possess higher pK_a values than the corresponding free or glycine-conjugated bile acids. They should therefore be eluted ahead of the latter bile acids from an anion-exchange column, as already noted by Gordon *et al.* (2). On account of the extra sulfate group on the steroid nucleus of the sulfates of glycine and taurine conjugates (4), these esters would have lower pK_a values than their parent conjugates and would therefore be retained longer. From a partition column the ornithocholanic acids are eluted between the glyco- and taurocholanic acids (3).

When the separations of the acids within the free and glycine conjugated groups are considered, it is seen that on both columns and thin layers, the cholic acid derivatives are eluted ahead of the derivatives of the deoxycholic and chenodeoxycholic acids, and presumably other dihydroxy bile acids. The derivatives of the lithocholic acid are eluted last. Complete separations of the mono-, di-, and trihydroxy bile acids are not realized even on the thin-layer plates of ion exchangers, and there is no discernible resolution of the various taurine conjugates. This order of elution of the bile acids is opposite to that expected on the basis of their pK_a values (Table II). Free cholic acid (pK_a 5.29) would have been expected to be retained longer than the dihydroxy acids (pK_a 6.18–6.29) which should have been retained longer than the lithocholic acid (pK_a 6.73 in aqueous alcohol). Similarly, glycocholic acid (pK_a 3.75) ought to have been retained longer than the glycodeoxycholic acid (estimated pK_a 4.77) and the latter longer than glycolithocholic acid (estimated pK_a 5.0?). Curiously, Josephson (7) reports lower values for glycodeoxycholic (pK_a 3.81) than for glycocholic acid (pK_a 4.41) in contrast to the values of Small (5) cited above. Josephson also reports lower value than Small for taurodeoxycholate (pK_a 1.74) than for taurocholate (pK_a 1.83). It remains to be established if these discrepancies can be accounted for on the basis of differences in the surface charges of these molecules or their aggregates in the chromatographic systems.

In the meantime complicating factors must be invoked to account for this unexplained deviation. Thus, Gordon *et al.* (2) suggested a sorption effect to explain this abnormality on the basis of diffusion in either the ion-exchanger particle or in adherent liquid film. The more hydrophobic the steroid molecule (fewer hydroxyl groups), the more strongly it would be retained by the hydrophobic resin surface. Such an explanation was considered likely by Blanchard and Nairn (19) who noted that the Dowex anion-exchange resins bound more cholate than glycocholate. Hydrophilicity and molecular size, however, appeared to be the governing factors in the relative uptake of these anions in their study. Whether or not the cholate and glycocholate anions were released from the resin in the order of their extent of binding was not established by Blanchard and Nairn (19). In this connection, it is interesting to note that during gel filtration of bile acids on Sephadex G-25, Norman (44) observed that both molecular filtration and absorption chromatography occurred during the procedure. The absorption was more pronounced when gel filtration was performed with saline than with water. Conjugates of monohydroxycholanic acid were absorbed to the greatest extent, followed by dihydroxycholanic acid, and then trihydroxycholanic acid. The rates of elution of taurine and glycine conjugates were about the same. The sodium salts of unconjugated bile acids, such as cholate, deoxycholate, and chenodeoxycholate were eluted at the same rate as the corresponding glycine and taurine conjugates.

It is obvious that even a simple resolution of the bile acids on the basis of their overall charge is likely to become extremely complicated as new components are being identified. Furthermore, a positive test for a specific bile acid may not be sufficient for rigid identification of the conjugate in any one group of conjugates, because the sulfates of the glycine conjugates, for example, may readily associate with the taurine conjugates. Neither ion-exchange columns nor thin layers appear to be capable of resolving the 5α and 5β series of cholanic acids, or mixtures of axially and equatorially substituted isomers, such as is possible by gas chromatography and chromatography on thin layers of silica gel. There is reason to believe that ion exchangers of greater selectivity might possibly be prepared in the future.

Hofmann and Small (6) have suggested that an examination of the oleophilic anion-exchange polymers recently prepared by Gregor and his colleagues (45) would be of interest as these resins might at least show greater affinity for bile salts in aqueous solutions. Such polymers, however, might also provide a more intimate interaction with the steroid nucleus in which steric factors would play a significant role.

V. APPLICATIONS

Most methods of detailed bile acid analysis require a preliminary isolation and purification of the extract to be analyzed. In several instances these steps have been effectively accomplished by ion-exchange chromatography. However, ion-exchange chromatography can also resolve effectively the glycine and taurine conjugates. Furthermore, both isolation and purification as well as conjugate resolution can readily be accomplished on the same ion-exchange column. Compared to thin-layer chromatography, the anion-exchange columns can handle much larger amounts of material and do not require a preliminary lipid extraction. Also anion-exchange chromatography is much more discriminative than liquid–liquid extraction for the resolution of bile acid conjugates and yields purer and better-defined fractions.

A special application of anion exchangers in bile acid work has resulted from the observations that many resins bind bile acids under physiological conditions (neutral *p*H and aqueous solutions).

A. Isolation and Purification

Ion-exchange chromatography as a means of purifying bile acids in biological extracts was introduced by Lambiotte (46). For this purpose he used the anion-exchange resin Amberlite IRA-400 in the hydroxyl form. The original method, which called for the initial absorption of the bile acids from a water-saturated ether solution, has since been modified (33). The crude bile acid mixture (0.5–5 mg) is now dissolved in 2–3 ml of 50% aqueous ethanol and then applied to a 0.8 × 10 cm column of IRA-400 in the hydroxyl form. The free and glycine-conjugated bile acids are then recovered with 30 ml of 1 *M* formic acid in 50% ethanol and the taurine conjugates are recovered by elution with 75 ml of a solution of sodium chloride–water–ethanol (23 g:100 ml:100 ml). Quantitative recoveries were reported for cholic and deoxycholic acids and their conjugates.

Ion-exchange chromatography as a means of purifying fecal bile acids was described by Kuron and Tennent (30). An alcoholic solution of fecal bile acids was first passed through a column of Dowex 50 × 1 in the hydrogen form (wet weight of the resin about 50 times the weight of solids in this extract). The purification was completed by passing the eluate through a column of Dowex 1 × 2 resin in the hydroxyl form which retained the acids. After washing the column with aqueous solutions of electrolytes, elution was achieved with ammonium carbonate in aqueous ethanol. Nearly quantitative recoveries were reported. Apparently satisfactory results with the method of Kuron and Tennent were also obtained by Roels and Hashim (47), who employed anion-exchange chromatography for the purification of bile acids before attempting quantitative chromatography on silicic acid-impregnated glass paper.

Purification of fecal bile acids by ion-exchange chromatography according to the method of Kuron and Tennent has not given satisfactory results in other laboratories (33). Some of the bile acid metabolites found in the feces were observed to be strongly absorbed to the resin and large volumes of solvent were required for quantitative elution even when the more open anion-exchanger DEAE-Sephadex was used.

The suitability of anion-exchange resins for the isolation of fecal bile acids has been investigated also in the author's laboratory (48) using the method of Hirshfeld *et al.* (49). Quantitative absorption of both free fatty acids and bile acids was obtained from water saturated petroleum ether–diethyl ether solutions. After washing the resin free of neutral fat with petroleum ether, the acids were converted into the methyl esters by a treatment with hydrochloric acid in anhydrous methanol. The methyl esters of mixed bile acids and fatty acids were recovered by extracting the resin with methanol and petroleum ether. Pure bile acids were isolated by absorption chromatography of the methyl esters on silicic acid (50). Although radioactive bile acids were not used to determine the completeness of recovery of the fecal bile acids, no distortion in the proportions of the common bile acids was observed when standard mixtures were added to the resin and removed as described. In order to absorb and recover the conjugated bile acids by this method, more polar solvents such as 50% ethanol had to be used. Other investigators have used alumina columns for the purification of bile acids isolated by ion exchange (20).

Apparently quantitative elutions of standard bile acids and bile acid conjugates from ion-exchange resins have also been obtained by Sandberg *et al.* (20), who determined the recovery of radioactive standards from anion-exchange columns used to isolate natural bile acids from dilute plasma. For the isolation of the bile acids from blood, the serum is diluted with an equal volume of water and passed through a column of the anion-exchanger Amberlyst XN-1006. After successive washings with water and aqueous alcohol, the acids are recovered by elution with 0.2 *M* $(NH_4)_2CO_3$ in 80% ethanol. The effectiveness of anion-exchange resins for the isolation of blood bile acids has been confirmed in subsequent work (51), and has remained unsurpassed despite great advances made in other chromatographic techniques.

Effective isolations of both free and conjugated bile acids by means of anion-exchange resins have also been obtained from extracts of subcellular fractions of rat liver (52) as well as from crude rat-liver homogenates (53). In the latter case the bile acids were first extracted from freeze-dried tissue homogenates with 95% ethanol containing 0.1% of aqueous ammonia. The pooled extracts were evaporated to dryness and the residue dissolved in 3–7 ml of 0.1 *N* NaOH (*p*H 11) and transferred to a column of Amberlyst A-26 (formerly XN-1006) anion-exchange resin. The transfer was completed with the addition of a total of 35 ml of water. All subsequent steps in the ion-exchange purification of the biliary bile acids were performed as described by Sandberg *et al.* (20). Although the overall recoveries of the bile acids from the entire procedure were only of the order of 80%, there was no reason to suspect that any selective losses of the acids occurred on the resin, as the discrepancies could be attributed to incomplete hydrolysis of the conjugates by the bacterial enzyme used or to partial degradation during alkaline hydrolysis of the conjugates.

The use of ion-exchange chromatography for the purification of commercial samples of glycocholate and taurocholate on a preparative scale was described by Gordon *et al.* (2). For this purpose the conjugates (0.1–1.0 g) were dissolved in 50% ethanol (20 ml) and applied to a column (2 × 30 cm) of Dowex-50 cation-exchange resin and the column washed with 50% ethanol until all the Pettenkofer-positive material had been recovered. Under these conditions, the inorganic cations, most of the pigment, and all of the amino acids except taurine were retained on the resin. A more sensitive and specific test of the elution of the bile acid conjugates would have been thin-layer chromatography. The purification was completed by reducing the eluate of the cation-exchange column to about 20 ml and applying it to an anion-exchange resin column (2 × 30 cm) in 25% ethanol. After washing this column with 300 to 400 ml of 25% ethanol, the glycine conjugates were eluted with 0.05 *N* HCl in 25% ethanol. The taurine conjugates were purified by washing the anion-exchange resin successively with 25% ethanol and 0.05 *N* HCl in 25% ethanol, and then recovered by elution with 0.5 *N* HCl in 25% ethanol. The elutions were continued in both cases until no more glyco- or taurocholate, respectively, was displaced from the column, as indicated by spot tests on paper chromatography. The pure bile acid conjugates were recovered by solvent extraction after neutralizing the eluates and reducing them to dryness. It was observed that the use of higher alcohol concentrations allowed the recovery of the conjugates by elution with solutions of acetic or formic acid, or dilute ammonia.

B. Separation

Although both isolation and purification of the bile acids on ion-exchange columns must of necessity involve certain elements of separation, in order to be fully effective, in only a few cases have the experimental conditions been deliberately designed for optimum separation of the bile acids or conjugates from each other. Because of the rather small differences in the pK_a values between the free and the glycine conjugated acids, ion-exchange chromatography holds little potential for the more refined separations, such as possible on thin-layer absorption-chromatography. Certain group separations, however, can be readily accomplished by ion-exchange methods, and the importance of this technique may grow as interest increases in the resolution of the various conjugate groups, as well as their sulfated derivatives.

1. Conjugates

A resolution of glycine and taurine conjugates of bile acids by ion-exchange chromatography (see above) was first achieved by Lambiotte (54). The general applicability of this method to the isolation and resolution of the bile acid conjugates in animal biles was independently reported by Gordon *et al.* (2). The methods developed for this purpose were identical to those elaborated for the purification of commercial glyco- and taurocholate described above. Considerable reduction of column size and effluent volume for analytical work was possible, which allowed a much more efficient operation all around. Furthermore, the Pettenkofer reaction was replaced by thin-layer spot-testing as a means of following the progress of the conjugate elution. Using this general procedure, effective resolutions of glycine and taurine conjugates were obtained of the gallbladder bile of dog, ox, and man (2) and rabbit, sheep, hog, and man (55). In the latter study the column size was reduced to 30 × 1 cm and the taurine conjugates were eluted with 0.05 *N* HCl in 50% ethanol. It was demonstrated that the bile of ox and man contained an unidentified Pettenkofer-positive bile acid conjugate. This conjugate was eluted as a separate peak with the 25% alcohol solution, and behaved on paper chromatography (Sjövall's system G_D) as glycocholic acid. This conjugate was eluted as an earlier and a separate peak from glycocholate, also with the 0.05 *N* HCl in 25% ethanol used to recover the glycine conjugates. In both ox and human bile it made up about 3.5% of the total Pettenkofer-positive material. The bile acid moiety of this conjugate was identified as cholic acid in several paper chromatographic systems (33). The amino acid part, however, was not glycine, since on paper chromatography (56) it showed an R_f value lower than that of standard glycine. In this system both ornithine and lysine showed R_f values comparable to those of the unknown ninhydrin-positive amino acid. The higher pK_a

exhibited on the ion-exchange chromatography also suggested that the amino acid moiety of this conjugate was a more basic amino acid than glycine. In this connection it must be pointed out that ornithocholanic acids have since been detected in human bile by other techniques (3), as well as in guinea pigs subjected to hepatic injury of unspecified nature (57). The ornithine conjugates have a free NH_2 and COOH group. A positive ninhydrin reaction is given by ornithocholanic acids from the bile of guinea pig, rat, ox, and man, and from liver incubated *in vitro*. On partition chromatography, the ornithocholanic acids appear in the eluate between the taurine and the glycine conjugates (3).

Although the conditions for the optimum separation of the dihydroxy and the trihydroxy derivatives in each conjugate group were not fully explored, in these studies (2, 55) it was noted that in both conjugate groups, the elution sequence was the same. The cholic acid derivative was eluted first. This was followed by the conjugates of the cheno- and deoxycholic acids. Any lithocholic acid conjugates were eluted last. Whether this elution order resulted from a greater absorption of the less polar bile acid derivatives to the hydrophobic resin matrix, or from a difference in the hydration of the molecular species in the inter- and outer-phases of the chromatographic system remained to be established. Figure 4 shows the ion-exchange chromatographic resolution of the bile acid conjugates of ox bile.

Although successful separations of standard bile acid conjugates have also been obtained using thin layers of anion exchangers, no practical applications have been reported. Preliminary investigations have shown (39) that this method may have advantages over conventional thin-layer systems in the separation of the glycine, taurine, and ornithine conjugates as well as their sulfated derivatives.

Both standard acids and conjugates from natural sources have been resolved by Briggs *et al.* (40) and Biserte *et al.* (41) using electrophoresis on paper, and the latter authors have reported applications to the analysis of the acids in animal biles. Norman (42) used paper electrophoresis to study the bile acids in the gallbladder bile and hepatic bile of man.

2. *Bile Acids*

The common ion-exchange chromatographic systems (columns, paper) are not sufficiently efficient to provide practical resolutions of mixtures of free bile acids or even mixtures of free bile acids and their glycine conjugates. Practical resolutions of mono-, di-, and trihydroxy bile acids, however, can be obtained (39) on thin layers of anion exchangers of the type described by Randerath (17). Since these separations do not represent an improvement over the existing thin-layer techniques they are of interest only as far as the purification or resolution of the bile acid conjugates in these systems is con-

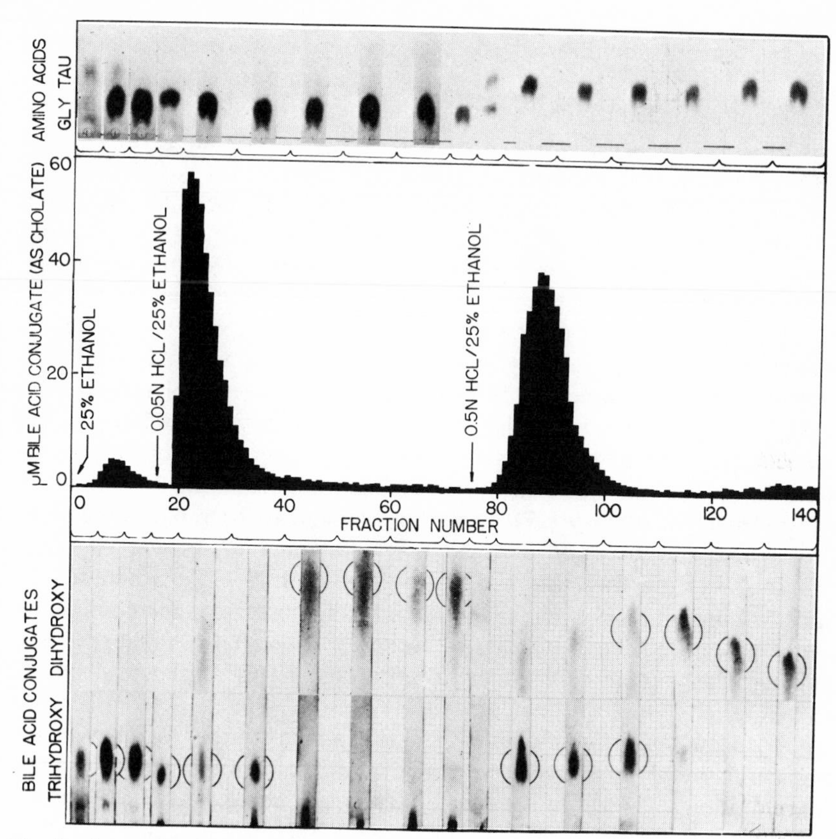

Fig. 4. Ion-exchange chromatography of the bile acids of ox bile. Column: 2 × 30 cm Dowex-1 (50–100 mesh) in acetate form.

A. *Upper Third*—paper chromatography (56) of amino acids, released from the conjugates on acid hydrolysis.

B. *Middle Third*—trace of the elution of Pettenkofer-positive material: OC, ornithocholanic acid; GC, glycocholic acid; TC, taurocholic acid. The glycine and taurine conjugates of the dihydroxy acids also present in ox bile do not yield sufficient chromogen to be measured effectively by this color reaction.

C. *Lower Third*—paper chromatography of the conjugates appearing in the eluate (Sjövall's system G_D). Redrawn with the permission of the authors and publishers of *Can. J. Biochem. Physiol.* **41,** 77 (1963).

cerned.

C. Ion-Exchange Resins as Therapeutic Bile Acid Sequestrants

The feeding of the bile acid-binding polymeric amines, cholacrylamine resin and cholestyramine resin, causes lowering of blood cholesterol in experimental animals (22), and in human subjects (58). These substances are neither absorbed nor digested and they produce no toxic effects on short-term administration. These resins bind and decrease the resorption of bile acids from the intestine, thereby stimulating the rate of oxidation of cholesterol in order to replenish the bile acid supply (59). Cholestyramine is now used for therapeutic purposes (60).

The cholacrylamine resin, or MK-325 (Merck, Sharpe & Dohme, Inc.), is a water-soluble polymer having quaternary amino groups attached to a polyacrylic skeleton by ester linkages (22). Its equivalent weight is about 325, and its molecular weight is about 2 million. The stock material was a clear, viscous, 12.5% aqueous solution. The cholestyramine resin, or MK-135, is an insoluble, solid quaternary ammonium anion-exchange resin, in which the basic groups are attached to a styrene-divinylbenzene copolymer skeleton by carbon–carbon bonds (similar to Dowex-1). Its equivalent weight is about 230. This material as used contains about 75% water. Both of these resins are used as their chloride salts, which are neutral in reaction.

In an *in vitro* assay (27), cholate was found to be rapidly bound by cholestyramine (Questran, a brand of cholestyramine supplied by Mead Johnson and Company). Most of the binding occurred within a minute, and after 20 min, binding of cholate was not further increased. Increasing temperature from 25 °C to 37 °C had no effect on binding after 10 min or more of shaking. The amount of cholate bound at *p*H 6.0 in 0.02 *M* phosphate buffer at 25 °C after 30 min was increased as the resin was increased from 10 to 80 mg. With 0.02 *M* phosphate buffer, cholestyramine bound progressively less cholate as the *p*H was increased from 5.0 to 8.0. At the higher phosphate concentrations, changes in *p*H did not influence binding. At *p*H 6.0, cholate binding was significantly less with 0.15 *M* or 0.3 *M* phosphate than with 0.02 *M* phosphate; at *p*H 8.0, phosphate concentration did not affect binding. Other anion exchange resins, such as Dowex 2X1 and preparations of Amberlite resin, had much lower activity than cholestyramine in the *in vitro* binding according to Whiteside *et al.* (27).

Cholestyramine resin in the form of glycocholate or taurocholate salts did not inhibit cholesterol rise in plasma in cholesterol-fed cockerels (22), presumably because these forms of resin were already saturated with bile acids and could not take up more. The stearate salts of cholestyramine resins, however, were fully active. The therapeutic effectiveness of the resin depends on the selectivity between the bile salt anions and the chloride anion as well as the capacity for the organic ion. Studies have therefore been made (18, 19) to determine the separation factors for these ions on various resins in order to investigate some of the properties of the resin that are responsible for the high affinity of the large organic ions for the ion-exchange resins. Since the number of equivalents of cholate anion bound were equal to the number of equivalents of chloride ion released by the resin, within the limits of experimental error, it was concluded (19) that the binding process was due to ion exchange only. As yet we have no precise data on the affinity of cholestyramine for different bile salt conjugates.

Although the resin-bound bile acids can be recovered by extraction with alcoholic solutions of ammonium carbonate following the excretion of the resin from the body, no detailed studies appear to have been made of their composition. Thus it is not known whether or not the resin-bound acids are subject to microbial modification in the large intestine. Carey and Williams (61) fed 10 g of cholestyramine resin per day to a healthy human volunteer and found that his fecal bile acid excretion was increased eightfold by this treatment. In patients with biliary cirrhosis, resin dosing caused a fall in the serum levels of bile acids, total lipid, and cholesterol, and relieved pruritis. The subjects exhibit steatorrhea of varying degrees because the absorption of triglycerides is also impaired by this regimen. The possible usefulness of substances of this general type in the treatment of hypercholesterolemia requires further investigation.

With advance in our knowledge of the physical chemistry of bile acids and lipids, it may be possible to synthesize resins with still greater selectivity *in vivo* than cholestyramine, should this method prove to be an effective long-term therapeutic measure. The administration of cholestyramine to guinea pigs induces cholelithiasis (62) and isolated incidents of adverse affects have been reported in the clinical literature following long-term treatment with the resin (63).

In addition to clinical use, cholestyramine has been employed in metabolic research to increase the turnover rate of the bile acid pool in the rat (64).

VI. SUMMARY AND CONCLUSIONS

Since ion exchange and electrochromatography are basically limited to ionogenic compounds, comparatively little use has been made of these separation methods in the steroid field. Although the bile salts possess acid functions, the differences in their pK_a values are not sufficient for complete resolution of the components differing also in the number of hydroxyl groups on the steroid nucleus. Moreover, the general theories of ion-exchange chromatography are not directly applicable to separations of bile acids because of the need of including organic solvents in the chromatographic systems, and the detergent properties of the bile acid molecules. The separation of bile acids into discrete conjugate groups by ion-exchange chromatography, nevertheless, can be accomplished with relative ease. With the identification of new conjugates and sulfates of conjugated bile acids, the importance of this method is likely to grow. The ion-exchange columns, however, will probably be replaced by thin layers of anion exchangers in these separations. This will also eliminate the need for recovering small amounts of bile acid from aqueous solutions of salt, which is one of the major drawbacks of ion-exchange chromatography on columns.

By far the most important use of ion exchangers in bile acid analyses at present is in the isolation of the free and conjugated bile acids from natural sources. The capability of absorbing the acids quantitatively from a dilute aqueous solution of a biological fluid or tissue extract is a distinct advantage not rivalled by solvent extractions or thin-layer chromatography on silica gel. Sufficient work appears to have been done to conclude that at least the common bile acids and their conjugates can be quantitatively recovered from ion exchangers, although additional purification may be necessary before these acids can be subjected to gas chromatography.

In view of the apparently successful therapeutic applications of anion-exchange resins, studies on bile acid binding and release by ion exchangers will probably be expanded, as will the efforts at synthesis of more selective resins. This should lead to a better understanding of the equilibria and rates of exchange operating in these chromatographic systems and to the development of better resins and more effective methods of bile acid separation by ion-exchange chromatography.

VII. ACKNOWLEDGMENTS

The author wishes to thank Drs. B.A. Gordon and A.F. Hofmann for help in locating certain pertinent references, and Dr. D.M. Small for making available selected experimental data prior to publication.

Special indebtedness is acknowledged to Dr. J.G. Nairn, who read the entire manuscript and commented upon it.

The studies of the author and his collaborators referred to in the manuscript were supported by the Medical Research Council of Canada, the Ontario Heart Foundation, Toronto, Canada, and the Eli Lilly Company, Indianapolis, Indiana.

REFERENCES

1. D. Kritchevsky and P. P. Nair, *in* "The Bile Acids" (D. Kritchevsky and P. P. Nair, eds.), Vol. 1, Chapter 1, Plenum Press, New York (1971).
2. B. A. Gordon, A. Kuksis, and J. M. R. Beveridge, *Can. J. Biochem. Physiol.* **41**, 77 (1963).
3. L. Peric-Golia and R. S. Jones, *Science* **142**, 245 (1963).
4. R. H. Palmer, *Proc. Natl. Acad. Sci. U.S.A.* **58**, 1047 (1967).
5. D. M. Small, *in* "The Bile Acids" (P. P. Nair and D. Kritchevsky, eds.), Vol. 1,

Chapter 8, Plenum Press, New York (1971).
6. A. F. Hofmann and D. M. Small, *Ann. Revs. Med.* **18**, 333 (1967).
7. B. A. Josephson, *Biochem. Z.* **263**, 428 (1933).
8. W. D. Kumler and I. F. Halverstadt, *J. Biol. Chem.* **137**, 765 (1941)
9. P. Ekwall, T. Rosendahl, and N. Lofman, *Acta Chem. Scand.* **11**, 590 (1957).
10. O. M. Henriques, *Acta Pathol. Microbiol. Scand.* Suppl. 111, 149 (1930).
11. F. Helfferich, "Ion Exchange," McGraw-Hill, New York (1962).
12. C. J. O. R. Morris and P. Morris, "Separation Methods in Biochemistry," Chapters 8, 9, and 10, Pitman, London (1963).
13. F. Helfferich, *in* "Advances in Chromatography" (J. Calvin Giddings and R. A. Keller, eds.), Vol. 1, Chapter 1, Marcel Dekker, New York (1965).
14. "Dowex-Ion Exchange," Second Ed., Dow Chemical Company, Midland, Michigan (1959)
15. D. F. Zinkel and J. W. Rowe, *Anal Chem.* **36**, 1160 (1964).
16. B. A. Gordon, "The Analysis of Bile Acid Conjugates by Ion Exchange Chromatography," M.Sc. thesis, Queen's University, Kingston, Canada (1961).
17. K. Randerath, "Thin Layer Chromatography" (Translated by D. D. Libman), Second Ed., Academic Press, New York (1966).
18. J. Blanchard, "A Study of Ion Exchange Resins and Certain Bile Acids, M.Sc thesis, University of Toronto, Toronto, Canada (1966).
19. J. Blanchard and J. G. Nairn, *J. Phys. Chem.* **72**, 1204 (1968).
20. D. H. Sandberg, J. Sjövall, K. Sjövall, and D. A.Turner, *J. Lipid Res.* **6**, 182 (1965).
21. B. A. Gordon, A. Kuksis, and J. M. R. Beveridge, *Can. J. Biochem.* **42**, 897 (1964).
22. D. M. Tennent, H. Siegel, M. E. Zanetti, G. W. Kuron, W. H. Ott, and F. J. Wolf, *J. Lipid Res.* **1**, 469 (1960).
23. D. M. Tennent, S. A. Hashim, and T. B. Van Itallie, *Fed. Proc.* **21** (Suppl. 11), 77 (1962).
24. C. H. Whiteside, H. B. Fluckiger, and H. P. Sarett, *Proc. Soc. Exptl. Biol. Med.* **121**, 153 (1966).
25. W. E. Cohn, *J. Am. Chem. Soc.* **72**, 1471 (1950).
26. C. W. Davies, *Biochem. J.* **45**, 38 (1949).
27. G. W. Bodamer and R. Kunin, *Indust. Engng. Chem. (Industr.)* **45**, 2577 (1953).
28. C. W. Davies and B. D. R. Owen, *J. Chem. Soc.* **1956**, 1676.
29. C. W. Davies and B. D. R. Owen, *J. Chem. Soc.* **1956**, 1681.
30. G. W. Kuron and D. M. Tennent, *Fed. Proc.* **20**, 268 (1961).
31. A Kuksis, unpublished results.
32. B. A. Gordon, A. Kuksis, and J. M. R. Beveridge, *Fed. Proc.* **20**, 248 (1961).
33. J. Sjövall, *in* "Methods of Biochemical Analysis" (D. Glick, ed.), Vol. 12, page 97, Wiley, New York (1964).
34. G. Rouser, G. Kritchevsky, D. Heller, and E. Lieber, *J. Am. Oil Chemists' Soc.* **40**, 425 (1963).
35. E. Hogfeldt, *in* "Ion Exchange" (J. A. Marinsky, ed.), Vol. 1, Chapter 4, Marcel Dekker, New York (1966).
36. A. F. Hofmann, *J. Lipid Res.* **8**, 55 (1967).
37. R. M. Smillie, *Arch. Biochem. Biophys.* **85**, 557 (1959).
38. B. A. Gordon, unpublished results.
39. M. T. Subbiah and A. Kuksis, manuscript in preparation.
40. T. Briggs, M. W. Whitehouse, and E. Staple, *Nature* **182**, 394 (1958).
41. G. Biserte, J. Vanlerenberghe, and F. Guerrin, *Compt. Rend. Soc. Biol.*, **153**, 618 (1959).
42. A. Norman, *Proc. Soc. Exptl. Biol. Med.* **115**, 936 (1964).
43. H. Shimura, *Hirosaki Igaku* **8**, 483 (1957); in *C. A.* **52**, 3269g (1958).
44. A. Norman, *Proc. Soc. Exptl. Biol. Med.* **116**, 902 (1964).
45. H. P. Gregor, G. K. Hoeschele, J. Potenza, A. G. Tsuk, R. Feinland, M. Shoda, and P. Teyssie, *J. Am. Chem. Soc.* **87**, 5525 (1965).

46. M. Lambiotte, *Bull. Soc. Chim. Biol.* **37**, 1023 (1955).
47. O. A. Roels and S. A. Hashim, *Fed. Proc.* **21** (Suppl. 11), 71 (1962).
48. S. S. Ali and A. Kuksis, unpublished results.
49. I. Hornstein, J. A. Alford, L. E. Elliott, and P. F. Crowe, *Anal. Chem.* **32**, 540 (1960).
50. S. S. Ali, A. Kuksis, and J. M. R. Beveridge, *Can. J. Biochem.* **44**, 957 (1966).
51. K. Sjövall and J. Sjövall, *Clin. Chim. Acta* **13**, 207 (1966).
52. T. Okishio, P. P. Nair, and M. Gordon, *Biochem. J.* **102**, 654 (1967).
53. P. P. Nair, M. Gordon, S. A. Tepper, and D. Kritchevsky, *J. Biol. Chem.* **243**, 4034, (1968).
54. M. Lambiotte, *Rev. Intern. Hepatol.* **7**, 521 (1957).
55. B. A. Gordon, "Studies on Bile Acid Metabolism," Ph.D. thesis, Queen's University, Kingston, Canada (1963).
56. R. R. Redfield, *Biochim. Biophys. Acta* **10**, 344 (1953).
57. L. Peric-Golia and R. S. Jones, *Proc. Soc. Exptl. Biol. Med.* **110**, 327 (1962).
58. S. S. Bergen, Jr., T. B. Van Itallie, D. M. Tennent, and W. H. Sebrell, *Proc. Soc. Exptl. Biol. Med.* **102**, 676 (1959).
59. T. B. Van Itallie and S. A. Hashim, *Med. Clin. N. Amer.* **47**, 629 (1963).
60. S. A. Hashim and T. B. Van Itallie, *J. Am. Med. Assoc.* **192**, 289 (1965).
61. J. B. Carey, Jr., and G. Williams, *J. Am. Med. Assoc.* **176**, 432 (1961).
62. L. J. Schoenfield and J. Sjövall, *Am. J. Physiol.* **211**, 1069 (1966).
63. Anon., *J. Am. Med. Assoc.* **197**, 261 (1966).
64. W. T. Beher, M. E. Beher, and B. Rao, *Proc. Soc. Exptl. Biol. Med.* **122**, 881 (1966).

Chapter 7

MASS SPECTRA OF BILE ACIDS*

J. Sjövall, P. Eneroth, and R. Ryhage

Department of Chemistry and Laboratory for Mass Spectrometry
Karolinska Institutet
Stockholm, Sweden

I. INTRODUCTION

In the last ten years mass spectrometry has grown to become one of the most valuable techniques for analysis and structure determination of bile acids. A very important reason for this development is the construction of combined gas chromatography–mass spectrometry instruments capable of dealing with complex biological mixtures at a high sensitivity level. Since the biochemist and clinical chemist are interested mainly in the analysis of biological materials, the aim of this chapter is to provide information on the use of this instrument combination, i.e., the use of a gas chromatograph as an inlet system or the use of a mass spectrometer as a gas chromatographic detector. Emphasis has been put on practical considerations and on correlations between mass spectra and structure rather than on mechanistic interpretations of the spectra. For details on the latter aspect the reader is referred to the books by Budzikiewicz, Djerassi, and Williams (1, 2).

Early studies of bile acid mass spectra were made by Bergström, Ryhage, and Stenhagen (3–5) and the results were used to determine the structure of Hammarsten's α-phocaecholic acid (6). Other groups applied mass spectrometry to the elucidation of structures of the closely related bile alcohols (7). At about this time combined gas chromatography–mass spectrometry instruments were developed that could be used in studies of compounds with molecular weights as high as those of bile acid derivatives (8–13). The early

*This work was supported by grants from the Swedish Medical Research Council (13X-219), the Wallenberg Foundation, and the Bank of Sweden Tercentenary Fund.

studies as well as subsequent studies with gas chromatography–mass spectrometry instruments required high temperatures to volatilize the bile acid esters. Partly due to thermal decomposition the spectra often failed to show a molecular ion. In an attempt to avoid this difficulty, direct inlet systems were used to extend the results of the early studies (14, 15). One may assume that, in biological studies, introduction through a gas chromatography column (at high temperatures) or through a direct inlet system (at low temperatures) will be the two common methods to introduce bile acid derivatives into the ion source. For reasons mentioned in the beginning, the following description will be concerned mainly with the former mode of operation and all figures show spectra obtained using a gas chromatographic inlet system.

II. CONDITIONS FOR GAS CHROMATOGRAPHY–MASS SPECTROMETRY

Our personal experience with gas chromatographic–mass spectrometric analysis of bile acids is limited to the use of a modified Atlas CH-4 mass spectrometer connected to the gas chromatography column via a molecule separator of the jet type (8, 16), and to the use of prototypes and production models of the LKB 9000 instrument.

When the conditions of analysis are chosen, the requirements of both the gas chromatographic and the mass spectrometric parts must be considered. The combination instrument differs from a conventional gas chromatograph since there is a vacuum at the exit of the column. The gas flow rate must be chosen so that the pressure in the column does not get too low. If this happens peak distortion and loss of resolution occurs. Flow rates of 30–60 ml/min are suitable when a two-stage jet separator is used with helium as carrier gas, and when the columns, 3–4 mm i.d., are packed with 80–100 or 100–120 mesh packing. Under these conditions 99–99.5% of the helium is removed in the separator and about 60–70% of the sample reaches the ion source (16). The pressure in the analyzer tube is about $1\text{–}3 \times 10^{-6}$ mm Hg.

The same column temperatures can be used as in conventional gas chromatography. Both the molecule separator and the ion source must be kept at the same or higher temperature than the column, and a constant even temperature of these parts is important to prevent tailing or memory effects. Since bile acid derivatives are usually separated at 200–250°C, thermal cracking in the mass spectrometer cannot be avoided. This is in contrast to the case when a direct inlet system is used. However, the thermal effects are less disturbing than one would at first anticipate. Although relative intensities will differ between spectra obtained with gas chromatographic and direct inlet systems, the general fragmentation patterns are very similar.

Another factor influencing the fragmentation is the energy of the bombarding electrons. Although most workers use 70 eV to ensure maximal ionization we have kept the energy at 22–23 eV to avoid ionization of helium. This was important in the first combination instrument and since most of the reference spectra were run in 1964 the same energy has been used later to get comparable results. Furthermore, we have found that low electron energy gives more valuable information in work with biological mixtures since high electron energy gives rise to many ions with low mass and little diagnostic value.

Some workers use a splitter before the mass spectrometer inlet to be able to attach other detection or collection devices. In routine analyses part of the total ion current produced in the mass spectrometer is used to follow the appearance of organic material from the column. A gas chromatographic analysis should, however, always be made prior to the analysis in the combination instrument to avoid introduction of grossly contaminated samples.

The major factor limiting sensitivity is the background bleed from the column. The properties of the silanized support and the injection membranes may also influence the background. Only stable phases can be used, e.g., silicones GC-grade SE-30 (methyl), OV-1 (methyl), OV-17 (phenyl), QF-1 (trifluoropropyl), OV-210 (trifluoropropyl), and OV-225 (cyanoethyl). A well-washed and siliconized support of the Gas-Chrom P, Q, or Chromosorb W types, carrying the lowest percentage of stationary phase compatible with good separation characteristics and efficiency should be used (1–3 %). Although use of little packing material (i.e., small-diameter columns) reduces background, biological extracts are often impure so that good column capacity (i.e., large diameter) is required. Columns of 3–4-mm diameter have given good efficiency coupled with a long usable lifetime. Of the three latter phases mentioned above, only QF-1 has been tested by us. According to the manufacturer OV-210 and OV-225 are more stable than the QF-1 phase. Silicones with trifluoropropyl groups, although not particularly thermostable, are very valuable for separation of bile acid derivatives because of the selectivity for carbonyl groups.

When a direct inlet system is used, spectra can be obtained of the free bile acids. This is not possible with a gas chromatographic inlet. The simplest derivative which can be analyzed with the combination instrument is the methyl ester. Valuable information on the nature of the fragment ions can be obtained by analysis of both the methyl and ethyl esters. Particularly for gas chromatographic reasons it is better, however, to protect hydroxyl groups by acetylation, trifluoroacetylation, or trimethylsilylation.

The choice of bile acid derivative depends on the nature and complexity of the bile acid mixture to be analyzed. Derivative formation is described in the chapter on gas chromatography. Generally it is of advantage to use derivatives which can be separated at low temperatures so that the background intensity can be kept low. Preferably the derivatives should have a high thermal stability, not be adsorbed to column support, glass, or metal, have a low molecular weight (simplifies evaluation and handling of spectra even when computers are used), and be simple to make in quantitative yield on a microscale. It is obviously difficult to meet all these requirements. Trifluoroacetates and trimethylsilyl (silyl) ethers have been the most commonly used derivatives of hydroxy bile acids since they are easily and rapidly made. Trifluoroacetates of common biliary bile acid methyl esters separate very well on QF-1 columns but not on SE-30, whereas the silyl ethers of these common compounds separate poorly on both phases but are very suitable for mass spectrometry. Depending on position and orientation, hydroxyl groups are trifluoroacetylated or silylated at different rates, and partial derivatives can be prepared that may aid in the interpretation of the mass spectra.

Although keto groups do not have to be derivatized this is sometimes of advantage for mass spectrometric reasons. For example, the presence of a 3-keto group can be established by conversion into a 1,1-dimethylhydrazone. O-methyloximes are simple to prepare and reaction rates are different depending on the position of the keto groups.

Certain potential sources of error must be kept in mind both when complex biological mixtures and supposedly pure compounds are analyzed. Poor gas chromatographic columns or use of derivatives that adsorb to the support may give rise to tailing and memory effects. If the molecule separator and ion source and housing are adequately and evenly heated, this memory effect always originates in the gas chromatographic part of the system and not in the mass spectrometer. An effect which may be considered as the opposite of a memory—an anticipation effect—can be observed if thermolabile derivatives are analyzed. Trifluoroacetates exemplify this situation. Thermal loss of trifluoroacetic acid on the column results in a continuous elution of decomposition product(s) before the actual compound appears. The mass spectra of the compound and its degradation product(s) may be very similar (see Section IV) since loss of trifluoroacetic acid on electron impact occurs in the ion source. The masses of the molecular ions are of course different but at high temperatures the abundance of molecular ions may be low for the parent compound. Problems of this nature are likely to be serious when compounds are present in widely varying concentrations in the mixture analyzed.

Another source of error is the formation of multiple products in the preparation of derivatives. Diazomethane may react with keto groups unless freshly distilled diazomethane and short reaction times are employed. Keto groups may react with trifluoroacetic anhydride to give enol esters, or in silylation reactions to give enol ethers. Under more vigorous conditions even substitution reactions may take place during silylation (17). However, the conditions needed to derivatize hydroxyl groups in bile acids can usually be kept such that no side reactions occur. The 3-keto-Δ^4 structure is an exception and gives enol derivatives with relative ease. However, mass spectra of enol silyl ethers may yield valuable structural information.

As a fourth source of error one may mention the introduction of contaminants during the workup of a biological sample. The contaminant may come from solvents, reagents, glassware, or the skin surface. Solvent residues, e.g., dioctylphthalate, are most common, but bile acids may also be introduced. It is important to make a strict separation of glass, solvents, etc., between laboratories for synthetic and analytical work.

All the factors discussed in this section influence the sensitivity of a gas chromatographic–mass spectrometric analysis. Ion yield in the mass spectrometer, number of ion species and their relative abundance, and slit settings for a desired resolution are also very important. In our experience the four major factors determining the amount that has to be injected to give full mass spectral information are: retention time, losses on the column, column bleed, and distribution of ion species. Usually this amount is 0.1–2.0 μg. However, less material (e.g., 10 ng) may be sufficient to show the major ions. It is clear that gas chromatography–mass spectrometry gives more structural information with ultramicro amounts than any other method.

III. GENERAL FRAGMENTATION OF BILE ACID DERIVATIVES

Many of the common major fragmentations were discussed by Bergström, Ryhage, and Stenhagen (5). Their results have been extended, with more comprehensive discussions of bile acid mass spectra (1, 12, 15, 18, 19). The fragmentation patterns of sterol derivatives are in many respects similar to those of the bile acids; discussions of sterol mass spectra are given in (1, 11, 20–23). General reviews of steroid mass spectra may be consulted (1, 24, 25).

The following discussion is based on spectra of the bile acid derivatives listed in Table I. This is not a complete list of all published spectra but the results are representative of general fragmentation patterns. The molecular weights of bile acid derivatives can be calculated from the group contributions listed in Table II. This table also shows how the molecular weight changes when a hydroxy and/or keto bile acid methyl ester is converted into the derivatives.

Figure 1 and Table III give a schematic summary of the fragmentations of bile acid derivatives. It should be stressed that detailed mechanistic studies have not been made, i.e. with bile acids labeled with isotopes. Thus, the

TABLE I. Bile Acid Derivatives Analyzed by Mass Spectrometry

Configuration at C-5	Substituents[a]	Derivative[b]	Side chain[c]	Reference[d]
5α,5β	3β	OH(Me)	C_{20}	26
5β	None	Me	C_{24}	5
5β	3α	Acid; OH(Me); OAc(Me)	C_{24}	A,5, 15, 27
5β	3α	TFA(Me; Et)	C_{24}	A
5β	3α	TMS(Me; Et)	C_{24}	A
5β	3α	OAc(Me); TFA(Me)	C_{27}	A, 5
5β	3β	TFA(Me)	C_{24}	A
5β	3α-Δ^{11}	TFA(Me; Et)	C_{24}	A
-	3β-Δ^{5}	TFA(Me; Et)	C_{24}	A
5α,5β	3-keto	Me	C_{20}	28
5β	3-keto	Me; Et; DMH(Me)	C_{24}	A
-	3-keto-Δ^{4}	Me	C_{24}	27
5β	7α	OH(Me); TFA(Me)	C_{24}	A
5β	7-keto	Me	C_{24}	A
5β	12α	OH(Me); TFA(Me)	C_{24}	A, 5
5β	12-keto	Me	C_{24}	A, 29
5β	3α,6α	Acid; OH(Me)	C_{24}	A, 15
5β	3α,6α	TFA(Me); TMS(Me; Et)	C_{24}	A
5α,5β	3,6-keto	MO(Me)	C_{24}	30
5α	3,6-keto	MO(Me)	C_{20}	30
5β	3α,7α	Acid; OH(Me)	C_{24}	A, 14, 15, 27
5β	3α,7α	OAc(Me); TFA(Me; Et)	C_{24}	A, 5, 31, 32
5β	3α,7α	TMS(Me; Et)	C_{24}	A
5β	3α,7α	OH(Me)	C_{24}	33
5β	3α,7α	OH(Me)	C_{27}	14
5β	3β,7α	TFA(Me)	C_{24}	A
5β	3β,7α	OH(Me); TFA(Me)	C_{24}	32, 33
5β	3α,7β	Acid; OH(Me); OAc(Me); TFA(Me)	C_{24}	A, 5, 14, 27, 32
5β	3α,7-keto	OH(Me); TFA(Me); TMS(Me)	C_{24}	A, 27
5β	3β,7-keto	TFA(Me)	C_{24}	A
5β	3-keto,7α	OH(Me); TMS(Me)	C_{24}	A, 27, 34, 35
5α	3-keto,7α	OH(Me)	C_{24}	34, 35
5β	3,7-keto	Acid; Me; DMH(Me)	C_{24}	A, 14, 27
5α	3,7-keto	Me	C_{24}	36
5β	3α,12α	Acid; OH(Me); OAc(Me)	C_{24}	A, 5, 14, 15
5β	3α,12α	TFA(Me; Et); TMS(Me; Et)	C_{24}	A, 31, 32
5α	3α,12α	TFA(Me); TMS(Me)	C_{24}	A
5β	3β,12α	TFA(Me)	C_{24}	A
5α	3β,12α	TMS(Me)	C_{24}	A
5β	3α,12β	TFA(Me)	C_{24}	A
5β	3β,12β	OH(Me)	C_{24}	A
5β	3α,12α-Δ^{6}	TFA(Me)	C_{24}	A
5β	3α,12α-Δ^{7}	TFA(Me)	C_{24}	31
5β	3α,12α-$\Delta^{8(14)}$	TFA(Me)	C_{24}	31
-	3,12α-$\Delta^{3,5}$	TFA(Me); TMS(Me)	C_{24}	A
5β	3α,12-keto	OH(Me); TFA(Me); TMS (Me)	C_{24}	A

TABLE I. (Continued)

Configuration at C-5	Substituents[a]	Derivative[b]	Side chain[c]	Reference[d]
5β	3β,12-keto	TFA(Me)	C_{24}	A
5β	3-keto,12α	OH(Me); TFA(Me); TMS(Me)	C_{24}	A, 37
5α	3-keto,12α	OH(Me)	C_{24}	37
-	3-keto-Δ^4-12α	OH(Me); TFA(Me); TMS (Me)	C_{24}	A, 27
5β	3,12-keto	Acid; Me; DMH(Me)	C_{24}	A, 14, 15
5β	12-keto-3,4-seco[e]	Me; Et	C_{24}	A
5β	7α,12α	TFA(Me)	C_{24}	A
5β	7,12-keto	Acid; Me	C_{24}	15
5β	3α,6α,7α	OH(Me); TMS(Me)	C_{24}	A
5β	3α,6α,7β	TFA(Me)[f]; TMS(Me)	C_{24}	A
5β	3α,6β,7α	TFA(Me)[f]; TMS(Me)	C_{24}	A
5β	3α,6β,7β	TMS(Me)	C_{24}	A
5β	3α,7α,12α	OH(Me); OAc(Me); TFA(Me)	C_{22}	A, 5
5β	3α,7α,12α	OH(Me); OAc(Me); TFA(Me)	C_{23}	A, 5
5β	3α,7α,12α	Acid; Cath(Me) OAc(Me)	C_{24}	5, 14, 15
5β	3α,7α,12α	OH(Me); TFA(Me; Et); TMS(Me; Et)	C_{24}	A, 15, 32
5α	3α,7α,12α	TFA(Me); TMS(Me)	C_{24}	A
5β	3α,7α,12α	OH(Me); OAc(Me); TFA(Me)	C_{25}	A, 5
5β	3α,7α,12α	Acid; OH(Me); OAc(Me); TFA(Me)	C_{27}	A, 15, 14
5β	3α,7α,12α	OAc(Me)	C_{28}	5
5β	3α,7β,12α	TFA(Me); TMS(Me)	C_{24}	A, 32
5β	3β,7β,12α	TFA(Me)	C_{24}	A
5β	3α,7α,12-keto	OAc(Me); TFA(Me); TMS(Me)	C_{24}	A, 15
5β	3α,12α,7-keto	OH(Et); Cath(Me); OAc(Me)	C_{24}	5, 15
5β	3α,12α,7-keto	TFA(Me); TMS(Me)	C_{24}	A
5β	3-keto,7α,12α	OH(Me); TMS(Me)	C_{24}	A, 34, 38
5α	3-keto,7α,12α	OH(Me); TMS(Me)	C_{24}	34, 38
5β	3α,7,12-keto	OAc(Me); Cath(Me);TMS (Me)	C_{24}	A, 15
5β	3,7-keto,12α	OH(Me); OAc(Me); TMS	C_{24}	A, 15, 18
5α	3,7-keto,12α	OH(Me)	C_{24}	36
5β	3,7,12-keto	Acid; Me; DMH(Me)	C_{24}	A, 15
5β	3α,7α,23ξ	OAc(Me)	C_{24}	5, 6

[a] Greek letters denote orientation of hydroxyl groups.
[b] Acid, free bile acid, no protecting groups; OH, free hydroxyl groups; Me and Et, methyl and ethyl esters, respectively; OAc, acetate; TFA, trifluoroacetate; TMS, trimethylsilyl ether; Cath, 3-cathylate; DMH, 3-dimethylhydrazone; MO, O-methyloxime.
[c] Number of carbon atoms in entire bile acid molecule is given.
[d] "A" refers to published and unpublished work by the authors and includes (12, 13, 18, 19).
[e] 12-Keto-3,4-seco-5β-cholane-3,4,24-trioic acid.
[f] Decomposition products formed on gas chromatography column.

TABLE II. Molecular Weights of Common Derivatives of Bile Acid Methyl Esters, and Change of Molecular Weights on Formation of Derivatives of Hydroxyl and Keto Groups

Contributing groups	Mass increase per substituent	Mass change on derivative formation
Methyl cholanoate	374	—
Methyl(ene) group	14	—
Keto group	14	—
Dimethylhydrazone group	56	42
O-methyloxime group	43	29
Hydroxyl group	16	—
Methoxy group	30	14
Acetoxy group	58	42
Trimethylsiloxy group	88	72
Trifluoroacetoxy group	112	96
Double bond	−2	—

TABLE III. Schematic Representation of the Tentative Origin of Ions Formed in the Fragmentation of Substituted Methyl Cholanoates

Fragmentation[a]	Carbon atoms[b] in		Mass[c] of	
	Ion	Lost fragment	Ion	Lost fragment
a	—	H_2O	—	18
a_1	—	CF_3COOH	—	114
a_2	—	$(CH_3)_3SiOH$	—	90
a_3	—	$(CH_3)_3SiO$	—	89
b	—	CH_3	—	15
c	ABCD–20–24	OCH_3	—	31
c_1	ABCD–20–24	CH_3OH	—	32

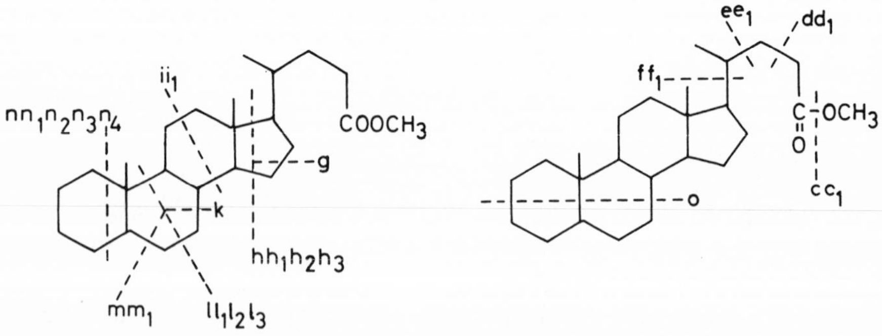

Fig. 1. Schematic illustration of diagnostically significant fragmentations of the carbon skeleton of methyl cholanoates. The masses of different fragments are given in Table III.

TABLE III. (Continued)

Fragmentation[a]	Carbon atoms[b] in		Mass[c] of	
	Ion	Lost fragment	Ion	Lost fragment
d	ABCD–20–22	23–24	—	73
d_1	23–24	ABCD–20–22	74	—
e	ABCD–20–21	22–24	—	87
e_1	22–24	ABCD–20–21	87	—
f	ABCD	20–24	—	115
f_1	20–24	ABCD	115	—
g	ABC–15	16–17;20–24	—	142
h	ABC	15–17;20–24	—	157
h_1	ABC	15–17;20–24	—	156
h_2	ABC	15–17;20–24	—	155
h_3	15–27;20–24	ABC	154	—
i	12–D–20–24	AB–11	208[d]	—
i_1	AB–11	12–D–20–24	—	210
k	CD–20–24	A–6–7	249[e]	—
l	7–CD–20–24	A–6	262[e]	—
l_1	7–CD–20–24	A–6	261[f]	—
l_2	7–CD–20–24	A–6	263[g]	—
l_3	A–6	7–CD–20–24	285[h]	—
m	6–7–CD–20–24	A	—	96[i]
m_1	A	6–7–CD–20–24	94[k]	—
n	BCD–20–24	1–4	—	54[k]
n_1	BCD–20–24	1–4	—	70[l]
n_2	BCD–20–24	1–4	—	145[m]
n_3	1–4	BCD–20–24	142–143[n]	—
n_4	BCD–20–24	1–4	—	142[n]
o	3–7	—	243[h]	—

[a] See Fig. 1.
[b] A, B, C, and D refer to entire rings with six (or five) carbons and angular methyl group(s). Individual carbon atoms are referred to by respective numbers.
[c] The mass will depend on presence or absence of substituents. This is evident from the footnotes and from the mass spectra shown in the figures.
[d] With no substituents in ring D and side chain.
[e] With no substituents in rings CD and side chain.
[f] After loss of hydroxyl function at C-12.
[g] In some 3,6- and 3,7-dihydroxycholanoate derivatives.
[h] Containing two trimethylsiloxy groups.
[i] With no substituents in ring A.
[k] With loss of hydroxyl function at C-3.
[l] With 3-keto group.
[m] Including one trimethylsiloxy group.
[n] Ring A of an enol trimethylsilyl ether of a 3-keto bile acid.

origins of several ions are hypothetical. It is sufficient to read the paper by Tökés *et al.* (39) about the course of ring D fragmentation of steroids to realize the complex nature of the fragmentation processes. An ion may arise by alternative reactions and a peak of a particular mass may be due to several ion species with different structures. However, in spite of this complexity, the peaks produced in bile acid fragmentations are of considerable diagnostic significance. By comparing the spectra of different derivatives of different bile acids the origin of the peaks can be established with reasonable certainty and structural information can be gained by simple arithmetic considerations. Detailed information regarding structure and formation of ions can be obtained by high-resolution mass spectrometry and deuterium labeling.

In substituted methyl cholanoates (Table I) loss of water or its equivalents (acetic acid, trifluoroacetic acid, trimethylsilanol, etc.) is pronounced. In gas chromatography–mass spectrometry this process is partly thermal, partly due to electron impact; when a direct probe is used, thermal elimination can be avoided. Since the mechanisms of the two types of elimination are different (26) the spectra will differ to some extent. However, in work with biological materials, the complexity of the mixtures and the small amounts of bile acids available usually make it necessary to use the former method and to accept the thermal component in the fragmentation process. When methyl esters of di- and trihydroxy bile acids and their acetates or trifluoroacetates are analyzed by gas chromatography–mass spectrometry, a molecular ion is usually not seen. This is partly due to the high temperatures used. Trimethylsilyl ethers usually give a molecular ion peak but it may be quite small.

Although spectra of different derivatives of the same bile acid usually show the same fragmentation pattern, there are often large differences in the relative intensities of the peaks. It is often of considerable help in identifications to analyze several derivatives. The statement that spectra of steroids with free hydroxyl groups are more informative than those of derivatives (25) is therefore too generalized in the case of bile acids.

Other fission reactions that give rise to prominent peaks are those involving loss of side chain with and without carbons of the D-ring. The latter reactions have been investigated in detail with steroid hydrocarbons (39). The structure of the side chain can be deduced from these fragmentations and from those leading to ions containing part of or the entire side chain. For the characterization of the latter ions it is of value to analyze both the methyl and ethyl esters since the mass of these ions will differ by 14 mass units. Analysis of both methyl and ethyl esters is also valuable when the presence of more than one carboxyl group is suspected. In a similar way the presence of a trimethylsiloxy group in an ion can be established by analysis of the derivative obtained after reaction with perdeuterotrimethylsilyl chloride in pyridine (40). Ions containing this group will differ by 9 mass units between the two spectra. The peaks at m/e 73 and 75 seen in all spectra of trimethylsilyl ethers (41) shift to m/e 82 and 81, respectively (40).

As will be seen in the mass spectra shown in the figures, many of the prominent peaks are due to the combined loss of the groups mentioned above. However, the fragmentation reactions are directed by the substituents and occur to a different extent so that spectra of isomeric bile acids can usually be distinguished. When silyl ethers are analyzed this is often the case even when the difference is only configurational. However, the gas chromatographic retention times are usually of greater importance for the determination of configurations.

In our experience with several different instruments (Atlas CH-4 and LKB 9000) over a period of six years the reproducibility in analyses of bile acid trimethylsilyl ethers is very good also with respect to differences between spectra of stereoisomers. Only once, when the ion source housing was contaminated, were the spectra quite different from normal. Spectra of bile acid methyl esters and particularly of their trifluoroacetates show somewhat larger variations; this is assumed to be due to the greater thermal sensitivity of these derivatives.

IV. FRAGMENTATION PATTERNS OF NONKETONIC BILE ACID DERIVATIVES

A. Monohydroxy Bile Acids

Three types of monohydroxycholanoates have been studied: derivatives of 3-, 7-, and 12-hydroxycholanoates. Spectra of some of these derivatives are shown in Figs. 2 and 3.

As seen in Fig. 2 the free, trifluoroacetylated, and trimethylsilylated methyl lithocholates give similar spectra but there are important differences with respect to preferred fissions. The molecular ion peak is small in all cases due to loss of the hydroxyl function. However, the peaks due to fragmentation through the D-ring (without loss of hydroxyl function) are much more prominent in the spectrum of the trifluoroacetate than in the other spectra (fragmentations h, h_1, h_3, and g). The masses of these ions indicate the number of hydroxyl functions (or double bonds) in the ABC-rings. Loss of the side chain (f; -115) is not seen without previous loss of the hydroxyl group (a, a_1, a_2,) but the mass of the ABCD ion then formed shows that the hydroxyl group was in the ring system. An ABCD ring ion is seen in almost all bile acid spectra and its mass is of considerable diagnostic importance.

Figure 2 also shows that spectra of trifluoroacetates of epimeric bile acids are very similar. Some examples given below indicate that trimethylsilyl ethers give more discriminating spectra in analyses of epimers. There are numerous examples of differences between spectra of trimethylsilyl ethers of

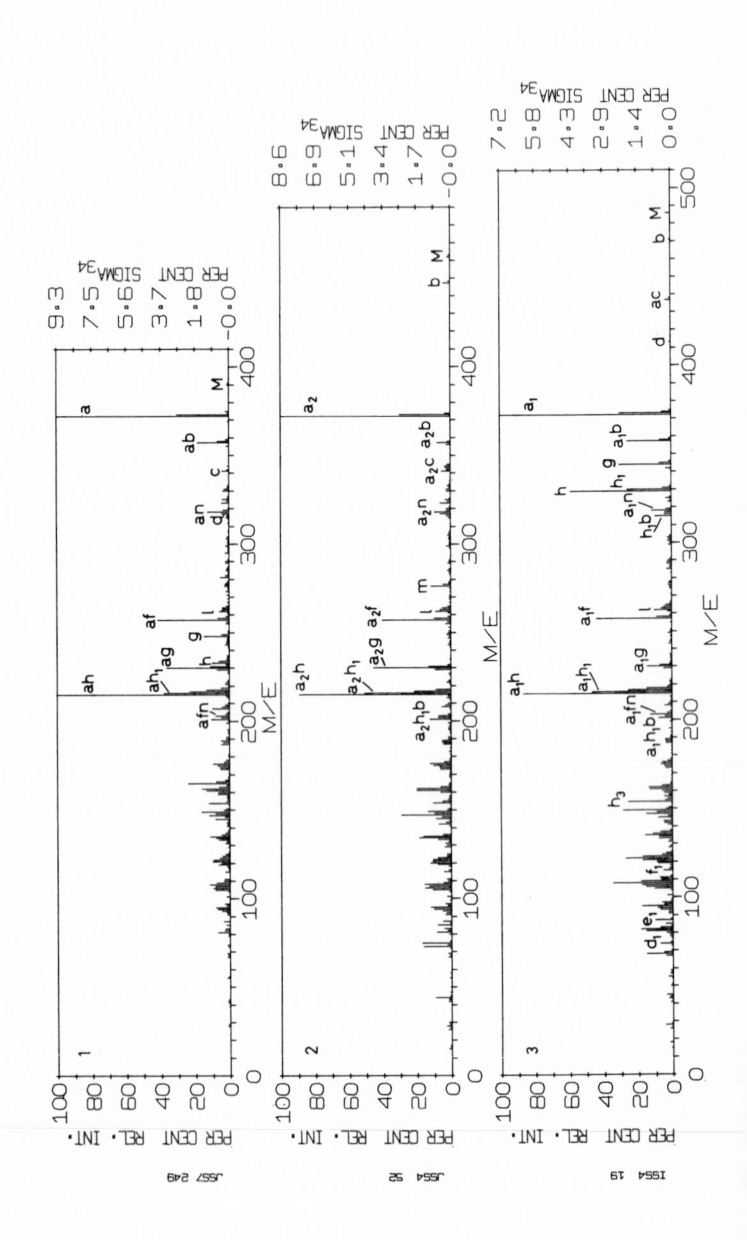
PER CENT REL. INT.
PER CENT SIGMA$_{34}$
M/E
JS57 249
JS54 52
IS54 19

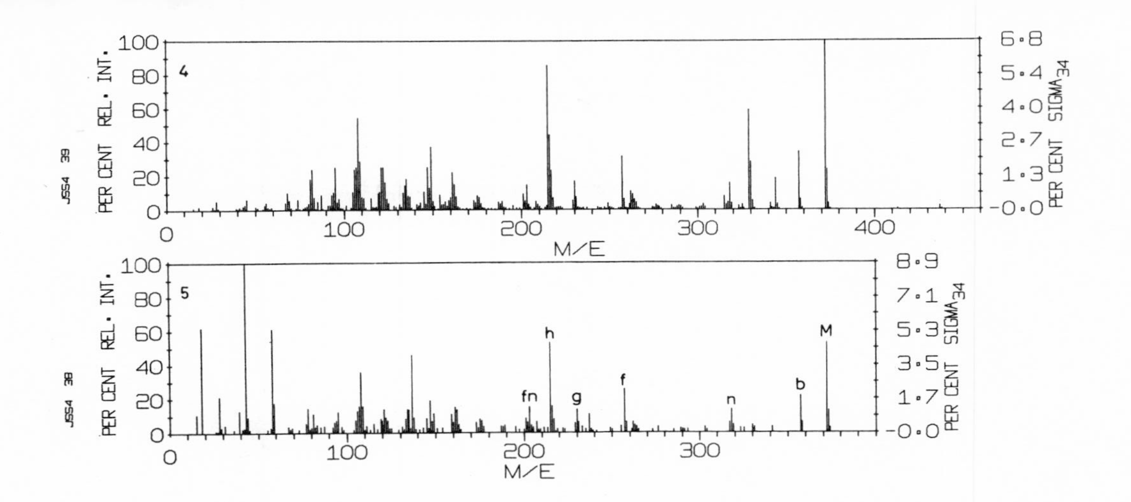

Fig. 2. Mass spectra of methyl lithocholate (1), its trifluoroacetate (3) and trimethylsilyl ether (2), and of the trifluoroacetate of the epimeric methyl 3β-hydroxy-5β-cholanoate (4) and the gas chromatographic decomposition product (5) of this compound. The letters in this and the following figures refer to the fragmentation modes illustrated in Fig. 1 and listed in Table III.

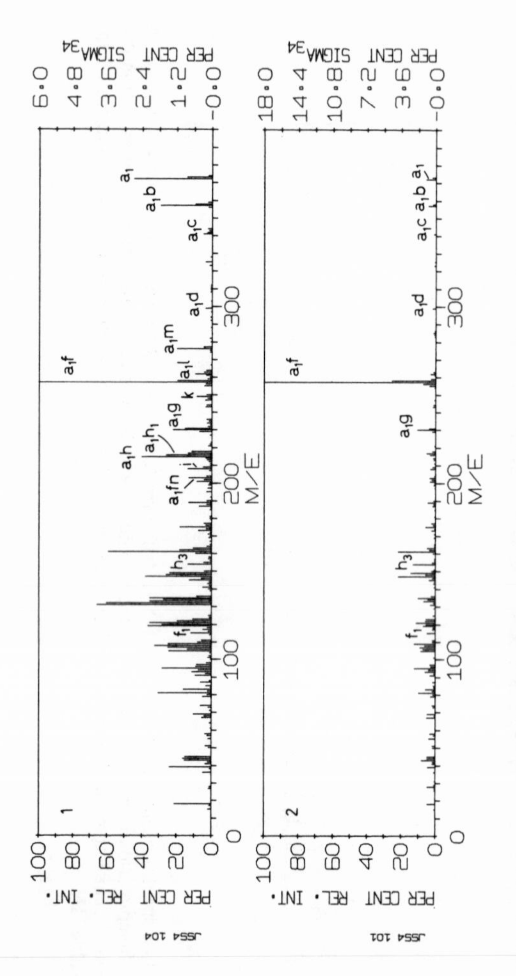

Fig. 3. Mass spectra of the trifluoroacetates of methyl 7α- (1), and 12α-hydroxy-5β-cholanoates (2).

epimeric C_{19} and C_{21} steroids. It is sufficient to mention that the spectra of eight epimeric 3-trimethylsiloxy-pregnan-20-ones all differed with respect to relative intensities (42), and these differences are very reproducible.

The similarities between spectra of a product formed by decomposition on the column and its parent compound are also seen in Fig. 2. It is evident from the spectra that loss of trifluoroacetic acid occurs during passage through the column. The extent of this degradation depends on the quality of the column and its temperature.

The position of the hydroxyl group (or its equivalent) influences the fragmentation markedly. Figure 3 shows spectra of the trifluoroacetates of methyl 7α- and 12α-hydroxy-5β-cholanoates which should be compared with the spectra in Fig. 2. Both the 7α- and the 12α-substituted compounds lose trifluoroacetic acid more readily than the 3-substituted ones. Whereas the latter show pronounced fragmentation through the D-ring the characteristic fragmentation of the 7α-hydroxy bile acid is through the B-ring (*k*, *l*, *m*). The mechanisms of formation of the ions of mass 262 (*l*) and 276 (*m*) are unknown. These ions contain the side chain, and analogous ions found in sterol spectra are discussed by Knights (21). The peak at *m/e* 249 (*k*) is of great diagnostic significance. It is found in bile acid derivatives having a hydroxyl group (or its equivalent) at C-6 or C-7, or a 5,6 or 6,7 double bond, and in some 7-keto bile acids. It is found at *m/e* 247 in C_{27} sterols with a 5,6 double bond (21, 43) and an analogous ion is formed in the fragmentation of Δ^4-3,6-diketosteroids (44). It is most likely an allylic carbonium ion of the CD-rings and side chain

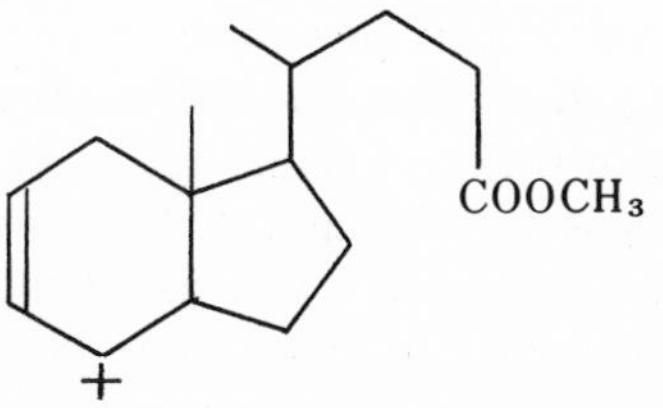

$COOCH_3$

+

and its mass gives information on the structure of that part of the molecule.

The 12-hydroxylated bile acids show a pronounced tendency to lose water (or its equivalent, partly thermally) followed by loss of the side chain. As a result, peaks due to loss of water (or its equivalent) + 115 mass units are always intense in spectra of these compounds, particularly at high temperatures. The importance of a preceding loss of the substituent at C-12 is indicated by the fact that methyl 3α-trifluoroacetoxy-5β-chol-11-enoate gives a base peak at *m/e* 369 (*M*-115) whereas the saturated compound (Fig. 2) shows a base peak at *m/e* 372 (*M*-114).

B. Dihydroxy Bile Acids

Examples of mass spectra of derivatives of 3,6-, 3,7- and 3,12-dihydroxy-cholanoates are shown in Figs. 4 and 5. At the temperatures used in gas chromatography–mass spectrometry, molecular ions are seen only with the free hydroxy compounds and the trimethylsilyl ether derivatives (intensity usually less than one percent of base peak). Molecular ions with a relative intensity of 5–10% are obtained from the methyl esters or acids when a direct probe is used (14, 15). Loss of 2 and 4 mass units from the molecular ions of compounds with free hydroxyl groups is most probably thermal and is not seen at low temperatures.

Loss of one water molecule (or its equivalent) is a major reaction followed by loss of the second hydroxyl function or the side chain. The latter pathway dominates in bile acids having a free or trifluoroacetylated 12-hydroxyl group whereas silyl ethers and derivatives of 3,6- and 3,7-dihydroxy bile acids preferably lose two hydroxyl functions. Thus, the ratio between the intensity of the peak formed by loss of one hydroxyl group plus the side chain and the intensity of the peak formed by loss of two hydroxyl groups is always much larger with dihydroxycholanoates having hydroxyl groups at C-3 and C-12 than in those having these groups at C-3 and C-6 or C-7 (see Figs. 4 and 5).

The mass spectrum of the trifluoroacetate of methyl 7α, 12α-dihydroxy-5β-cholanoate is dominated by the fragmentations induced by the presence of a 12-hydroxy group. Since the 7-trifluoroacetoxy group is also readily lost, the base peak is at *m/e* 255 and is 10 times more intense than the peak at *m/e* 370.

Trifluoroacetates show a relatively more pronounced fragmentation through the D-ring (*g, h,*) than other derivatives. The ion of mass 154 (h_3) is particularly prominent in trifluoroacetates of 3,6- and 3,7-dihydroxy bile acids. Its relative intensity is variable and its formation may be influenced by the temperature of the ion source. The mass changes with the length of the side chain and evidently this ion arises from that part of the molecule.

Fragmentation *g* through the D-ring is enhanced by the presence of an allylic 8,4 double bond. Thus, Kallner (31) found an unsaturated dihydroxy bile acid in bile from the teleost *Cottus quadricornus,* and the trifluoroacetate of this compound gave a mass spectrum typical of 3,12 substituents (intense peaks at *m/e* 367 and 253). When it was treated with hydrogen platinum catalyst the double bond was not saturated but the peak at *m/e* 340 (a_1, g) in the mass spectrum of the trifluoroacetate increased six times. This was interpreted as being due to migration of the double bond from the 7,8 to the 8,14 position and this proved to be correct on comparison with reference bile acid spectra.

The 6- and 7-hydroxylated compounds give ions due to fragmentation through the B-ring (*m/e* 249, 262). This fragmentation may take different courses for different derivatives and the cleavage of the 7,8 bond may be α to the oxygen substituent or allylic to a double bond formed after loss of the hydroxyl function. For comparisons, the spectrum of methyl 3β-trifluoroacetoxy-5-cholenoate is shown in Fig. 4. The intense ion at *m/e* 249 and the similarities between this spectrum and that of the silyl ether of methyl 3α,7α-dihydroxy-5β-cholenoate are evident. However, if the 3β-Δ^5 bile acid is analyzed as silyl ether the typical peaks at *m/e* 129 and *M*-129 (11, 23, 45) will appear, and the spectrum is very different from that of the disubstituted compound. This shows the importance of analyzing several different derivatives.

The spectra of the trifluoroacetates of 3,6- and 3,7-dihydroxy bile acids are similar but clearly different (Fig. 4). However, more pronounced differences are obtained by analysis of O-methyloximes of the diketones formed after oxidation (see Section V).

An ion of particular interest for the establishment of a 3,7-bis(trimethylsiloxy) structure is found at *m/e* 243 (*o*). It is also found in silyl ethers of 3,7,12-trihydroxycholanoates but it is insignificant in the 3α,6α-dihydroxy derivative. Its probable structure is

$$(CH_3)_3Si{-}O{-}CH{=}CH{-}CH{=}CH{-}CH{=}\overset{+}{O}{-}Si(CH_3)_3$$

and it might be formed by cleavage of the 2,3, 5,10, and 7,8 carbon bonds. When the compound is labeled with deuterium at C-3 and C-5 this ion contains both the deuterium atoms (Cronholm, Makino, and Sjövall, unpublished work).

Trimethylsilyl ethers of bile acids having a 12-hydroxy group yield a peak at *m/e* 208 (Fig. 5). This ion is less prominent in spectra of many other bile acid derivatives. It changes mass with the length of the side chain and it is assumed to arise by fragmentation through the C-ring and to contain the D-ring and side chain and one additional carbon.

All the derivatives of bile acids with a 3-hydroxy group discussed above show a peak (sometimes small) which arises through loss of carbons 1–4 (*n*) probably after prior loss of the 3-hydroxy group (or its equivalent). If the latter loss results in a 2,3 double bond, carbons 1–4 are lost through a *retro* Diels–Alder reaction (1). The mass of the ion gives information about the structure of the BCD-rings and side chain. However, since 7- and 12-hydroxy groups are lost more readily than a 3-hydroxy group the ion appears at *m/e* 314, 316, or 318 depending on whether there is two, one, or no hydroxyl group in the BCD-rings. Loss of the A-ring is further discussed in the section on ketonic bile acids.

It is clear that spectra of positional isomers differ. As shown in Fig. 5

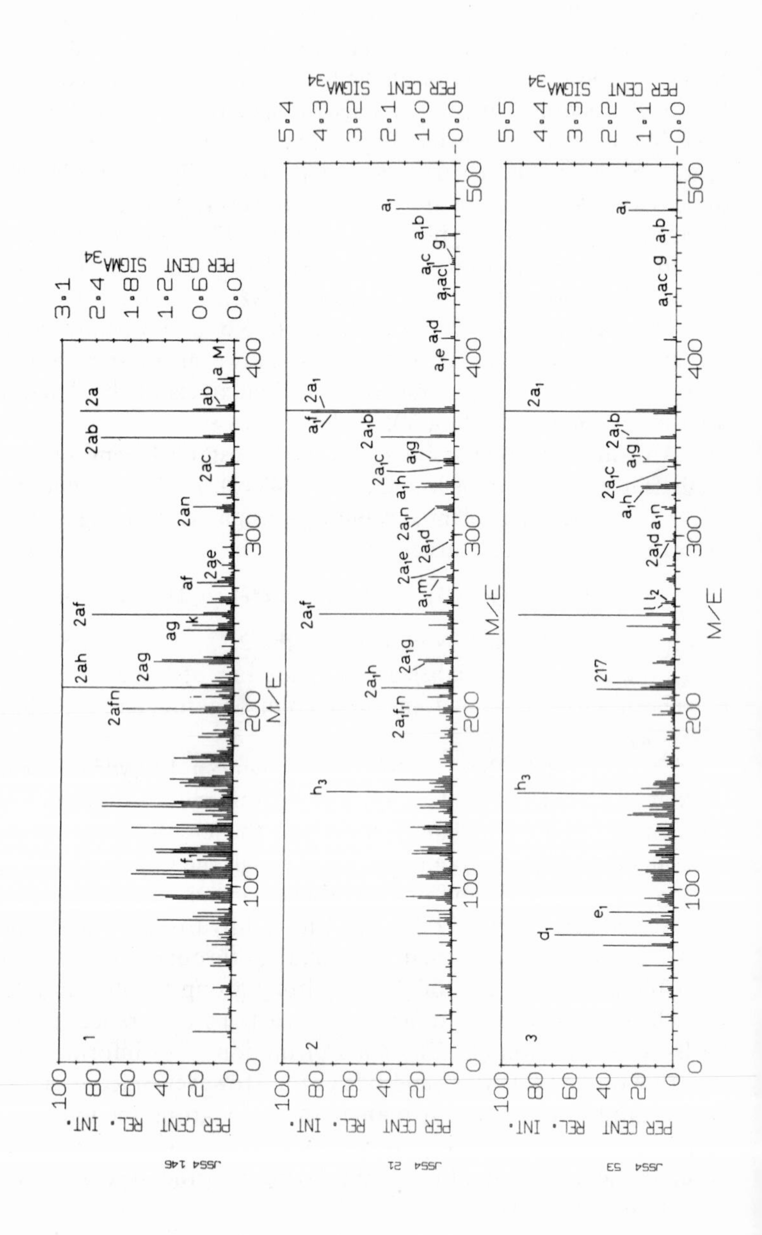
PER CENT REL. INT.
PER CENT SIGMA34
M/E
JS54 146
JS54 21
JS54 53

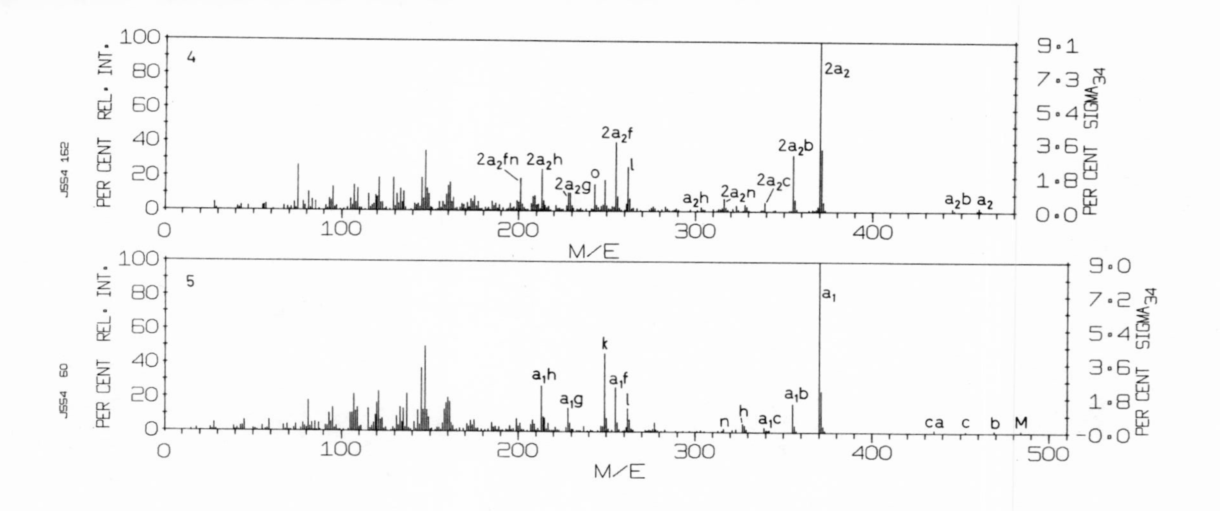

Fig. 4. Mass spectra of methyl 3α,7α-dihydroxy-5β-cholanoate (1), its trifluoroacetate (2) and trimethylsilyl ether (4), and of the trifluoroacetates of methyl 3α,6α-dihydroxy-5β-cholanoate (3) and 3β-hydroxy-5-cholenoate (5).

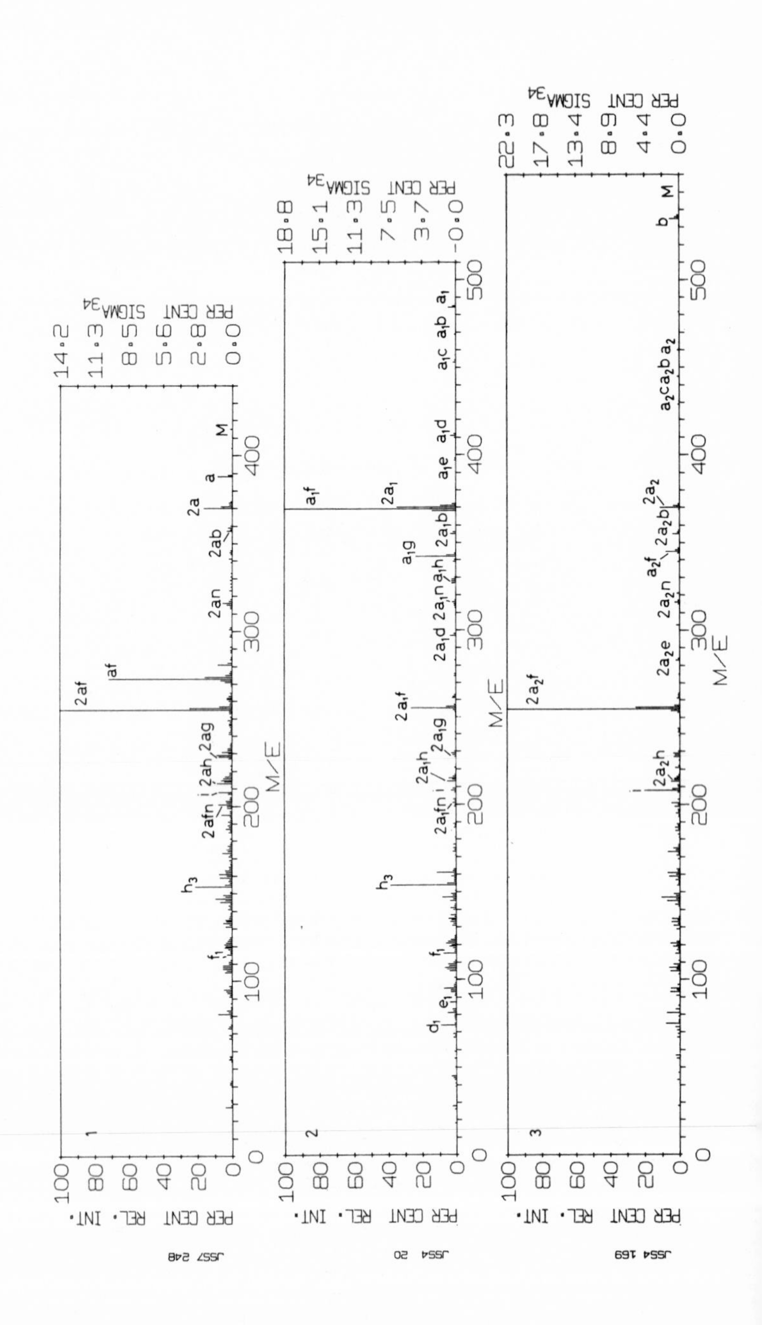
PER CENT REL. INT.
M/E
PER CENT SIGMA34
1
2
3

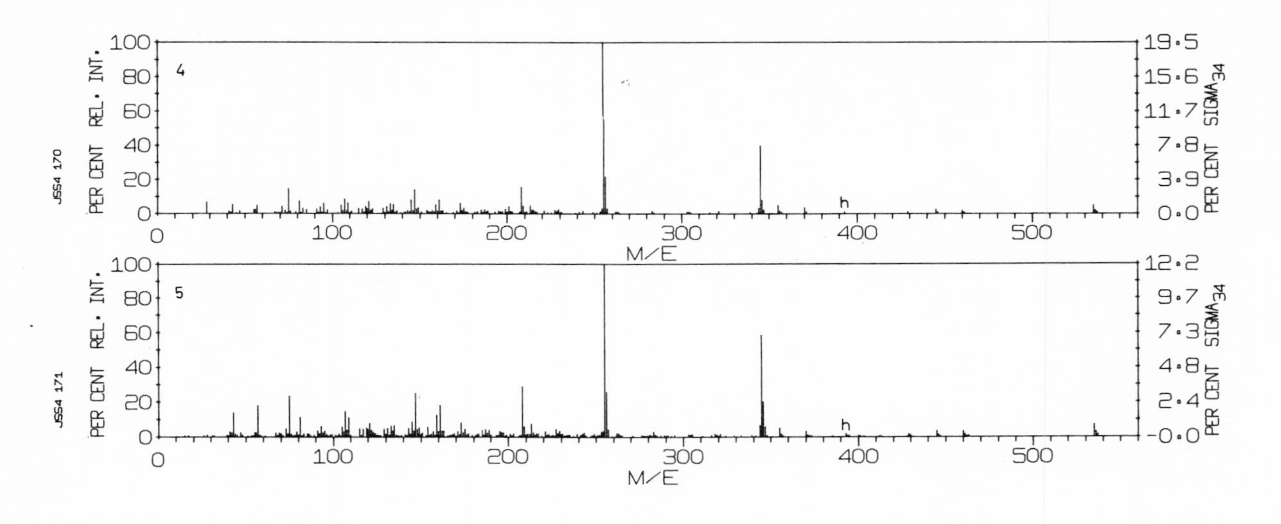

Fig. 5. Mass spectra of methyl 3α,12α-dihydroxy-5β-cholanoate (1) and its trifluoroacetate (2) and trimethylsilyl ether (3) and of the trimethylsilyl ethers of methyl 3α,12α- (4) and 3β,12α-dihydroxy-5α-cholanoates (5).

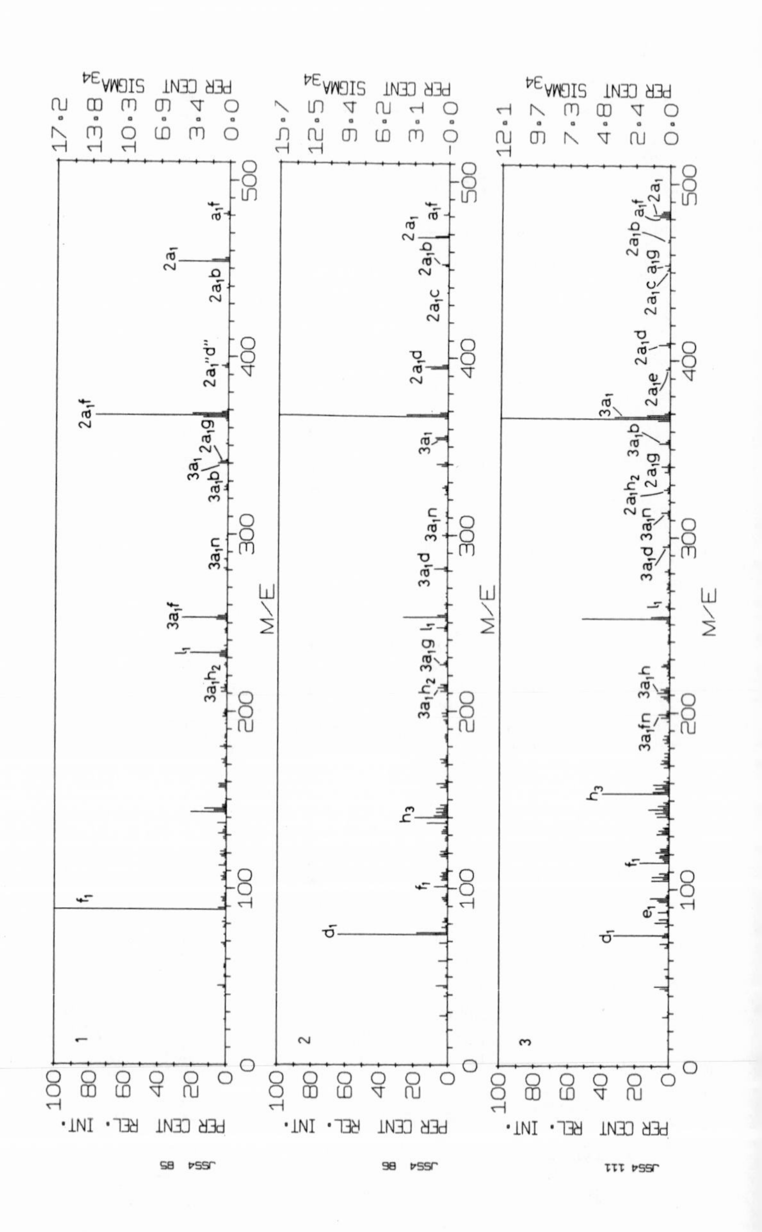
PER CENT REL. INT.
PER CENT SIGMA34
M/E
JSS4 85
JSS4 86
JSS4 111
1
2
3

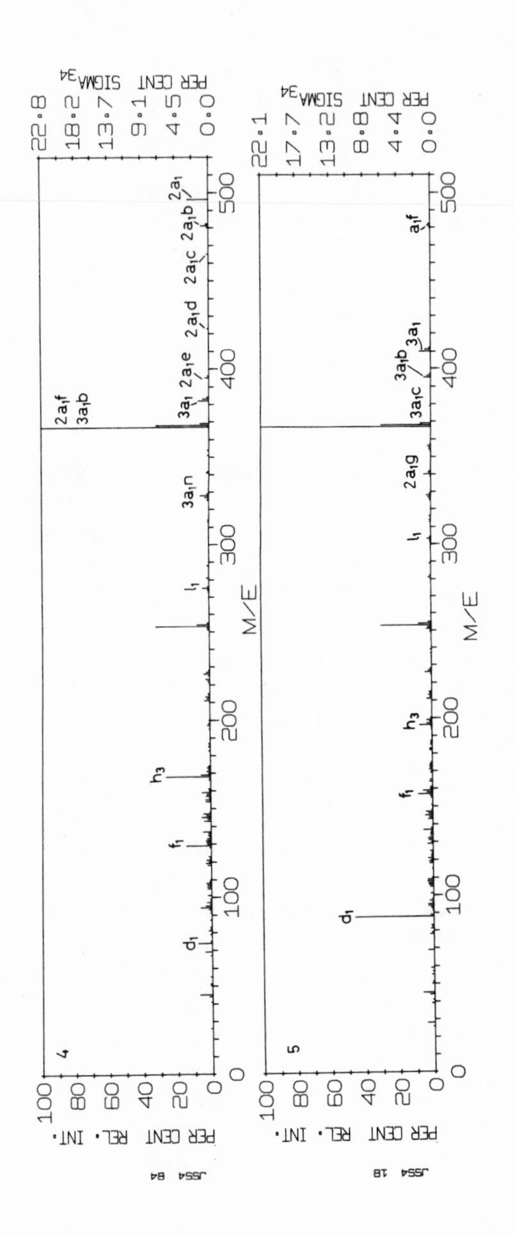

Fig. 6. Mass spectra of the trifluoroacetates of methyl 3α,7α,12α-trihydroxy-5β-bisnor-(1), -5β-nor- (2), -5β- (3), -5β-homo- (4) cholanoates and -5β-cholestanoate (5).

the relative intensities are also different in spectra of epimeric 3,12-dihydroxycholanoate silyl ethers. Loss of the 3α-trimethylsiloxy group in the 5β bile acid occurs more readily than loss of the 3α and 3β substituents in the 5α bile acids. This leads to a much higher ratio between the peaks at *m/e* M-(90+115) and M-2$\times$90 in the spectrum of the 5α compound. This is in agreement with the finding by Egger and Spiteller [using direct probe (26)] that the elimination of water from 3-hydroxysteroids depends more on the A/B configuration (*cis* or *trans*) than on the configuration of the hydroxyl groups (axial or equatorial) (see also 33, 34). However, with a high-temperature gas chromatographic inlet there may also be a slightly larger tendency to loss of axial than equatorial 3-hydroxy groups [see Figs. 2 and 5 and (25)] when comparisons are made within each of the 5α and 5β groups of compounds.

C. Trihydroxy Bile Acids

One of the bile acids first studied by mass spectrometry, α-phocaecholic acid, has hydroxyl groups at C-3, C-7, and C-23 and this type of substitution is seen from the ABC and ABCD ring ions (5). The most commonly occurring trisubstituted bile acids have substituents in the 3,7,12 or 3,6,7 positions. Figure 6 shows spectra of trifluoroacetates of 3,7,12-trihydroxy bile acid methyl esters with different side chains. The length of the side chain and the presence of a 12-trifluoroacetoxy group determine the character of the spectra (intense peak at M-(2$\times$114+side chain); 2*a,f*) and only a small peak at M-3$\times$114 is seen. The last trifluoroacetoxy group lost is that at C-3 (14).

Ions arising from the side chain are clearly seen and the fragmentation corresponds to that of fatty acid esters (46, 47). The base peak in the spectrum of the bisnorcholic acid derivative is the side chain ion at *m/e* 88, and the three-carbon end of the C_{27} acid derivative yields a prominent peak at *m/e* 88. The peak corresponding to *m/e* 154 in the C_{24} bile acids (h_3) is seen in the spectra of the derivatives of the C_{23}–C_{27} acids. The ABCD and ABC ring ions are seen at *m/e* 253 and 211, respectively.

Fragmentation through the D-ring (g, h, h_3) gives rise to ions discussed previously. The peak at *m/e* 226, due to loss of three trifluoroacetic acid molecules, the side chain, and C-16, C-17 ($3a_1$, g) is accompanied by a peak at *m/e* 227. The analogous peaks in dihydroxy derivatives are at *m/e* 228 and 229. These ions most probably contain the same carbon atoms (21) but the source of the additional hydrogen in the heavier ion is unknown. It is possible that one of the trifluoroacetoxy groups is lost without a ring hydrogen.

Fragmentation l_1 gives rise to a diagnostically very important peak. The fragmentation mechanism is not known. The peak is found at *m/e* 261

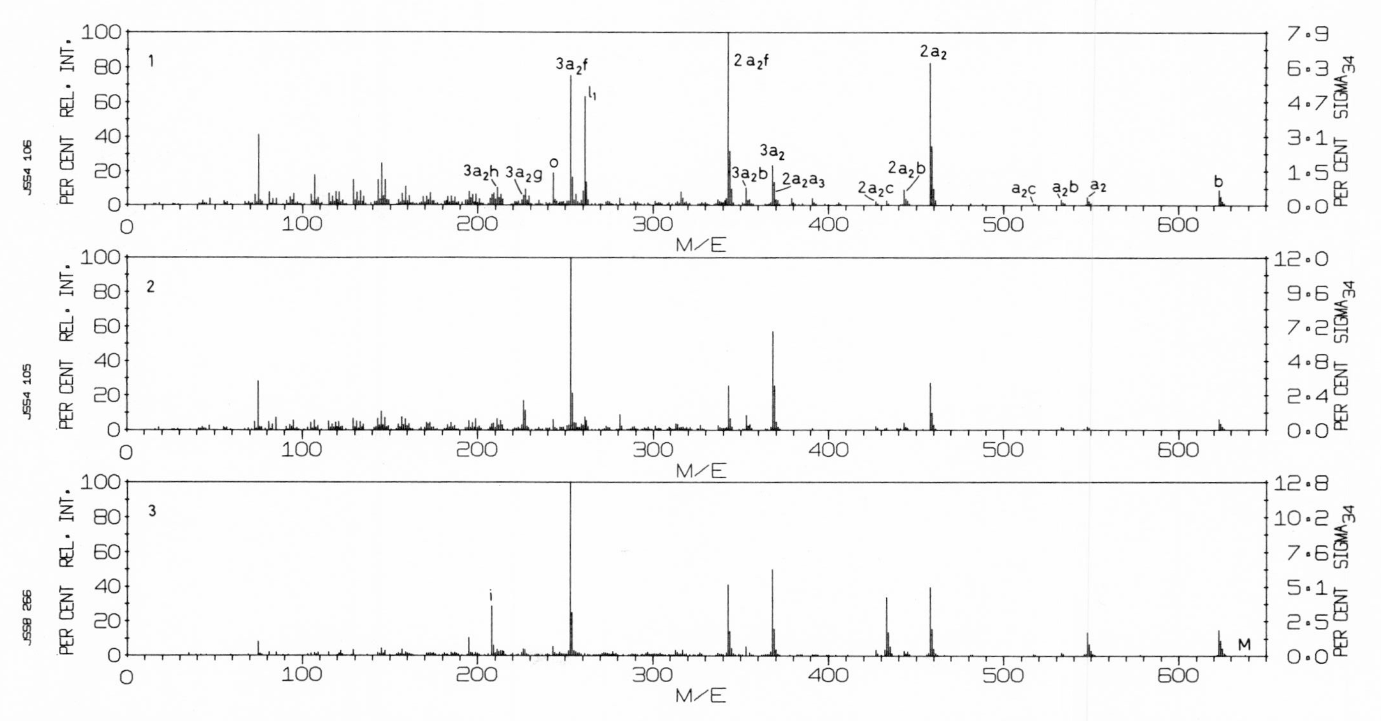

Fig. 7. Mass spectra of the trimethylsilyl ethers of methyl 3α,7α,12α-trihydroxy-5α- (1), 3α,7α,12α-trihydroxy-5β- (2), and 3α,7β,12α-trihydroxy-5β-cholanoates (3).

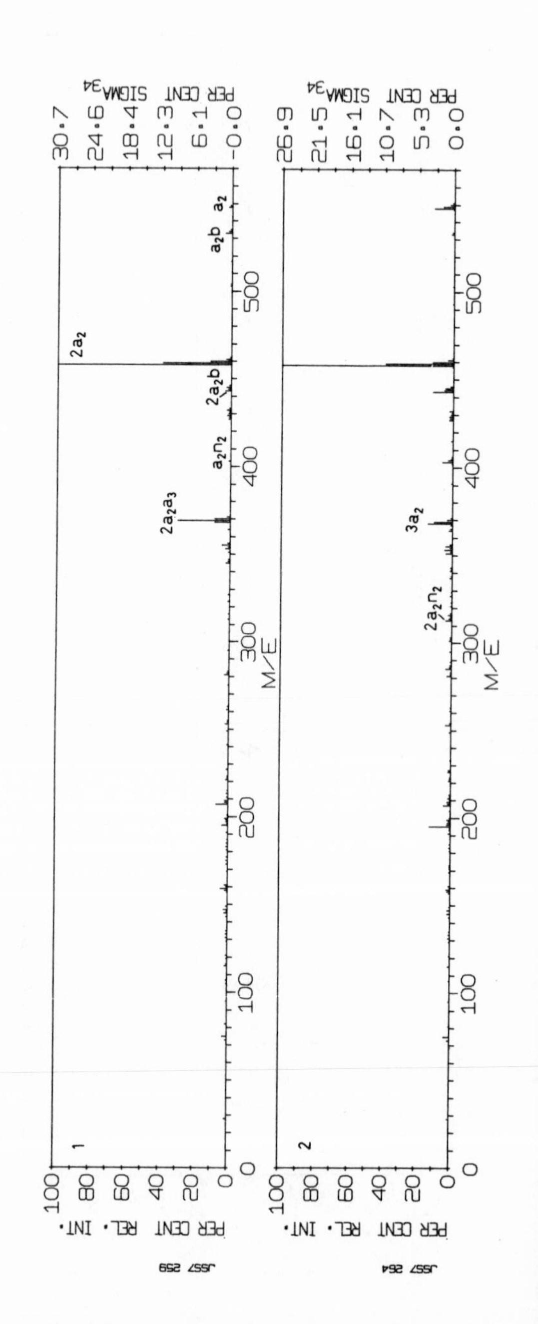
1
2
PER CENT REL. INT.
100 80 60 40 20 0
0 100 200 300 400 500
M/E
PER CENT SIGMA34
30.7 24.6 18.4 12.3 6.1 -0.0
26.9 21.5 16.1 10.7 5.3 0.0
2a2a3
a2n2
2a2b
2a2
a2b
a2
2a2n2
3a2
JS57 259
JS57 264

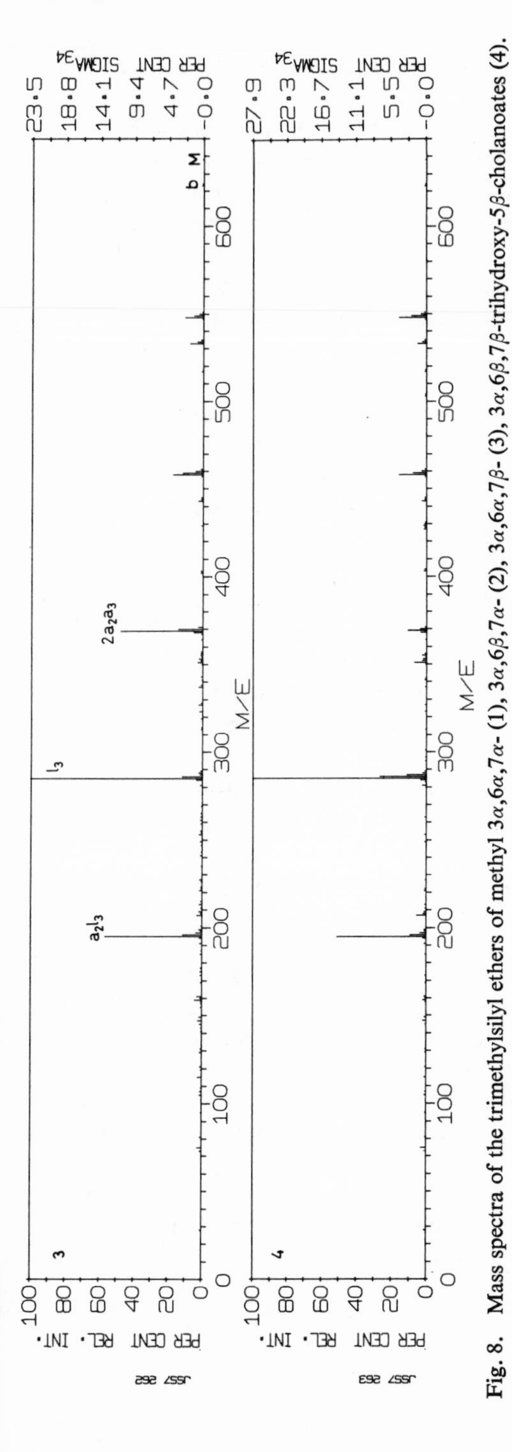

Fig. 8. Mass spectra of the trimethylsilyl ethers of methyl 3α,6α,7α- (1), 3α,6β,7α- (2), 3α,6α,7β- (3), 3α,6β,7β-trihydroxy-5β-cholanoates (4).

in derivatives of the C_{24} acid and this ion contains the CD-rings, side chain, and probably C-7. As shown in Fig. 7 it is prominent in the spectrum of the silyl ether of methyl $3\alpha,7\alpha,12\alpha$-trihydroxy-5α-cholanoate but it has a low intensity in the 5β-epimer. This figure also shows that the axial 7α-trimethylsiloxy group is lost more readily than the equatorial 7β group (the peak at M-(90+115) is much more intense in the spectrum of the 7β-substituted compound) and that the 3α-trimethylsiloxy group is lost less readily in the 5α (*trans*) than in the 5β (*cis*) compounds. In contrast, spectra of trifluoroacetates do not show significant differences between isomers.

The ion of mass 243, indicating a 3,7-bis(trimethylsiloxy) structure, is seen in all spectra in Fig. 7.

Spectra of 3,6,7-trihydroxycholanoic acid derivatives are quite interesting. The spectra of methyl esters are typical of cholanoic acids with three hydroxyl groups in the ABCD-rings. Trifluoroacetates are thermally unstable, give one or several decomposition products during gas chromatography, and give uninformative spectra with ions of low mass. The trimethylsilyl ethers on the other hand give very typical spectra, some of which can hardly be recognized as mass spectra of bile acid derivatives (Fig. 8). The isomers with a 7β-trimethylsiloxy group give a base peak at m/e 285 and this ion loses 90 mass units to form an ion of mass 195 (attested by metastable ion). The base peak is most likely due to fissions between carbon atoms 9 and 10 and 6 and 7, giving an ion consisting of the A-ring and C-6, with the trimethylsiloxy groups at C-3 and C-6. This is supported by the spectrum of the $3\alpha,6\beta,7\beta$-epimer with deuterium at C-3 and C-5 (Cronholm, Makino, and Sjövall, unpublished results). The spectra of the 7α-epimers are quite different, showing a base peak at m/e M-2×90 (m/e 458). The ABCD ring ion is of very low intensity in the four 3,6,7-substituted silyl ethers and the methyl esters should be analyzed if the ions common for trihydroxy bile acids are to be studied.

The spectra of the silyl ethers also show a fragmentation indicating the presence of vicinal trimethylsiloxy groups: M-($2\times90+89$) ($2a_2a_3$). The last silyl ether group in silylated trihydroxysteroids is often lost both as trimethylsilanol and as a trimethylsiloxy group (see Fig. 7). The latter mode of elimination is particularly prominent in steroids with vicinal trimethylsiloxy groups (Fig. 8). The spectra shown in Figs. 7 and 8 also illustrate the general value of using silyl ethers in mass spectrometric analyses of bile acids in order to obtain information about configuration of substituents. A drawback is that silyl ethers of bile acids in biological mixtures often do not separate well on the most thermostable stationary phases.

V. FRAGMENTATION PATTERNS OF KETONIC BILE ACIDS

The presence of a keto group in a bile acid has a profound effect on the mass spectral fragmentation pattern. Early studies of mass spectra of steroid ketones were made by Budzikiewicz and Djerassi (29) and the directing effect of keto groups on fragmentation reactions are discussed in (1) and in later papers by these research groups. In the bile acid series, keto groups are usually found at C-3, C-6, C-7, or C-12. The position of a keto group is usually recognized from the mass spectrum of a suitable derivative, although the simultaneous presence of several other substituents may sometimes change the directing effect of the keto group considerably. For purposes of comparison, mass spectra of compounds with the same number of substituents but different location of the keto group(s) are shown together in the following Figs. 9–12. The discussion of the spectra, however, is based on the position of the keto group, since this is usually the important problem when bile acids from biological materials are being investigated. It should be mentioned that comparisons have been made only between 3-, 6-, 7-, and 12-keto bile acids and it is possible that keto groups in other positions could give rise to spectra similar to those of bile acids studied [spectra of 5α-cholestan-2-one and -3-one are almost identical (29)]. However, simple microchemical reactions such as preparation of different derivatives and partial O-methyloximes, sodium borohydride reduction, chromic acid and periodic acid oxidations, etc., should give sufficient information to permit a tentative mass spectrometric localization of the substituents so that appropriate reference compounds can be made.

Common to spectra of ketonic bile acid methyl esters lacking hydroxyl groups is that the peak at *m/e* *M*-73 (*d*) has a considerably higher relative intensity than in spectra of derivatives of hydroxylated bile acids. The influence of keto groups on fragmentations in the side chain is also seen in the relatively pronounced loss of 32 mass units (c_1) from the molecular ions of 3-keto- and 7-ketocholanoates (Fig. 9). Derivatives of hydroxylated bile acids usually give an ion at *m/e* *M*-31 by loss of the ester methoxyl group.

A. 3-Keto Bile Acids

When a 3-keto group is the only substituent in a cholanoic acid structure the mass spectrum clearly indicates its position. The mass spectrum of the ethyl ester of 3-keto-5β-cholanoic acid is shown in Fig. 9 and it is seen that a major fragmentation is loss of carbons 1–4 with the carbonyl group (n_1, *M*-70). The ion formed is found at *m/e* 332 and 318 in spectra of the ethyl and methyl esters, respectively. Budzikiewicz and Djerassi noted that cleavage of ring A in 5β- and 5α-cholestan-3-one was favored in the 5β-epimer. The same difference is seen in the bile acid series. When 7α-, 12α-, or 7α,12α-hydroxy groups are also present the difference is slight but significant (34, 35, 37, 38). In these cases a peak at *m/e* 316 and 314, respectively (additional loss of hydroxyl functions), is seen only in spectra of the 5β-isomers (see Fig. 10). The peak at *m/e* 314 is found only in the spectrum of the methyl ester, not in that of its silyl ether (cf. Fig. 11). Simultaneous presence of 7- and/or 12-keto groups influences the fragmentation pattern markedly and the 3-keto group does not give rise to a peak at *m/e* *M*-70 in spectra of such bile acid derivatives (cf. Figs. 9 and 12). However, the peak at *m/e* 177 in the spectrum of methyl 3,12-diketo-5β-cholanoate may be due to loss of the side chain and 70 mass units.

One way to study the presence of a 3-keto group is to convert it to an enol trimethylsilyl ether which gives pronounced peaks at *m/e* 142 and 143 (cf. 23). The latter ion is the base peak in the spectrum of an enol silyl ether of methyl 3-keto-12α-hydroxy-5β-cholanoate. An intense peak is also seen at *m/e* 316, i.e., *M*-(90+142). In view of the preferred loss of the side chain following loss of a 12-trimethylsiloxy function the fragment of mass 142 is unlikely to represent the side chain with C-16 and C-17. A prominent peak at *m/e* 201 (*M*-[90+142+115]) indicates that it represents carbon atoms 1–4.

A 3-keto group can also be converted into a 1,1-dimethylhydrazone under mild conditions (48). Spectra of compounds derivatized in this way show large molecular ion peaks (often the base peak), thermal loss of 2 and 4 hydrogens, and peaks at *m/e* 42, 69, and 70.

B. 6-Keto Bile Acids

Spectra of O-methyloximes of methyl 3,6-diketo-5α- and 3,6-diketo-5β-cholanoates have been reported by Allen *et al.* (30). The O-methyloximes, prepared as described by Fales and Luukkainen (49), were found to be particularly useful derivatives since the 5α- and 5β-isomers could be separated by gas chromatography, which was not the case with the free ketones. Considerable differences in relative intensities of the peaks were noted between the spectra of the two isomers. The 5β compound gave a base peak at *m/e* 138 (probably ring A) whereas the molecular ion was base peak in the spectrum of the 5α-epimer. These results show the value of O-methyloximes in mass spectrometry of ketonic bile acids. It should be recognized that partial derivative formation may occur, and that formation of *syn*- and *anti*-isomers may result in a complex mixture of products from a pure compound.

C. 7-Keto Bile Acids

The presence of a 7-keto group in a bile acid derivative often results in the appearance of several diagnostically important peaks. The spectrum of methyl 7-keto-5β-cholanoate (Fig. 9) shows an ion at *m/e* 292 which is typical and is formed by fission of the 9,10 and 5,6 carbon bonds (1). The same peak, accompanied by a peak at *m/e* 293, is seen with different derivatives of 3-hydroxy-7-keto-5β-cholanoates (Fig. 10) and analogous peaks are also seen with some trisubstituted bile acids having a 7-keto group. In the latter cases these peaks (when present) have low intensities (Figs. 11 and 12).

The intensity of the peak at *m/e* 292 depends on the stereochemistry of A/B ring junction. It is very low in the spectrum of methyl 3,7-diketo-5α-cholanoate (36) and is about 30% of the base peak in that of the 5β-isomer. The same difference is seen for the peak at *m/e* 177. Comparisons of spectra of different 7-keto bile acid derivatives indicate that the latter ion may represent loss of side chain from the ion of mass 292.

Bile acid derivatives with a 7-keto group also give ions due to fragmentation through the C-ring. The ions of mass 178 (*M*-210) and 192 (*M*-210) in the spectra of methyl 7-keto- and 3,7-diketo-5β-cholanoates, respectively, are probably formed in this way (Fig. 9). Cleavage of the 8,14 and 11,12 bonds has been found to give analogous peaks in other 7-ketosteroids (29). The relative intensity of the peak at *m/e* 192 is 90% in the spectrum of the 5α epimer of 3,7-ketocholanoate (36) but only 30% in that of the 5β-epimer (Fig. 9). Other ions which may arise through cleavage of the C-ring include those at *m/e* 191 and 205 (fission of the 8,14 and 12,13 bonds) in the 7-keto and 3,7-diketo compound, respectively [Fig. 9 (29)]. Analogous peaks in spectra of the trifluoroacetates of 3-hydroxy-7-ketocholanoates (at *m/e* 303, trifluoroacetoxy group retained) have relative intensities of 10–20%. The ion of mass 165 in 7-ketocholanoate (Fig. 9) may also represent the A,B-rings with the carbonyl group (29).

Dean and Aplin suggested that the ion of mass 150 in 3,7-diketo-5β-cholanoic acid contained the A-ring, C-6, C-7, and the two oxygen atoms (14). It should be noted, however, that an ion of mass 150 is prominent in the spectrum of methyl 7-ketocholanoate (Fig. 9) and 5α-cholestan-7-one (29).

Loss of water from ketonic bile acid derivatives is often seen and this loss is more pronounced with 7-keto than with other ketocholanoates. This is true also with a low ion source temperature (14).

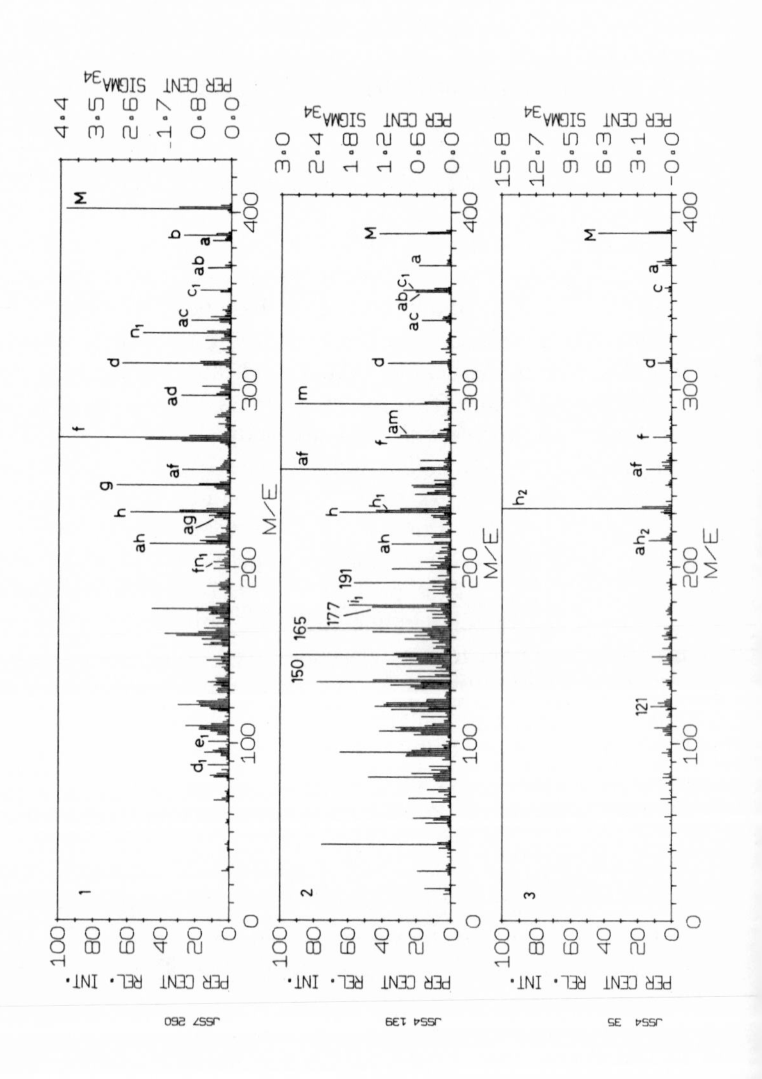
1
2
3
PER CENT REL. INT.
PER CENT SIGMA34
M/E
M
f
g
h
ah
d
150
165
177
191
121
JSS7 260
JSS4 139
JSS4 35

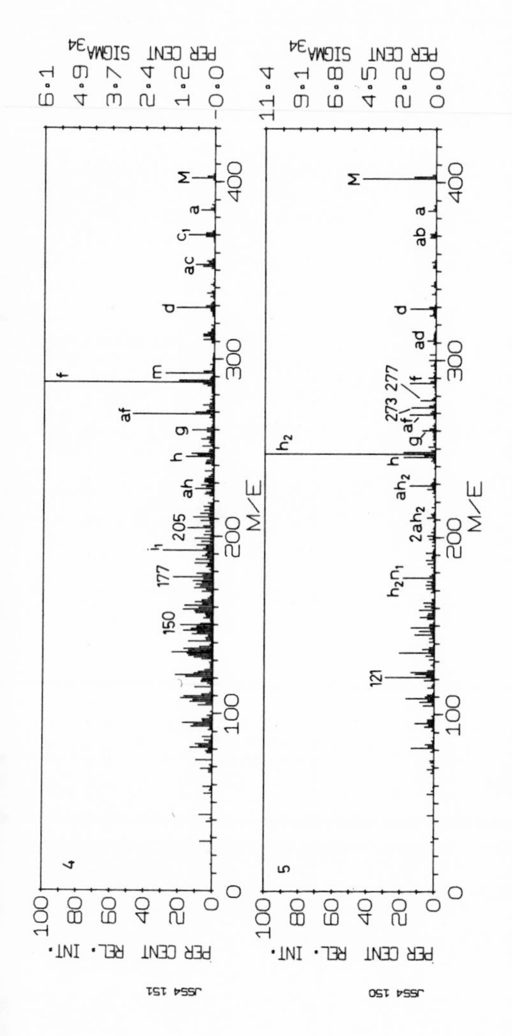

Fig. 9. Mass spectra of ethyl 3-keto-5β-cholanoate (1), and methyl 7-keto- (2), 12-keto- (3), 3,7-diketo- (4), and 3,12-diketo-5β-cholanoates (5).

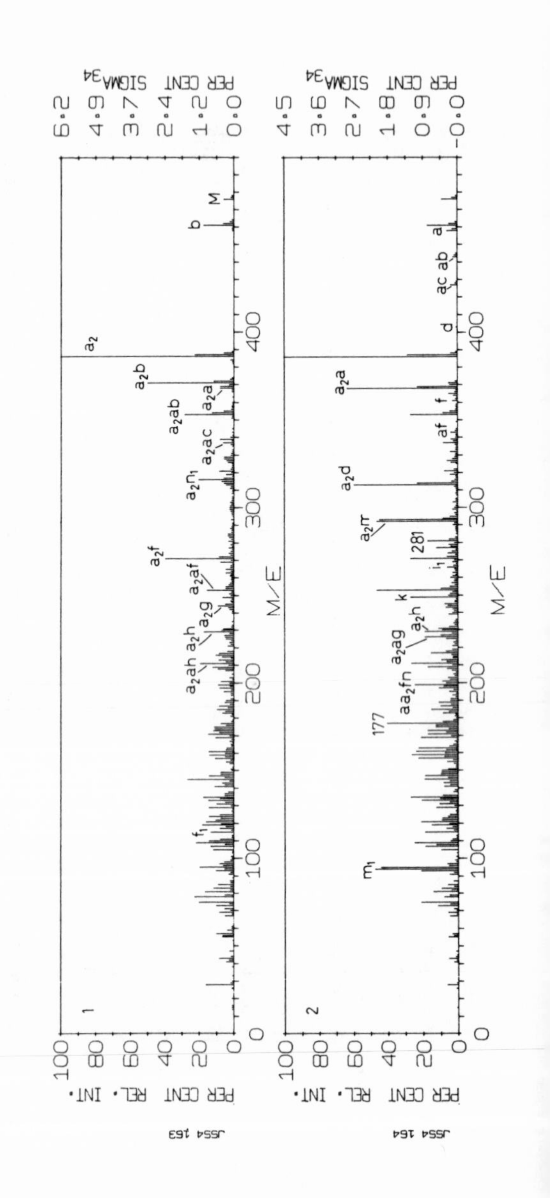
PER CENT REL. INT.
PER CENT SIGMA34
M/E
JSS4 163
JSS4 164

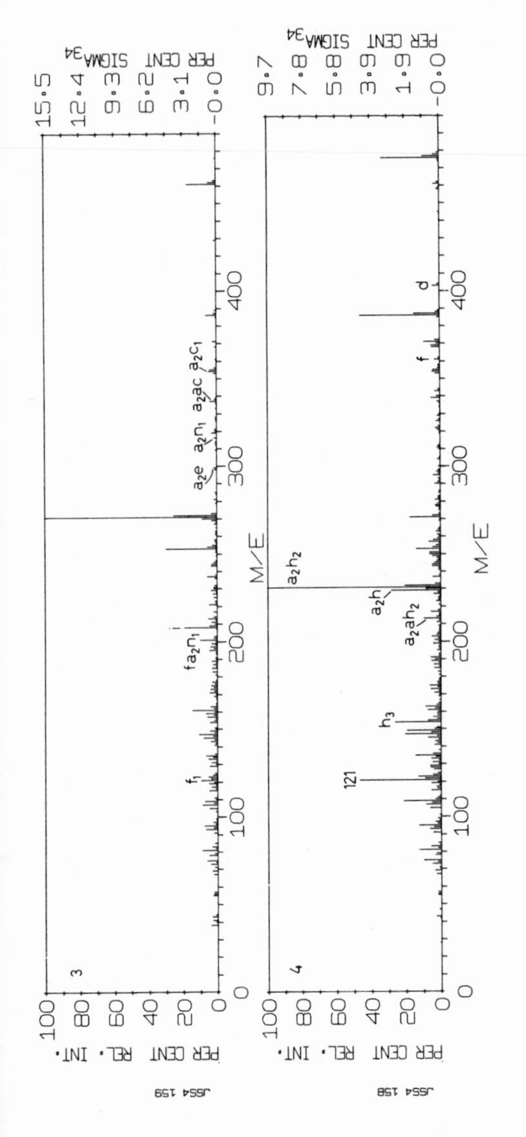

Fig. 10. Mass spectra of silyl ethers of methyl 3-keto-7α-hydroxy- (1), 3α-hydroxy-7-keto- (2), 3-keto-12α-hydroxy- (3), and 3α-hydroxy-12-keto-5β-cholanoates (4).

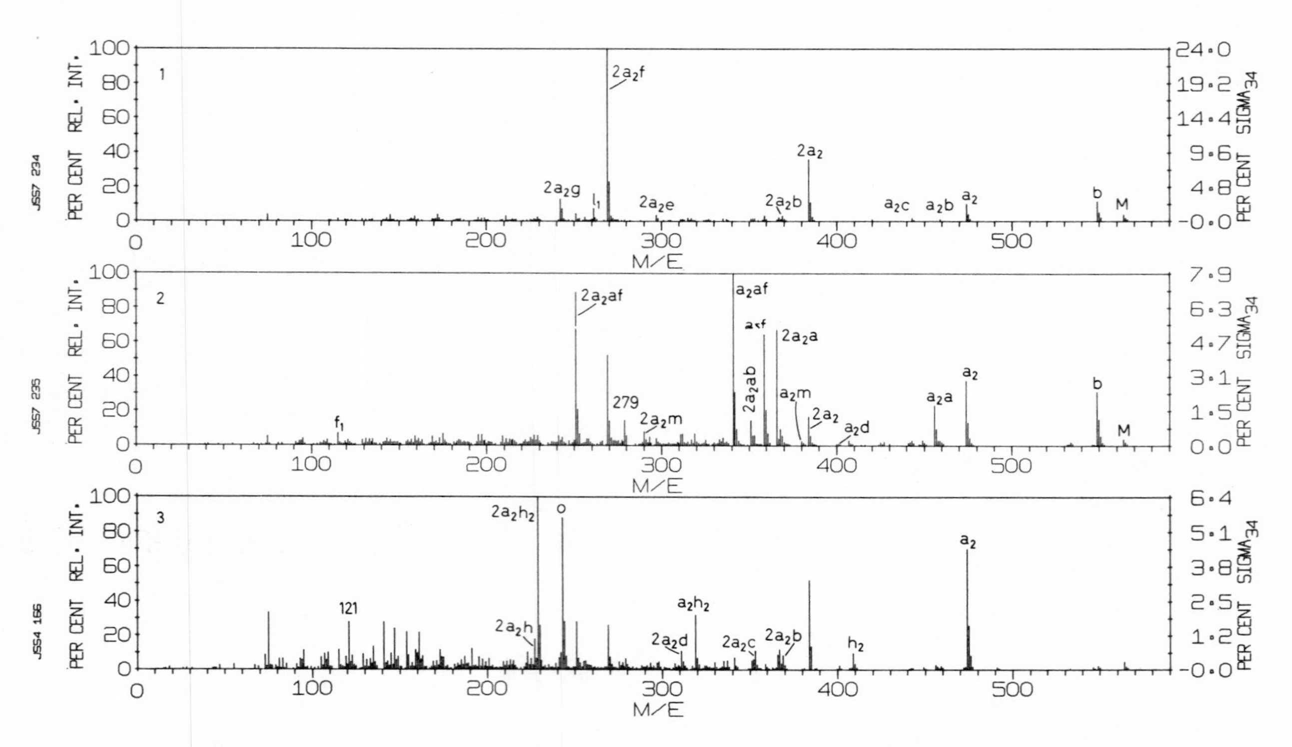

Fig. 11. Mass spectra of silyl ethers of methyl 7α,12α-dihydroxy-3-keto- (1), 3α,12α-dihydroxy-7-keto- (2), and 3α,7α-dihydroxy-12-keto-5β-cholanoates (3).

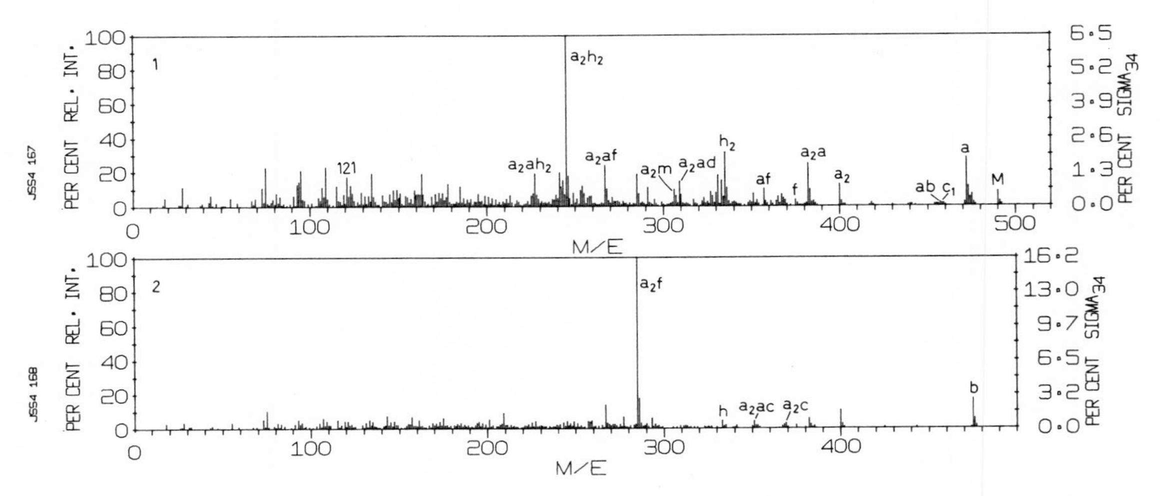

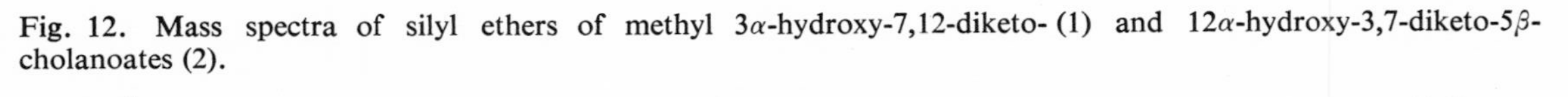
Fig. 12. Mass spectra of silyl ethers of methyl 3α-hydroxy-7,12-diketo- (1) and 12α-hydroxy-3,7-diketo-5β-cholanoates (2).

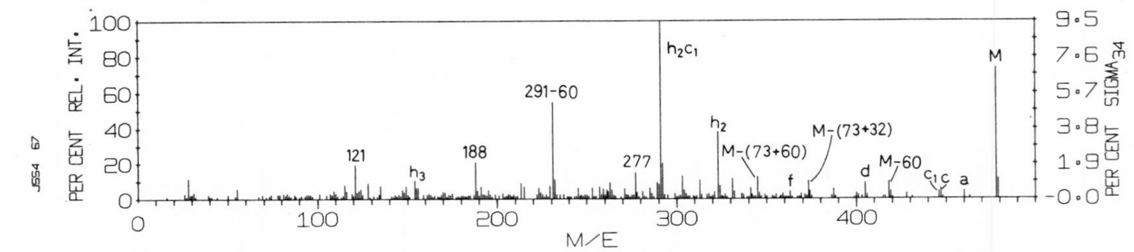

Fig. 13. Mass spectrum of the methyl ester of 12-keto-3,4-seco-5β-cholane-3,4,24,-trioic acid.

D. 12-Keto Bile Acids

Bile acids with a 12-keto group give most characteristic mass spectra. The side chain and D-ring are lost with capture of one hydrogen by the oxygen-containing fragment (1,29). Thus, a peak at *m/e* *M*-155 (or *M*-[18 or its equivalent)+155]) appears in spectra of all methyl cholanoates having a 12-keto group, irrespective of the presence of other substituents in the molecule (Figs. 9–12). This is also the case in the spectrum of 12-keto-3,4-seco-5β-cholane-3,4,24-trioic acid methyl ester (Fig. 13).

The spectra of 12-ketocholanoates also show a peak at *m/e* 121 in almost all cases studied. The nature of this ion is not known; it may be asumed to contain the C-ring with the carbonyl group and two additional carbon atoms. The presence of this ion is also support for a 12-keto group, at least when naturally occurring bile acids are studied.

A few ions in the spectra of 12-ketocholanoates shown in Figs. 9–12 deserve mentioning. The ions of mass 277 and 273 in the spectrum of methyl 3,12-diketo-5β-cholanoate are found at *m/e* 263 and 273, respectively, in the spectrum of the free acid (14). The latter ion evidently does not contain the side chain and the peak at *m/e* 277 may be due to fragmentation through the B-ring. It is of interest that the ion of mass 243, typical of a 3,7-bis(trimethylsiloxy) structure, is particularly intense in the spectrum of the silyl ether of methyl 3α,7α-dihydroxy-12-keto-5β-cholanoate.

CONCLUSION

The combination of gas chromatography and mass spectrometry is already the most powerful single method for identification of bile acids. Its importance will increase as more mass spectral data on different derivatives are collected and more studies on the mechanisms of fragmentations are made. Mass spectrometry makes possible the study of bile acid metabolism using compounds labeled with stable isotopes; in many cases it is possible to determine both abundance and position(s) of the isotope with less than 1 μg of material. The mass spectrometer can also serve as a detector for specific compounds appearing in the effluent from a gas chromatography column by being focused on specific ions formed in the fragmentation of these compounds (50). With advanced computer handling of data, the results of mass spectrometric analyses are obtained rapidly (51, 52) and manual handling will no longer limit the amount of information that can be collected.

REFERENCES

1. H. Budzikiewicz, C. Djerassi, and D.H. Williams, "Structure elucidation of natural products by mass spectrometry. Vol. II. Steroids, terpenoids, sugars and miscellaneous compounds," Holden-Day, San Francisco (1964).
2. H. Budzikiewicz, C. Djerassi, and D.H. Williams, "Mass spectrometry of organic compounds," p. 417, Holden-Day, San Francisco (1967).
3. S. Bergström, R. Ryhage, and E. Stenhagen, *Acta Chem. Scand.* **12,** 1349 (1958).
4. R. Ryhage and E. Stenhagen, *J. Lipid Res.* **1,** 361 (1960).
5. S. Bergström, R. Ryhage, and E. Stenhagen, *Svensk Kem. Tidskr.* **73,** 566 (1961).
6. S. Bergström, L. Krabisch, and U.G. Lindeberg, *Acta Soc. Med. Uppsal.* **64,** 160 (1959).
7. A.D. Cross, *Biochem. J.* **90,** 314 (1964).
8. R. Ryhage, *Anal. Chem.* **36,** 759 (1964).
9. E. Stenhagen, *Z. Anal. Chem.* **205,** 109 (1964).
10. J.T. Watson and K. Biemann, *Anal. Chem.* **37,** 844 (1965).
11. P. Eneroth, K. Hellström, and R. Ryhage, *J. Lipid Res.* **5,** 245 (1964).
12. P. Eneroth, B. Gordon, R. Ryhage, and J. Sjovall, *J. Lipid Res.* **7,** 511 (1966).
13. P. Eneroth, B. Gordon, and J. Sjövall, *J. Lipid Res.* **7,** 527 (1966).
14. P.D.G. Dean and R.T. Aplin, *Steroids,* **8,** 565 (1966).
15. H. Egger, *Monatsh. Chem.* **99,** 1163 (1968).
16. R. Ryhage, *Arkiv Kemi* **26,** 305 (1967).
17. E.M. Chambaz, G. Maume, B. Maume, and E.C. Horning, *Anal. Letters* **1,** 749 (1968).
18. J. Sjövall, *in* "The Gas Liquid Chromatography of Steroids" (J. K. Grant, ed.), Mem. Soc. Endocrinol. No. 16, p. 243, Cambridge University Press, London (1967).
19. J. Sjövall, *in* "Bile Salt Metabolism" (L. Schiff, J. Carey, Jr., and J. Dietschy, ed.), p. 205, Charles C Thomas, Springfield, Illinois (1969).
20. P. Eneroth, K. Hellström, and R. Ryhage, *Steroids* **6,** 707 (1965).
21. B.A. Knights, *J. Gas Chromatog.* **5,** 273 (1967).
22. B.A. Knights, *Phytochemistry* **6,** 407 (1967).
23. C.J.W. Brooks, E.C. Horning, and J.S. Young, *Lipids* **3,** 389 (1968).
24. E.C. Horning, C.J.W. Brooks, and W.J.A. VandenHeuvel, *in* "Advances in Lipid Research" (R. Paoletti and D. Kritchevsky eds.), Vol. 6, p. 273, Academic Press, New York (1968).
25. M. Spiteller-Friedman and G. Spiteller, *Fortschr. Chem. Forsch.* **12,** 440 (1969).
26. H. Egger and G. Spiteller, *Monatsh. Chem.* **97,** 579 (1966).
27. W.H. Elliott, personal communication (1969).
28. J. Sjövall and K. Sjövall, *Steroids* **12,** 359 (1968).
29. H. Budzikiewicz and C. Djerassi, *J. Am. Chem. Soc.* **84,** 1430 (1962).
30. J.G. Allen, G.H. Thomas, C.J.W. Brooks, and B.A. Knights, *Steroids* **13,** 133 (1969).
31. A. Kallner, *Acta Chem. Scand.* **22,** 2353 (1968).
32. A. Kallner, *Acta Chem. Scand.* **22,** 2361 (1968).
33. S.A. Ziller, Jr., E.A. Doisy, Jr., and W.H. Elliott, *J. Biol. Chem.* **243,** 5280 (1968).
34. A. Kallner, *Acta Chem. Scand.* **21,** 322 (1967).
35. S.A. Ziller, Jr., M.N. Mitra, and W.H. Elliott, *Chem. Ind.* **1967,** 999.
36. M.N. Mitra and W.H. Elliott, *J. Org. Chem.* **33,** 2814 (1968).
37. H. Danielsson, A. Kallner, and J. Sjövall, *J. Biol. Chem.* **238,** 3846 (1963).
38. M.N. Mitra and W.H. Elliott, *J. Org. Chem.* **33,** 175 (1968).
39. L. Tökés, G. Jones, and C. Djerassi, *J. Am. Chem. Soc.* **90,** 5465 (1968).
40. J.A. McCloskey, R.N. Stillwell, and A.M. Lawson, *Anal. Chem.* **40,** 233 (1968).
41. A.G. Sharkey, Jr., R.A. Friedel, and S.H. Langer, *Anal. Chem.* **29,** 770 (1957).

42. H. Eriksson, J.-Å. Gustafsson, and J. Sjövall, *European J. Biochem.* **6,** 219 (1968).
43. S.S. Friedland, G.H. Lane, R.T. Longman, K.E. Train, and M.J. O'Neal, *Anal. Chem.* **31,** 169 (1959).
44. C. Djerassi, J. Karliner, and R.T. Aplin, *Steroids* **6,** 1 (1965).
45. J. Diekman and C. Djerassi, *J. Org. Chem.* **32,** 1005 (1967).
46. R. Ryhage and E. Stenhagen, *Arkiv Kemi* **13,** 523 (1959).
47. R. Ryhage and E. Stenhagen, *Arkiv Kemi* **15,** 291 (1960).
48. W.J.A. VandenHeuvel and E.C. Horning, *Biochim. Biophys. Acta* **74,** 560 (1963).
49. H.M. Fales and T. Luukkainen, *Anal. Chem.* **37,** 955 (1965).
50. C.C. Sweeley, W.H. Elliott, I. Fries, and R. Ryhage, *Anal. Chem.* **38,** 1549 (1966).
51. R.A. Hites and K. Biemann, *Anal. Chem.* **40,** 1217 (1968).
52. B. Hedfjäll, P.-Å. Jansson, Y. Mårde, R. Ryhage, and S. Wikström, *J. Scient. Instr.* **2,** 1031 (1969).

Chapter 8

THE PHYSICAL CHEMISTRY OF CHOLANIC ACIDS

Donald M. Small*

Section of Biophysics
Department of Medicine
Boston University School of Medicine

I. INTRODUCTION

The conjugated bile acids are biologically the most important detergent-like molecules (1). They solubilize the insoluble components of bile, such as lecithin and cholesterol (2–9), aid in digestion by removing the products of pancreatic hydrolysis (monoglycerides and fatty acids) (10–12), and probably control certain intracellular events such as the synthesis of cholesterol in the intestinal mucosa (13). In certain cases, secondary bile salts may be toxic to cells and may cause cholestasis and hepatic necrosis (14–26). It is the belief of this author that the physicochemical characteristics of bile salts and their molecular interactions with other lipids, proteins, and complex carbohydrate molecules probably account for not only the physiological properties of bile salts but also the pathological conditions attributed to them.

The purpose of this chapter is to cover the physicochemical properties of cholanic acids and their salts. Some 150 cholanic acids are mentioned in Elsevier's Encyclopedia of Organic Chemistry (27) and about 440 monocarboxylic, unsubstituted hydroxyl, and carbonyl-substituted bile acids are cataloged by Sobotka (28). Obviously, it would be impossible to deal specifically with each one. Many of these acids are by-products of chemical reactions and are not naturally occurring compounds. Further, meaningful physicochemical data are not available for the great majority of bile acids.

*Markle Scholar in Academic Medicine.

Therefore, I shall concentrate on only those bile acids that have been reasonably well studied from a physicochemical point of view and which have some relation to physiology and biochemistry of living things. Because the specific physical characteristics of the bile acids and their alkaline metal salts vary considerably with the number of hydroxyl groups present on the steroid nucleus, I will present a fairly detailed description of the physicochemical properties of cholanic acid (no hydroxyl groups), monohydroxy, dihydroxy, and trihydroxy bile acids. Since the triketo bile acid (dehydrocholic acid) has been used widely as a choleretic, its properties will also be discussed. Unfortunately, many interesting bile acids and bile alcohols isolated from a variety of vertebrates (29–32) have not been studied physicochemically. However, knowing their molecular structure, many of the properties of these compounds can be deduced by comparison with the known properties of bile acids discussed in this chapter.

Because some of the terms used in this chapter will be unfamiliar to readers in the biological sciences, and because some of these terms have ambiguous meanings, even in common use, a glossary of terms and definitions is given in an appendix at the end of the chapter. It should be stressed here that these definitions are those of the author and may at times not be exactly the same as those in general usage.

The topic of bile acids has been the subject of a number of general reviews and books (28, 29, 30, 33, 34). Certain reviews have dealt, in part, with physicochemical aspects of the bile salts (35–39) or have mentioned physicochemical properties only briefly (28, 30, 34, 40). A review dealing specifically with the physicochemical properties of bile salts and their relation to physiologic function was published in 1967 (1). This chapter will be limited to a discussion of the physicochemical properties of bile acids and their salts. (For physiologic correlations, see 1, 8, 9, 10, 41, and 197.)

Bile salts, the sodium and potassium salts of di- and trihydroxy bile acids, are amphiphilic detergent-like molecules whose properties differ considerably from ordinary aliphatic detergent molecules (42–47). These differences have been amplified by the comparison of phase diagrams and the physicochemical properties of ordinary soaps and bile salts (42). Briefly, bile salts alone in water form no liquid crystal phases, form very small micelles, have Krafft points below 0 °C, and have excellent solubilizing properties for various kinds of swelling amphiphilic compounds such as lecithin and monoglycerides. Typical long-chain aliphatic detergents form liquid crystalline phases, form fairly large micelles, often have high Krafft points, and have poor solubilizing capacities for swelling amphiphilic compounds.

In this review, I will cover molecular structure, crystalline structure, surface chemistry, solubility in organic solvents and in aqueous solutions, micelle formation by bile salts, and the bulk interactions with other lipids.

II. MOLECULAR STRUCTURE

A. Chemical Definition

Bile acids are C_{22} to C_{28} carboxylic acids with a cyclopentenophenanthrene nucleus containing a branched side chain of three to nine carbon atoms ending in the carboxyl group. Most naturally occurring bile acids are C_{24} saturated acids. The AB ring juncture is usually *cis* (5β hydrogen) in the bile acids of vertebrates, although AB *trans* (5α hydrogen) bile acids occur in lower vertebrates (29). Different representations of six common C_{24} 5β hydrogen bile acids are shown in Figs. 1–6.

B. Formulae and Molecular Models

These figures contrast the conventional chemical configuration, the perspective formula (similar in appearance to the Dreiding model), the Stuart–Breigleb space-filling models, and the author's diagrammatic representation of the molecule. While the Dreiding models are important in the chemistry of bile acids because they give accurate bond angles and center-to-center atomic distances, the Stuart–Breigleb space-filling model is more useful to the physical chemist who wishes to study molecular interaction between the molecules. Dreiding models show the molecular skeleton whereas space-filling models show the average van der Waals' radii of the atoms or, so to speak, the molecules with all their muscle and flesh. The pictures of the bile acids (bottom Figs. 1–6) represent crude space-filling pictures as longitudinal or cross sections of the molecule. These molecular representations will be used henceforth in the text.

In general, the free bile acid molecules are about 20–21 Å long. Contrary to popular belief, the cross section shows that the steroid nucleus is *not* flat, but nearly circular in cross section. For instance, cholanic acid is about 6×7 Å at narrowest and widest diameters. This gives a cross-sectional area of 42 Å^2 (figured as a rectangle) or 38.6 Å (figured as a circle with radius of 3.5 Å). A rough guess at the molecular volume may be made by assuming the bile salt occupies a cylinder of radius about 3.5 Å and height of 20 Å. This gives a molecular volume of about 770 Å^3. This is, of course, too large, because the narrow end of the molecule has a smaller radius, but as we shall see in Section III, Table III, the values are not much larger than those obtained experimentally.

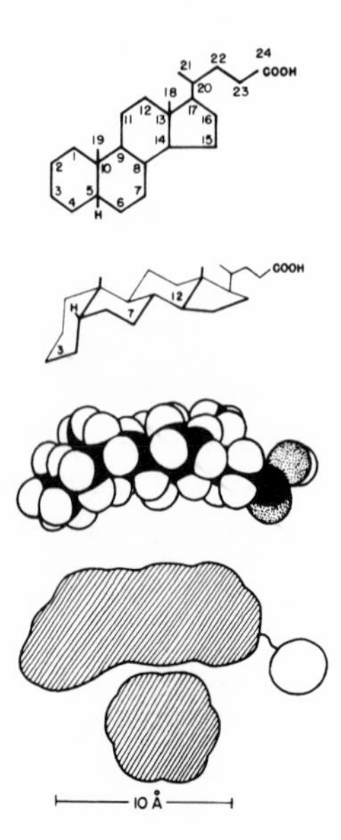

Fig. 1. Different representations of cholanic acid. Top—chemical formula; second from top—perspective formula; third from top—Stuart-Breigleb space-filling model; bottom—diagrammatic representation of the molecule in longitudinal and cross-section. Scale 10 Å.

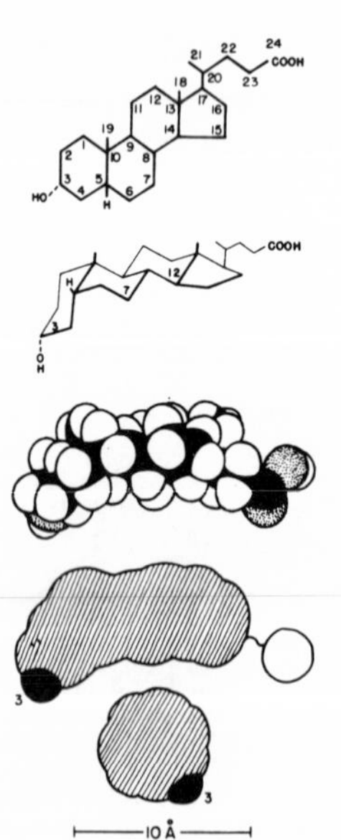

Fig. 2. Different representations of lithocholic acid—see legend for Fig. 1 for further explanation. The 3 refers to the hydroxyl group on the 3α position of lithocholic acid.

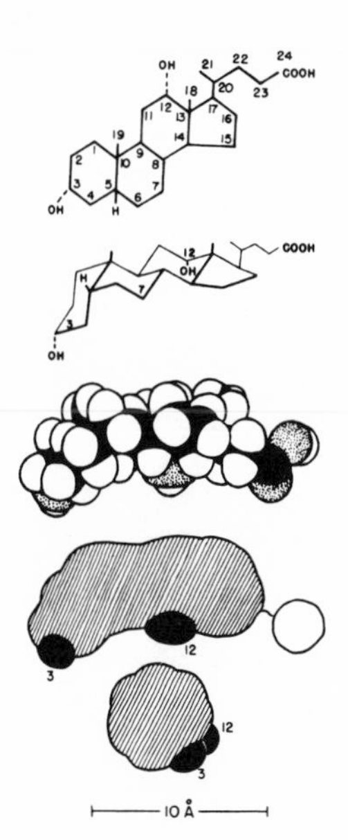

Fig. 3. Deoxycholic acid. Refer to legend of Fig. 1 for further explanation. The 3 and 12 refer to alpha hydroxyl groups at positions 3 and 12.

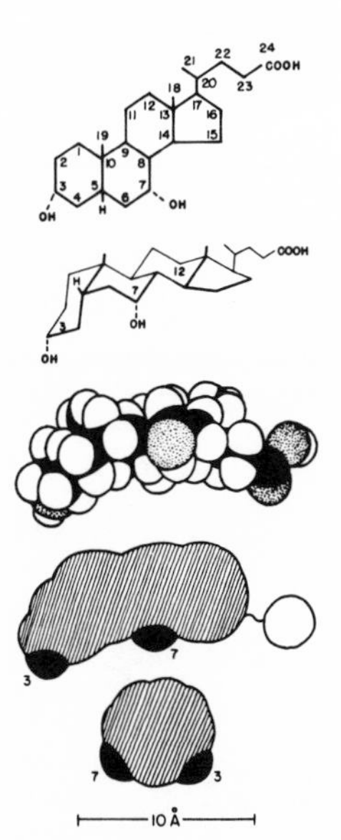

Fig. 4. Chenodeoxycholic acid. Refer to legend of Fig. 1 for further explanation. The 3 and 7 refer to alpha hydroxyl groups at positions 3 and 7.

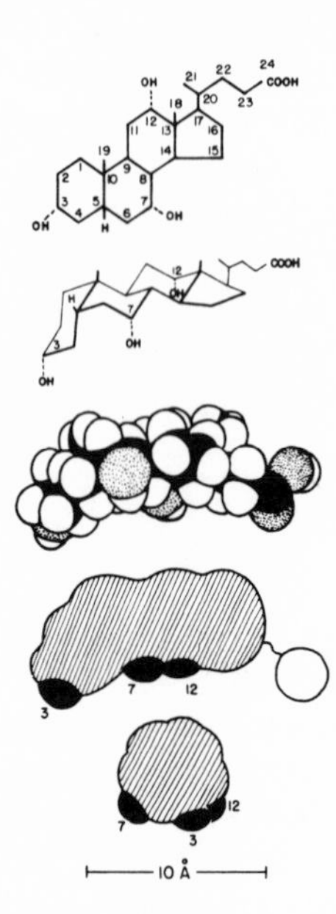

Fig. 5. Cholic acid. Refer to legend of Fig. 1 for further explanation. The 3, 7, and 12 refer to alpha hydroxyl groups at those positions.

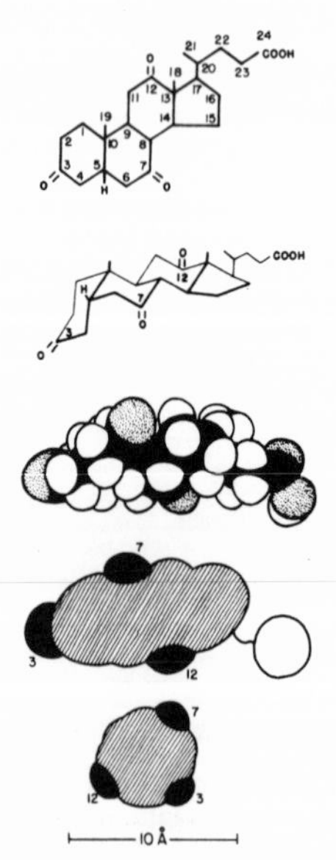

Fig. 6. Dehydrocholic acid. Refer to legend of Fig. 1 for further explanation. The 3, 7, and 12 refer to ketone groups at those positions.

Cholanic acid, the parent compound (Fig. 1), has no hydroxyl groups. It is not found in animals, but serves as a reference compound to which the physicochemical characteristics of the mono- and polyhydroxylated bile acids may be compared. It is virtually totally insoluble in water, either as the acid or as its sodium salt. Lithocholic acid (Fig. 2) has an equatorial hydroxyl in the alpha position at carbon number 3. Deoxycholic acid (Fig. 3) has hydroxyl groups at both the 3α position and at the 12α (axial) position. Chenodeoxycholic acid (Fig. 4) has the 3α hydroxyl and an axial 7α hydroxyl while cholic acid (Fig. 5) has alpha hydroxyl groups at the 3, 7, and 12 positions. These hydroxyl groups account for the extraordinarily different physicochemical properties of mono-, di-, and trihydroxy bile acids and their alkaline metal salts. The 3α hydroxyl group and the carboxyl group are at the opposite ends of the molecule. However, even though the 3α hydroxyl is an equatorial group, it tends to lie on one side of the molecule, that is, the side opposite the angular methyl groups at C_{18} and C_{19}. Thus, lithocholic acid has a potentially strong hydrophilic group at the carboxyl end and a moderately weak hydrophilic group at the other end separated by a large mass of water-insoluble lipophilic cyclic hydrocarbon. Chenodeoxycholic acid and deoxycholic acid have two hydroxyl groups. These hydroxyl groups are on the same side of the molecule. Cholic acid has three hydroxyls on the same side of the molecule. The distance between these three hydroxyl groups measured by either Dreiding or Stuart–Breigleb models is about 5 Å. They form a triangle on the hydrophilic side of the cholic acid steroid nucleus. Therefore, the di- and trihydroxy acids (the physiologically active bile acids) are molecules that have one water-soluble and one lipophilic water-insoluble side of the steroid nucleus ending in a small tail, which, when ionized, is highly water-soluble. Dehydrocholic acid, a triketo bile acid, has its three carbonyl groups spaced fairly evenly around the radial cross section of the bile acid (Fig. 6). This largely eliminates the hydrophobic side of the molecule. This structure bares directly on the fact that the sodium salt of the triketo acid is highly soluble as individual ions and does not form micelles readily (see Section VIII).

C. Infrared Spectra (IRS) of Bile Acids*

Infrared radiation, i.e., wavelengths from 1 to 100 μ (corresponding to the frequency range from about 10,000 to 100 cm^{-1}), is absorbed and converted by organic molecules into energy of molecular vibration. Although this absorption is quantitized, the vibrational spectra appear as bands rather

---

*The author is indebted to Mrs. Erika Smakula who obtained the IR spectra and who helped prepare this subsection.

than lines because a single vibration energy change is accompanied by a number of rotational energy changes. Conventional IR spectroscopy is carried out between the frequencies of 4000 cm^{-1} and 650 cm^{-1} (2.5 μ-15.4 μ). The band position can be presented either as wavelengths (μ) or wave numbers (cm^{-1}). I will use the wave number (frequency) in this section, although the spectra given in Figs. 7 and 8 will also show the wavelength. In general, there are two types of molecular vibration distinguished by IRS, stretching and bending. The stretching vibration occurs along the bond axis and results in rhythmic change in intra-atomic distance. When two or more like atoms are connected to another atom, such as occurs with $-CH_2-$ or $-CH_3$ groups, stretching may be symmetrical (all connected atoms vibrating in phase) or asymmetrical (the connected atoms vibrating out of phase). Bending vibrations result from changes of bond angle and may be of several types, i.e., in-plane bending (scissoring, rocking) and out-of-plane bending (twisting, wagging).

A nonlinear molecule, containing n atoms, possesses $3n-6$ (225 for cholanic acid) fundamental modes of stretching and bending vibrations, most of them giving rise to absorption bands in the wavelength region 2–25 μ or the wave number region 5000–400 cm^{-1}. The number of bands apparent is considerably less ($<$50) due to band overlap, poor intensity, or lack of required change in dipole character of the molecule.

1. *Intensities*

Incident radiation, which coincides in energy with a particular molecular vibrational level, is absorbed only if during the radiation a change in dipole moment occurs. Bonds of weak electric moment (e.g., C—H and C—C in saturated hydrocarbons) give rise to absorption bands of low intensity while polar bonds (C—O, C—OH) exhibit strong absorption bands, but also have an enhancing effect on dipole moments and thus absorption intensities of neighboring groups.

2. *Frequencies of Absorption*

The band positions associated with the various vibrations are primarily dependent on mass- and force-constant effects. They are sensitive to small, but highly specific perturbations, induced by effect of stereo chemistry, mechanical coupling, intramolecular electrical effects, and intermolecular interaction. Thus, the physical state of the compound, liquid, amorphous solid, crystalline solid (including polymorphism) may alter the frequency of absorption. When using band-by-band correlation with a standard compound for identification of an unknown, these factors should be considered. More detailed information appears in the textbooks on infrared spectroscopy (48–53).

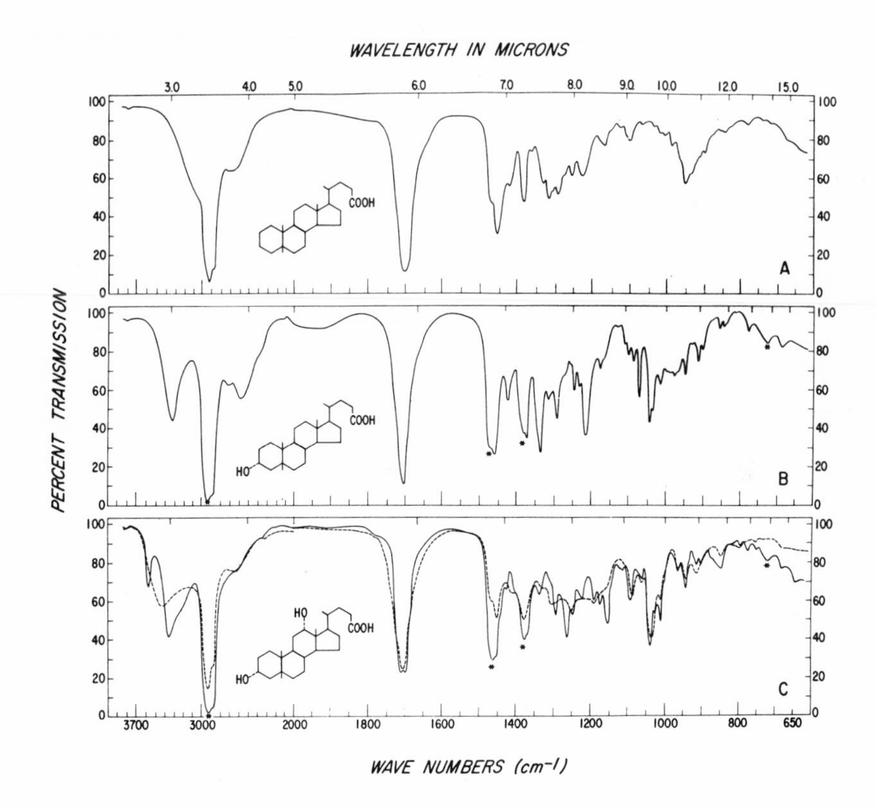

Fig. 7. The infrared spectra of bile acids. (A) Cholanic acid, crystallized from a melt. (B) Lithocholic acid, Nujol mull. (C) Deoxycholic acid, Nujol mull—glassy amorphous melt. Asterisks indicate Nujol bands superimposed on methyl and methylene bands of lithocholic and deoxycholic acid.

3. *Bile Acids*

A number of authors have reported infrared spectra of certain bile salts and related bile alcohols (30, 54–60). The lower frequency range of spectra of a number of ethyl and methyl esters of bile acids has been recorded (51). In general, infrared spectroscopy has been used in this field as a method to identify unknown compounds. In this section the IR spectra of the common bile acids (shown previously in Figs. 1–6) will be presented. The physical state of the bile acid was either crystalline (from a melt or in a Nujol mull) or a glassy supercooled amorphous melt. Spectra were obtained using a Perkin-Elmer Model 21 double-beam spectrometer at 25 °C. These spectra give some information on the intermolecular relations of bile acids in different physical states. Molecules capable of various discrete modes of hydrogen bonding such as the bile acids may exhibit polymorphism (but see Section IV). This

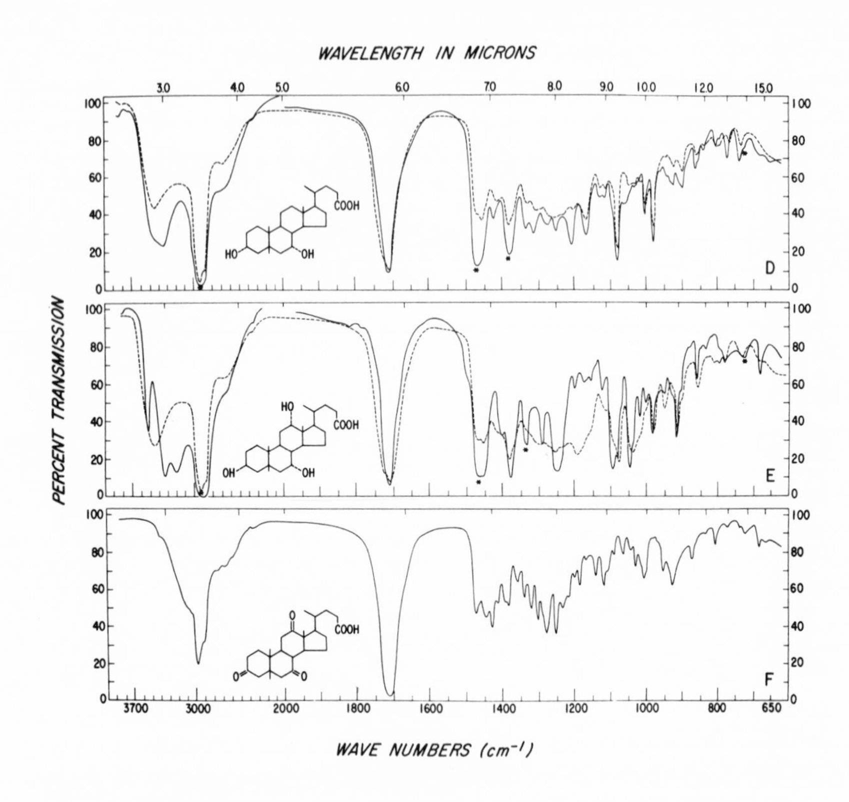

Fig. 8. The infrared spectra of the bile acids. (D) Chenodeoxycholic acid—Nujol mull—glassy amorphous melt. (E) Cholic acid—Nujol mull—glassy amorphous melt. (F) 5β Cholanic acid, 3,7,12-trione (dehydrocholic acid)—crystallized from a melt. Asterisks indicate Nujol bands, superimposed on methyl and methylene bands of chenodeoxycholic and cholic acids.

property should be considered when solid-state spectra of bile acids are to be used as references for the identification of unknown material. Usually solution spectra serve as unequivocal reference spectra. The amorphous phase serves well as a substitute for a solution, since intermolecular association effects are randomized in both.

The spectra of the bile acids are shown in Figs. 7 and 8. Band positions and assignments are given in Table I and spectra are discussed using letters *a* through *i* as indicated in Table I.

4. *The Carboxylic Acid Group (Bands b, d, f_3, g, and i)*

In carboxylic acids with no other proton-accepting or -donating groups present in the molecule or available in a solvent the predominant mode of

TABLE I. Infrared Data of Bile Acids[a]

	Band position, cm^{-1}	Assignment
a	3600–3200	OH stretch of alcohol groups
b	3000 very broad; 2700 shoulder; 2600 shoulder	OH stretch, overtones and combinations — carboxylic acid group hydrogen bonded dimer
c	2900	Overlapping integral of all unresolved CH, CH_2, CH_3 stretching vibrations
d	1700	C=O stretch of carboxylic acid group lowered in frequency by hydrogen bonding
e	1700	C=O stretch — Integral of 3 carbonyl groups, each substituted on a 6-membered ring, not hydrogen bonded
f	1500–1350	Band envelope, bending vibrations: CH_3 asymmetrical and symmetrical, and CH_2 scissoring
f_1	1470	Unperturbed methylene groups of ring system and side chain
f_2	1450	Unperturbed methylene groups of ring system and of angular methyl groups (asymmetrical)
f_3	1420	Methylene group at C_{23} lowered in frequency by adjacent carboxyl group
f_4	1430	Methylene groups adjacent to carbonyl groups in compound F only
f_5	1375	Both angular methyl groups (symmetrical)
g	1350–1150	C—OH of carboxylic acid group, coupled C—O stretch, and OH in plane bending
h	1100–950	C—OH of alcohol groups, C—O stretch
h_1	1042	3α hydroxy, rings A,B *cis*
h_2	1040	12α hydroxy
h_3	1080	two bands, characteristic for
	980	$3\alpha,7\alpha$ dihydroxy rings A,B *cis* only
i	ca. 950 broad	OH of carboxylic acid dimer out of plane bending

[a] See Figs. 7 and 8.

intermolecular association (persisting even in dilute solutions) is that of the hydrogen bonded dimer.

$$R-C\begin{matrix}\diagup O...H-O \diagdown \\ \diagdown O-H...O \diagup\end{matrix}C-R$$

This type of hydrogen bonding is quite strong and is characterized by a very broad absorption band of the OH stretching vibration centered around 3000 cm^{-1} with shoulders (presumably overtones) near 2700 and 2600 cm^{-1}.

A broad band near 950 cm^{-1}, assigned to the OH out of plane bending vibration, is characteristic of this dimer. The two broad bands *b* and *i*, though each partially overlapped by narrow bands of other origin, are most clearly seen in cholanic acid (A), lithocholic acid (B), and dehydrocholic acid (F).

The presence of another proton-donating and/or -accepting group (3α

OH) in lithocholic acid and the presence of the three proton-accepting carbonyl groups in dehydrocholic acid do not seem to interfere with the mode of dimerized hydrogen bonding of the carboxylic end group. This type of bonding is less clearly defined in the dihydroxy (C,D) and trihydroxy acids (E) though probably also present. Band *d*–the strongest band in each spectrum– and band f_3 are common to all 6 bile acids. The apparent relative intensity increase of band *d* in the triketo acid (F) is due to overlap with band *e*. Considerable differences between the crystal spectra of the various bile acids are present in the band pattern of region *g*. Some of the relatively strong bands in the crystal phase spectra are probably associated with twisting motions of the methylene groups of the flexible side chain, which probably has an ordered configuration in a crystal lattice. The random arrangement of rotational isomers expected in the amorphous phase is reflected in the decrease in structural detail of the bands in this region (C—O stretch as well as CH_2 twist).

5. The Hydroxyl Groups (Band in Regions a and h)

The stretching vibration of the OH group, when not involved in hydrogen bonding, gives rise to a sharp band near 3600 cm^{-1}. This band is lowered in position (usually not lower than about 3200 cm^{-1}) intensified and broadened in agreement with increasing strength of hydrogen bonding. In cholanic (A) and dehydrocholic acid (F) this band is absent as expected. The band of low intensity at 3400 cm^{-1} in F is due to an overtone of the strong carbonyl band (*d*) at 1700 cm^{-1}.

In the amorphous phase spectra of the dihydroxy compounds C and D and the trihydroxy compound E one strong band is seen near 3440 cm^{-1}, which is indicative of random hydrogen banding of all OH groups belonging to the same molecule. In the crystal phase spectra multiple peaks indicate that with increasing numbers of hydroxyl groups per molecule an increasing number of different modes of intermolecular hydrogen bonding prevails in the crystal. Lithocholic acid (B), the monohydroxy acid, shows one well-defined band at 3310 cm^{-1}, believed to arise from one polymeric mode of self-association.

```
(   R              R              R          )
    |              |              |
   /O\            /O\            /O\       ,H
  /    H      ,H/     H      ,H/     \   O/
        \    /         \    /          \ |
         O/              O/              R
         |               |
         R               R               )n
```

The other end of the molecule (carboxylic acid group) is engaged in dimerized H-bonding. One might expect that the crystals of lithocholic acid would be arranged in planes with alternating lamella of dimer H-bonds and polymeric

OH H-bonds. Of interest is a well-defined sharp band near 3550 cm^{-1} in the crystal phase spectra of deoxycholic acid (C) and cholic acid (E). Only these compounds have a 12α hydroxyl group. It would appear that this band arises from a free OH stretching vibration of the axial 12α hydroxyl group, which is sterically shielded from H-bonding, presumably by the position of the side chain. This hydroxyl group participates in random hydrogen bonding in the amorphous phase, where increased motion (rotational isomerism) of the flexible side chain is expected to be random. Region *h* (C—O stretch) is relatively free of strong bands in cholanic acid, as expected. As the number of hydroxyl groups is increased, more bands appear in this region. In lithocholic acid (B) the position of the strongest band is at *h* (1042 cm^{-1}). This band, together with the vicinal band pattern in region *h* is very nearly identical to that exhibited by several 3α hydroxy, AB *cis* steroids (51). Deoxycholic acid (C) shows an increase in intensity near 1040 cm^{-1} due to the additional OH group at carbon number 12 (h_2). Chenodeoxycholic acid (D) shows an unusual separation of the two C—O stretching bands of $<$100 wave numbers (h_3). Two strong bands appear at 1080 cm^{-1} and 980 cm^{-1}. This absorption feature is also exhibited by other 3α,7α dihydroxy steroids (51). The effect is maintained in the trihydroxy compound E of similar stereochemistry. This is not seen in structures with rings A and B in the *trans* configuration (51). Band separation by coupling of vibration is ordinarily found only in bands that not only possess two nearly alike vibrational frequencies but which also share the same atom. The unusual effect exhibited by 3α,7α dihydroxy steroids and bile acids may be due to multiple coupling, involving two nearly alike ring vibrations of rings A and B, which possess two nearly parallel C—O bonds.

6. *The Methyl and Methylene Groups (Band Regions c and f)*

The strong, poorly resolved summation band *c* is superimposed on the very broad band *b*. Except for an expected relative intensity variation due to the differing number of ring methylene groups in the bile acids, this band is common to all six compounds.

Assignments of the structurally useful band envelope *f* conform [with the proper allowance for the differing number of ring methylene groups in the differing bile acids to those established for other steroids (51)].

Other C—H bending vibrations of lower energy are more difficult to assign due to overlap and coupling with C—C skeletal stretching and bending vibrations. Band positions of twisting motions are expected in region *g*.

7. *Summary of the Structural Information Obtained from Infrared Spectra*

In crystals of cholanic (A), lithocholic (B), and dehydrocholic (F) acid, the predominant mode of intermolecular association is that of the carboxylic acid dimer. In lithocholic acid the equatorial 3α hydroxyl group appears to be engaged in one mode of polymeric self-association. The three carbonyl groups of dehydrocholic acid do not seem to precipitate greatly in hydrogen bonding effects.

In the di- and trihydroxy bile acids the dimeric hydrogen bond of the carboxyl acid group is less evident, though it appears to be maintained to a certain degree. The sterically unhindered equatorial 3α OH group and the axial 7α OH group, though sterically hindered, participate in various modes of hydrogen bonding in a crystal lattice. On the other band the axial 12α hydroxyl group in crystalline deoxycholic acid (C), as well as in cholic acid (E), appears to be protected from hydrogen bonding in the crystal lattice, possibly by steric shielding of a rotational isomer of the flexible side chain. This effect is not seen in the amorphous phase, where rotational isomerism of the side chain is expected to be random.

In the amorphous phase of all bile acids, hydrogen bonding of all hydroxyl groups is random, while the carboxylic acid dimer bond persists to a certain degree.

D. Nuclear Magnetic Resonance (NMR) Spectra of Bile Acids

Under appropriate conditions protons can be made to absorb electromagnetic radiation in the radio-frequency region. The frequency at which each proton absorbs energy is determined by the chemical and physical environment of the proton. For a given compound a plot of the frequency of the absorption against the intensity of absorption gives a proton NMR spectrum. The area under each absorption peak is proportional to the number of protons absorbing at that frequency. Thus, the proton NMR spectra of bile acids help not only with identification, but if the environment is altered, NMR can give valuable information on the motion of the protons in the molecule. For the theory and techniques of NMR spectroscopy, the reader is referred to the standard texts (61–66). Here we will consider the assignment of peaks to the bile acids and their sodium salts. Later (Sections VIII and IX) we will examine the use of NMR spectra of bile salts in deuterium oxide to give information about the structure of the bile salt micelle and the lecithin-bile salt micelle (67).

Figure 9 shows the NMR spectra of several bile acids in organic solvents (67). Table II contains information concerning the major peaks that could be assigned to the spectra of the bile salts and acids. Representative spectra in CD_3OD are given in Fig. 9. The chemical shift τ on the scale relative to tetramethyl silane (TMS) at ten ppm and the peak width at half peak height in cycles per second is recorded in Table II. Various solvents are used and referred to in the table. The assignment of chemical shifts referred to in the table was carried out by comparison of spectra for other steroids (64, 68, 69).

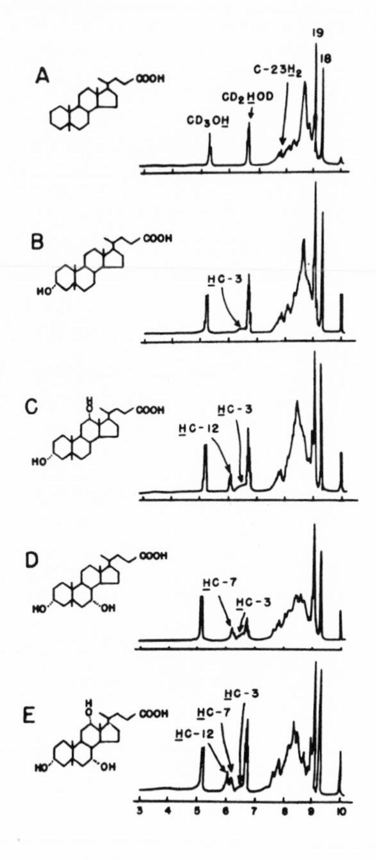

Fig. 9. Spectra of the common bile acids in $C^2H_3O^2H$. (A) Cholanic acid. 18 and 19 refer to signals from protons of C-18 and C-19 angular methyl groups. The large peak between 7.5 and 9 ppm is due to the protons of the steroid nucleus. Protons of the C-23 H_2 can be identified at about 7.8 ppm. Contaminating methanol protons are seen at 6.7 and 5.2 ppm in all spectra. (B) Lithocholic acid. The small broad peak at 6.5 is due to the proton at C-3. (C) Deoxycholic acid. Small peaks due to the protons attached to the rings at the C-12 and -3 position are noted at about 6.05 and 6.5 ppm. (D) Chenodeoxycholic acid. Small peaks due to the protons attached to the steroid nucleus at position C-7 and C-3 are noted at 6.23 and 6.6 ppm. (E) Cholic acid. Peaks are due to the protons attached to the nucleus at position C-3, -7, and -12 are noted at 6.6, 6.2, and about 6 ppm. All spectra were taken at 33.4°C (67).

1. Cholanic Acid

Cholanic acid was studied by MeOD and $CDCl_3$. In these spectra one is able to identify the two protons associated with the C-23 carbon, a large peak due to the steroid nucleus, and discrete peaks due to the C-21, C-19, and C-18 methyl groups. The C-21 methyl group is a doublet and merges in part with the C-19 methyl group.

2. Lithocholic Acid

The spectrum from lithocholic acid is very similar to that of cholanic acid, except that a rather broad peak appears at about 6.5 ppm which corresponds to the proton attached in the β position to the C-3 carbon. The signal from this proton is broadened by multiple spin–spin splittings.

3. Deoxycholic Acid

The spectrum of deoxycholic acid was studied in CD_3OD and Na deoxycholate in D_2O. Peaks due to the protons attached to the steroid nucleus at positions C_{12} (about 6.0 ppm) and C_3 (about 6.4 ppm) are present.

TABLE II. Proton, NMR, Data of Bile Acids (67)

Bile acids	Chemical groups	Peak width at 1/2 height in CD_3OD or $CDCl_3$ (cycles/sec)	Chemical shift ppm relative to tetramethylsilane)		
			D_2O[b]	C_2O_3OD	$CDCl_3$
A. Cholanic acid (5β-cholanoic)	C-23 H_2	18		7.8	7.75
	Steroid nucleus	30		8.66	8.70
	C-12 H_3	3(doublet)		9.05	9.07
	C-19 H_3	2.7		9.06	9.07
	C-18 H_3	2.2		9.32	9.33
B. Lithocholic acid (3α-hydroxy-5β-cholanoic acid)	C-3 H	20		6.50[c]	6.4
	C-23 H_2	18		7.8	7.8
	Steroid nucleus	30		8.65	8.68
	C-21 H_3	3(doublet)		9.07	9.07
	C-19 H_3	2.7		9.07	9.07
	C-18 H_3	2.2		9.32	9.34
	(C-3 OH[a]				
	C-24 OOH[d])	4.6			5.15
C. Deoxycholic acid (3α,12α-dihydroxy-5β-cholanoic acid)	C-12 H	7	5.95	6.05	
	C-3 H	20	6.35	6.5	
	C-23 H_2	18	7.85	7.8	
	Steroid nucleus	30	8.5	8.45	
	C-21 H_3	3(doublet)	9.08	9.00	
	C-19 H_3	2.7	9.08	9.08	
	C-18 H_3	2.2	9.30	9.30	
D. Chenodeoxycholic acid (3α,7α-dihydroxy-5β-cholanoic acid)	C-7 H	8	6.12	6.23	6.1
	C-3 H	20	6.5	6.5	6.5
	C-23 H_2	18	7.9	7.85	7.8
	Steroid nucleus	30	8.7	8.55	8.6
	C-21 H_3	3(doublet)	9.08	9.08	9.08
	C-19 H_3	2.7	9.08	9.10	9.08
	C-18 H_3	2.2	9.32	9.33	9.31
	(C-3 OH[a]				
	C-7 OH[a]				
	C-24 OOH[d])	6.0			5.26
E. Cholic acid (3α,7α,12α-trihydroxy-5β-cholanoic acid)	C-12 H	7	6.0	6.04	
	C-7 H	8	6.15	6.20	
	C-3 H	20	6.5	6.55	
	C-23 H_2	18	7.8	7.8	
	Steroid nucleus	30	8.35	8.32	
	C-21 H_3	3(doublet)	9.05	8.99	
	C-19 H_3	2.7	9.08	9.08	
	C-18 H_3	2.2	9.30	9.29	

[a] The number after the C refers to the number of the carbon atom in the chemical formula (standard numbering of carbon atoms is shown in Fig. 1).

[b] Studies as the sodium salt of the bile acid in $2H_2O$.

[c] $C^2H_3O^2H$ contained 10% C^2HCl_3 to increase solubility of cholanic and lithocholic acids.

[d] This peak disappeared when sample was shaken with a drop of $2H_2O$.

4. Chenodeoxycholic Acid

Chenodeoxycholic acid shows a spectrum very similar to that of lithocholic acid, except that a peak at about 6.2 ppm due to the proton attached in the β position to C-7 carbon is present. This peak is somewhat broader than the sharp peak at about 6.0, corresponding to the proton of the C-12 carbon seen in deoxycholic acid.

5. Cholic Acid

The spectrum of cholic acid shows peaks of the protons attached to the nucleus corresponding to the 12, 7, and 3 positions. They have the following respective assignments in CD_3OD: 6.04 ppm, 6.20 ppm, and 6.55 ppm.

The sharp peaks due to the C-18 and C-19 methyl groups are common to most steroids (64, 68, 69). The positions of these two methyl groups are quite constant in all of the bile acids examined. They vary only by about 0.05 ppm. The line widths in organic solvents are very narrow, being about 2.7 and 2.2 cps, respectively. The absence of any spin–spin coupling makes these peaks useful for the study of molecular motion since broadening of the signals should be due to motional effects only. The signal from the C-21 methyl group is a doublet due to spin–spin coupling with the single proton at C-20. Since the chemical shift is nearly identical to the signal from the methyl protons of the C-19, accurate measurement of line widths of these two chemical groups are difficult. The signal from the C-23 protons (next to the carboxyl group) can be identified in all the bile acids as a broad hump at about 7.8 ppm. The peak of a large broad signal due to the remaining protons is found at about 8.7 ppm in cholanic acid, which has no hydroxyl groups. This signal is shifted downfield as hydroxyl groups are added to the steroid nucleus such that the signal from cholic acid (the trihydroxy acid) is found at about 8.35 ppm. In pure $CDCl_3$, lithocholic acid (Fig. 9B) gave a peak corresponding to two protons at 5.15 ppm. Chenodeoxycholic acid (Fig. 9D) showed a similar peak at 5.26 ppm, corresponding to three protons. Since these peaks were not present in CD_3OD or D_2O and could be abolished in $CDCl_3$ by shaking the sample with a drop of D_2O, they most probably represent the protons of the OH and COOH groups (64). These protons, of course, are not seen in CD_3OD or D_2O because of rapid exchange with solvent deuterium.

III. PHYSICAL CONSTANTS

A. Molecular Weight and Melting Point

Molecular weights (MW), melting points (m.p.), densities (*d*), apparent

TABLE III. Physical Constants

Common name	Abbreviation	Systematic name	Molecular weight	Melting point (°C)	by pycnometry in water (20°C)
			1	2	3
Cholanic acid	—	(5β-cholanoic acid)	360.56	163–164	1.16
Lithocholic acid	LA	(3α-hydroxy-5β-cholanoic acid)	376.56	184–186	
Chenodeoxycholic acid	CDCA	(3α,7α-dihydroxy-cholanoic acid)	392.56	143	1.20
Deoxycholic acid	DCA	(3α,12α-dihydroxy-5β-cholanoic acid)	392.56	176–178	1.19
Cholic acid	CA	(3α,7α,12α-trihydroxy-5β-cholanoic acid)	408.56	198	1.39
Dehydrocholic acid	DHCA	(3,7,12-triketo-cholanoic acid)	402.51	237	

[a] Methyl lithocholate: C_2H_5OH.
[b] 3α,7β,-dihydroxy-5β-cholanoic acid.
[c] DCA: C_2H_5OH.
[d] CA: $4H_2O$.
[e] Sodium cholate.
[f] Calculated by the equation $MV = \frac{MW \times 10^{24}}{d \times N}$, where MW is the molecular weight of sodium salt, d is the density of sodium salt in aqueous solution, and N is Avogadro's number.
[g] Density of Na-lithocholate calculated to be 1.30.

partial specific volumes (V_{ap}), and apparent molecular volume (MV) of the common bile acids or their salts are given in Table III. The melting points were taken from older standard references concerning bile acids (27). Similar melting points have been obtained in our laboratory by differential scanning calorimetry (DSC) and conventional methods. Different solvents can give different melting points. Although melting point alone is not adequate data for identification of a specific bile acid, it is helpful in determining purity of the bile acid, since organic impurities of reasonably high molecular weight will lower the melting point of a bile acid in proportion to the amount of the impurity present. Mass spectroscopy, the best method of analyzing the structure of pure organic compounds, has recently been used with gas chromatography to study the bile acids present in feces (70, 71). Very accurate molecular weights may be obtained by this method.

of Bile Acids

Density, g/cm³				
by flotation (72)	calculated from partial crystal data (72)	by pycnometry of Na salt in aq. solution (43)	V_{ap}, cm³/g Apparent partial specific volume of Na salt in aq. solution (43)	MV, Å³/molecule[f] Apparent anhydrous molecular of Na slat in aq. solution
4	5	6	7	8
1.163[a]	1.217[a]			508[g]
1.174[b]	1.185[b]	1.28	0.780	538
1.160[c]	1.152[c]	1.31	0.765	526
1.156[d]	1.153[d]	1.33	0.75	539
1.167[e]	1.191[e]	1.30	0.77	546

B. Density and Apparent Specific Volume

The density of a substance depends on its physical state. Solids are usually more dense than liquids (ice and water are obvious exceptions). Thus, the density depends on the type of packing of molecules in a crystal or a liquid. If the density is measured in solution, the relation of the molecules of the solvent to the molecule of solute may influence density. In Table III are recorded densities obtained by several different methods. The variable values reflect not only technical problems of measuring density, but also the differences obtained when crystal structure differs and when the molecule is in solution.

Column 3 of Table III gives the density of some of the crystalline bile acids in water obtained in my laboratory by pycnometry. The density increases with the number of OH groups, as might be expected. In column 4 are presented the density of a number of bile acids or their derivatives (several with solvent of crystallization) measured by flotation (72). Column 5 gives the density of those compounds calculated from partial crystallographic data (72). Some variability exists (due to method, impurities, or incorrect crystallographic calculation of unit cell volume) and it is not possible at the present time to be sure which value to choose, although the values calculated from crystallographic data are probably best. When a compound (solute) is dissolved in a solvent it may be very intimately surrounded by solvent molecules and thus have an apparent specific volume (V_{ap}) which is quite small. A density calculated in this way may, therefore, be higher than that found in the crystal of the compounds. Thus, when the sodium salts were dissolved in aqueous solution (water or dilute NaCl solutions) the densities (43) were considerably higher (see column 6)* than those of bile acids in crystalline form. Column 7 gives the partial specific volume of these bile salts (43) and column 8 the anhydrous apparent molecular volume (MV) of the sodium salts of the bile acids calculated from their density in aqueous solutions (d), their molecular weight (MW), and Avogadro's number (N) by $\mathrm{MV} = \mathrm{MW} \times 10^{24}/d \times N$. Since the molecular volume of a water molecule is about 29.9 $Å^3$ at 4 °C, the bile salts have a volume about 17 or 18 times as great. These values are about 20% lower than the 770 $Å^3$ estimated roughly from molecular dimensions in Section II.

C. Miscellaneous Physical Properties

The optical activity of bile acids and their sodium salts depends on their purity, the wavelength of light employed, the solvent, and the concentration. The values given in standard references (27, 74) are often not comparable. Josephson (75) showed that increased concentration of bile salts in water substantially decreases the optical activity. The concentration, however, made little difference if the salts were studied in alcohol. Since the bile salts do not form micelles in alcohol (unpublished observations by the author), the variations noted in water may be related to the formation of micelles. The optical rotation of the acids in other organic solvents is not affected appreciably by concentration. The reader is referred to the earlier work of Sobotka (28) and Josephson (75).

The older work concerning other physical properties such as molecular refractivity and ultraviolet absorption spectra are summarized in Sobotka (28).

---

*Densities of the sodium salts of the cholic acid in aqueous solution were given by Ekwall (36). Although no experimental points are given, his curve suggests that there is a slight increase in the density around the CMC. Thus, at 20 °C the density is about 1.31 g/cc in concentration below about 20 mM. Between 20 mM and 65 mM the density is about 1.36, and in concentrations above 65 mM decreases to about 1.34. The author has studied the density at 20 °C and 36 °C of nine common bile salts (43) in concentrations above the CMC, and found little or no variation of density of individual bile salts at constant temperatures. Further, increased counterion concentration (NaCl) did not affect the density of the bile salts in solution. The density of the common free bile salts, and their glycine and taurine conjugates ranges between 1.285 and 1.351 g/cc. In general, the trihydroxy bile salts are slightly more dense than the dihydroxy salts. Increasing the temperature from 20 °C to 36 °C decreases the density of all the bile salts slightly. Laurent and Persson (73) have found that the partial specific volume of NaTDC was constant at 0.76 g/cc when studied at different concentrations of bile salt (0.5 g%–4 g%) in 0.15 M phosphate buffer at 20 °C.

IV. CRYSTALLINE STRUCTURE

The crystalline structure of cholanic acids and their alkaline metal salts has unfortunately been neglected. The major crystallographic work was carried out by Kratky, Giacomello, and co-workers (76–79) 30 years ago and was oriented toward solving the structure of the "choleic acids." These substances, isolated by Wieland and Sorge (80), are mixed crystals of deoxycholic acids (or certain other bile acids such as α and β apocholic acids) obtained on crystallization from organic solvents. A summary of the work on choleic acid appears in Sobotka (28).

The structure of the space unit cell of deoxycholic acid (Fig. 10A) crystallized from acetone (79) has the following dimensions: $a = 25.8$ Å, $b = 13.5$ Å, and $c = 7.2$ Å. Since the lattice parameters did not change appreciably with various choleic acids of long-chain aliphatic substances such as alcohols, fatty acids, and esters (Fig. 10B), it was deduced that there were holes in the crystalline lattice of deoxycholic acids that could accommodate aliphatic compounds. The so-called coordination numbers or molecular ratios of the deoxycholic acid and its guest molecule are related to the number of carbons in the aliphatic chain of the guest compounds and therefore the chain length of the molecule.

The structures of certain uncommon bile acids and their ethyl esters have also been studied (79, 81), and incomplete crystal data of certain bile acid derivatives given in a brief note (72). The crystalline structures of most of the common bile acids and their alkaline salts have not yet been defined.

The powder patterns of some 17 bile acids have been recorded by Parsons, Beher, Wong, and Baker (82–85). Powder patterns may be used as a means of identifying these compounds, provided solvent and method of crystallization are known and the compound is relatively pure. Specific data

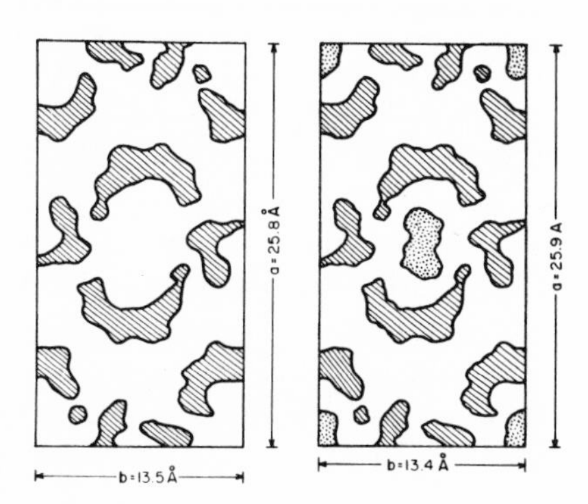

Fig. 10. Partial representations of the unit cell of deoxycholic acid (B) and the choleic acid formed from deoxycholic acid and a long-chain fatty acid (A) (79). Note that the unit cell dimensions do not change appreciably with the addition of the long-chain fatty acid. It appears to fit in holes in the deoxycholic acid lattice. Shaded areas are electron densities due to deoxycholic acid and stippled areas are those due to the long-chain fatty acid.

on crystalline structure of the bile acids, such as unit cell, is difficult to deduce from these patterns. Obviously this field needs further study.

V. SURFACE CHEMISTRY OF BILE ACIDS AND BILE SALTS

The study of the surface properties and characteristics of insoluble and soluble monolayers gives information concerning molecular dimensions, intermolecular interaction, and interaction with substrate.

A. Insoluble (Stable) Monolayers of Bile Acids

In this technique the molecules to be studied are dissolved in an appropriate organic solvent (e.g., benzene) and spread on the clean substrate surface (e.g., water, aqueous salt solutions) in a Langmuir trough (86–91). After the solvent has evaporated from the surface, the surface theoretically contains only the substrate (for instance water) and a given known number of molecules. The area of the surface can then be decreased by a movable barrier at one end of the trough, and by means of either a Wilhelmey plate or a movable barrier at the other end of the trough, the pressure exerted by the film can be measured. One then obtains a compression isotherm for the substance spread on the surface. This isotherm is plotted in terms of the average surface area per molecule in angstroms squared (Å^2) versus the surface pressure (π) exerted in dynes/cm. Figure 11 shows the compression isotherm of deoxycholic acid on 3 *M* NaCl, *p*H 1.5. The portion of the isotherm extending from 250 Å^2/molecule to 20 Å^2/molecule was obtained by the author on a continuously recording surface balance designed by Dervichian (92).

In order to understand this isotherm, it is worthwhile to apply Gibbs' phase rule (93) which, at *constant temperature* and *pressure,* states that the number of degrees of freedom (F) in a system at equilibrium is equal to the number of components (C) in the system minus the number of phases (solid, liquid, gas) present in the system, or $F = C - P$. Crisp (94) has derived a two-dimensional phase rule to apply to a single plane surface containing q surface phases. The rule predicts that the number of degrees of Freedom (F) will be $F = C - P_b - (q - 1)$, where C is the total number of components in the system, P_b is the number of bulk phases, and q is the number of surface phases. In the case of deoxycholic acid spread on aqueous substrate the number of components (C) can be considered to be two, the water of the aqueous phase and the deoxycholic acid. The number of bulk phases, that is the substrate, can be 1 or 2 and the number of surface phases can be 1 or 2. When the area per molecule is very large, for instance 10,000 Å^2 molecule (right side of Fig. 11), the surface pressure is very low ($>$0.1 dynes/cm) but

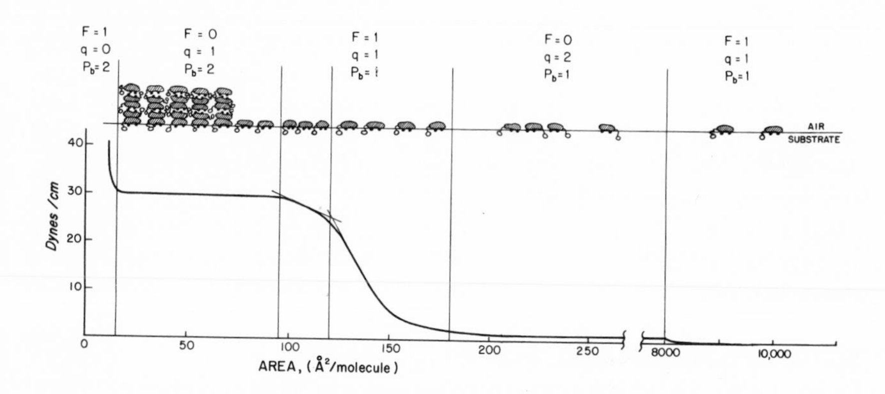

Fig. 11. Compression isotherm of deoxycholic acid spread on 3 *M* NaCl at *p*H 2. 20 °C/ Vertical axis surface pressure in dynes. cm; horizontal axis, area of the molecule in Å². Numbers and letters in the upper part of the figure are explained in the text. Diagrammatic representation of the surface is shown above the isotherm. The representation of the isotherm at very high molecular areas (right-hand side of figure) has been magnified for the purpose of explanation (see text).

does increase slightly with each decrease in area. The DCA molecules are very highly separated on the surface and act as a two-dimensional vapor. Decrease in surface area (analogous to volume in a bulk system) causes an increase in pressure. Since $q = 1$ (the vapor); $P_b = 1$; $C = 2$ the number of degrees of freedom (F) is $F = 2 - 1 - (1 - 1) = 1$. At a certain area [about 8000 Å²/molecule—for more specific details see (95)] the isotherm becomes flat (i.e., there is no increase in pressure with decrease in area). This shows that there are zero degrees of freedom ($F = 0$) and therefore we can conclude from the equation that there most be two surface phases ($q = 2$). This suggests the formation of islands of DCA packed much more closely together. These islands are analogous to a liquid phase. Therefore, in the flat portion of the curve (8000 Å² – 180 Å²/molecule) two surface phases coexist—a liquid phase in equilibrium with its two-dimensional vapor. As the area is decreased between 8000 Å²/molecule to about 180 Å²/molecule, more and more surface is covered with the two-dimensional liquid and less and less with the vapor. When the area is decreased below 180 Å²/molecule, the isotherm begins to rise steeply so that each change in area gives a change in pressure. Again a degree of freedom is present in the system and, therefore, we can conclude from the formula $F = C - P_b - (q - 1)$ that there is one surface phase, a two-dimensional liquid of bile acid molecules packed rather tightly in the surface. As the area is decreased the molecules pack more tightly. At about 120 Å²/molecule there is a change in the compressibility of the film, suggesting that there is a rearrangement or secondary phase transformation in the two-dimensional liquid phase covering the surface. At less than 90 Å²/molecule the isotherm again becomes flat. This again means that we have zero degrees of freedom. Since there can be only one surface phase the conclusion must be that a second bulk phase ($P_b = 2$) has been produced by the forcing out of the surface film a bulk phase of bile acid that piles up on the surface as diagrammatically suggested in the figure.

Finally, at very small surface areas there is again an increase in surface pressure which suggests that no surface phase (monolayer) remains and we are now dealing with the compression of two bulk phases, and F again equals one.

Monolayers of amphiphilic molecules may be gaseous, liquid, condensed, or solid, depending upon the surface viscosity. This can be judged simply by spreading clean talcum powder on the surface and with a small jet of air noting whether the talc granules move freely (gaseous or liquid), appear to have some viscous resistance (condensed), or do not move at all (solid). Liquid, condensed, solid transformations take place in certain films, while others remain liquid even though the isotherm is very steep (e.g., that of cholesterol). Dervichian (96, 97) has clearly described the various types of isotherms of aliphatic compounds. Using the talc test, bile acids appear to form gaseous and liquid monolayers but not condensed or solid films (see below).

Thus, the isotherm not only gives us the cross-sectional area per molecule at any given pressure, the collapse pressure of the monolayer in dynes/cm, and, knowing the density of the molecule, the height of the film in angstroms (Å) at any pressure, but also the number of surface and bulk phases present at any given area. The actual measuring of viscosity can be done simultaneously using the talc test or more sophisticated surface viscometers (98) and the measure of the potential difference between the top of the monolayers and the substrate phase beneath can also be obtained by using surface potential attachments (99). The molecule can be made to behave in different fashions by varying the substrate and temperature, thus affording much knowledge of not only the molecular dimensions, but also the interactions of the polar parts of the molecule with one another and with the substrate, and nonpolar parts with each other. General information concerning theory and technique will be found in texts on interface phenomena (86–90) and the articles of Langmuir (91), Dervichian (92), Gershfeld (100), and Reis (101).

Surface studies of insoluble monolayers of various bile acids have been studied by several groups, especially those of Ekwall (102–105) and Otero Aenlle (95, 106–110). In Fig. 12 are given the isotherms of some of the bile acids on 3 M NaCl, 20 °C, compiled from the work of Ekwall *et al.* (102–105). Figure 13 summarizes the important data derived from these isotherms. A diagrammatic representation of the surface structure of the bile acids at the

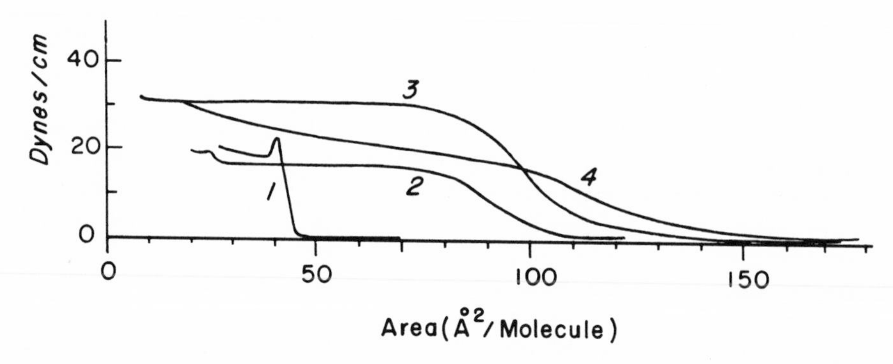

Fig. 12. Compression isotherms of bile acids [after Ekwall, 102–105]. (1) Cholanic acid; (2) lithocholic acid; (3) chenodeoxycholic acid (note similarity to Fig. 11); (4) cholic acid. 20°C, 3 *M* NaCl, *p*H "acidic."

interface is given at the top of Fig. 13. The area in Å²/molecule, the surface pressure at monolayer collapse in dynes/cm, and the rough viscosity of the film at collapse point are recorded. Similar studies have been carried out in the laboratory of Dervichian by the author, which corroborate the work of Ekwall.

Cholanic acid (Fig. 12) shows a very steep isotherm from 44 to 40 Å²/molecule. The viscosity of the film remains liquid up to the collapse point of the monolayer at about 20 dynes/cm. Lithocholic acid spread (Fig. 12) on water at *p*H 2 obviously has a different isotherm. At about 119 Å²/molecule there is a sudden rise in pressure to a first collapse point at 81 Å²/molecule. Thus, between 119 and 81 Å²/molecule the film is a single liquid phase. This range of surface areas corresponds to a monolayer of molecules lying flat

	Cholanic A.	Lithocholic A.			Deoxycholic A.	Cholic A.
AIR						
SUBSTRATE	pH 2	pH 2	pH 2 Collapsed Film	pH 10.75 3 M NaCl	pH 2 3 M NaCl	pH 2 3 M NaCl
AREA, Å²/mol max.	44	119	28 (3x28 = 84)	119	140	190
AREA, Å²/mol min.	40	81	24 (3x24 = 72)	44	85	105
SURFACE PRESSURE AT FILM COLLAPSE dynes/cm	20	12.3	16.5	29.4	30	14
STATE OF FILM	Liquid	Liquid	Solid	Liquid	Liquid	Liquid

Fig. 13. Diagrammatic representation of the surface configuration of the molecules of cholanic acid, lithocholic acid, deoxycholic acid, and cholic acid under varying conditions of substrate. Area per molecule at the first inflexion point of the curve (max) and at the collapse point (min), surface pressure at film collapse, and the state of the film are given.

on the surface with both the 3α OH and the carboxyl group in the aqueous surface. The decrease from 119 to 81 Å²/molecule represents closer and closer packing until at 81 Å²/molecule the liquid monolayer collapses, causing the formation of a second bulk phase. This extra phase results in zero degrees of freedom ($F=0$) and a flat isotherm. At about 27 Å²/molecule a second increase in pressure occurs from 14 to 16.5 dynes/cm with a decrease in area of 5 Å²/molecule. Since there is one degree of freedom and no monomolecular surface phase, the isotherm records the compression of two bulk phases. Because the area of 27 Å²/molecule is ⅓ of the minimum area of the lithocholate lying in a monolayer (81 Å²), Ekwall (102) suggested that between about 27 and 23 Å²/molecule the bile acid froms a single bulk phase made up of a trilayer of bile acid. This trilayer then collapses to form a multilayer bulk phase similar to deoxycholic acid (Fig. 11). Therefore, on dilute HCl, it is impossible to displace either the OH or the carboxyl group from the water separately. However, if one spreads lithocholic acid on 3 *M* NaCl at *p*H 10 (Fig. 14) there is only one collapse point, at about 44 Å²/molecule, which corresponds to the molecule standing up as pictured in Fig. 13. Here the increase in *p*H allows the carboxyl to ionize and therefore become much more water-soluble, and the high salt concentration tends to salt out (dehydrate) the hydroxyl group and allow it to be pushed out of the water.

On 3 *M* NaCl at an acid *p*H, deoxycholic (Fig. 11) and chenodeoxycholic

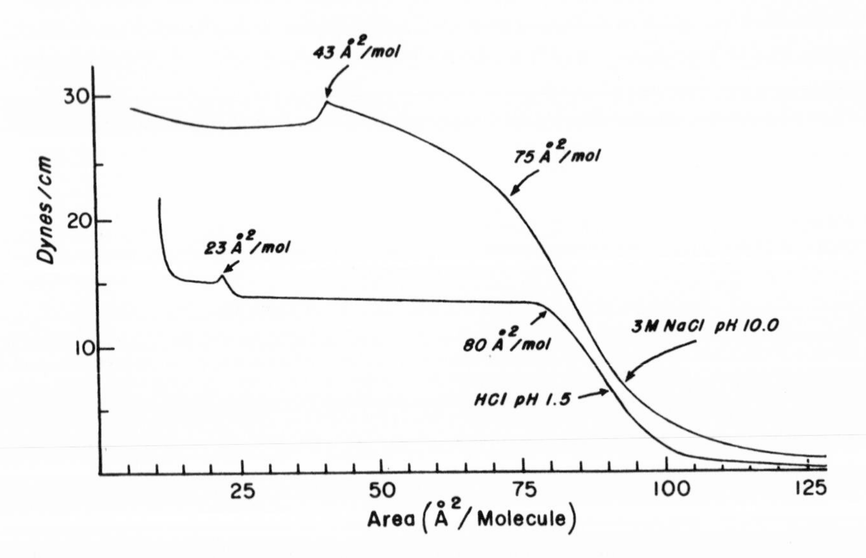

Fig. 14. Compression isotherm of lithocholic acid on two different substrates. Lower curve—dilute hydrochloric acid, *p*H 1.5. Upper curve—3 *M* sodium chloride, *p*H 10.0. For further explanation see text.

acid (Fig. 12) have nearly identical isotherms, with a monolayer collapse point of the liquid film at a surface pressure of 29–30 dynes/cm and a surface area of about 85–90 Å^2/molecule. This corresponds to molecules being tightly packed together, but lying flat with both hydroxyl groups and the carboxyl groups in the water, and demonstrates the tenacity with which the hydroxyl groups anchor to the aqueous phase even in the presence of 3 *M* NaCl. Cholic acid (Fig. 12), which has very slight solubility in the bulk at *p*H 2 and 3 *M* NaCl, shows an isotherm similar to dihydroxy acids. The collapse point occurs at only about 14 dynes and at somewhat greater surface area per molecule (105 Å^2/molecule). The flat portion of the curve after the collapse point so evident in the curves of the dihydroxy acids is replaced by a gradually rising curve which might be due to some solubility of cholic acid in the aqueous bulk phase.

B. Mixed Monolayers of Bile Acids and Other Amphiphiles

Two methods have been used to study molecular interaction in mixed films. In the first, binary mixtures of two insoluble components are spread on the surface of the substrate and compression isotherms obtained. In the second method (111), an insoluble component is spread and a soluble substance is injected into the aqueous phase. After allowing a given time for the soluble substance to equilibrate with the surface, an isotherm is obtained. The interpretation of the latter studies is often difficult and the scope of this review does not permit a thorough discussion of the studies that have been carried out (112).

The method of spreading binary mixtures of insoluble substances on substrate has been used to study the interrelations of a large number of compounds of biological interest. For instance, the interaction of triglyceride with phospholipid (113), triglycerides with fatty acids (114), lecithin with cholesterol (115, 116), sterols with hydrocarbons (117), bile acids with fatty acids (108,109), and bile acids with ovolecithin (118) have been investigated. Certain differences between the interaction of mono-, di-, and trihydroxy bile acids with lecithin can be shown by this technique. Figures 15, 16, and 17 give representative compression isotherms of lithocholic acid with lecithin; deoxycholic acid with lecithin; and cholic acid with lecithin, respectively. The deoxycholic acid and cholic acid mixtures were spread on acid surfaces of 3 *M* NaCl to decrease the solubility of the OH groups in the substrate. The pressures are given in dynes/cm and the curves are area per molecule of *lecithin*. The molar ratios of each mixture are given next to each curve. These studies were carried out using a continuous recording surface balance designed by Dervichian.

Representative curves for mixtures of lecithin and lithocholic acid spread

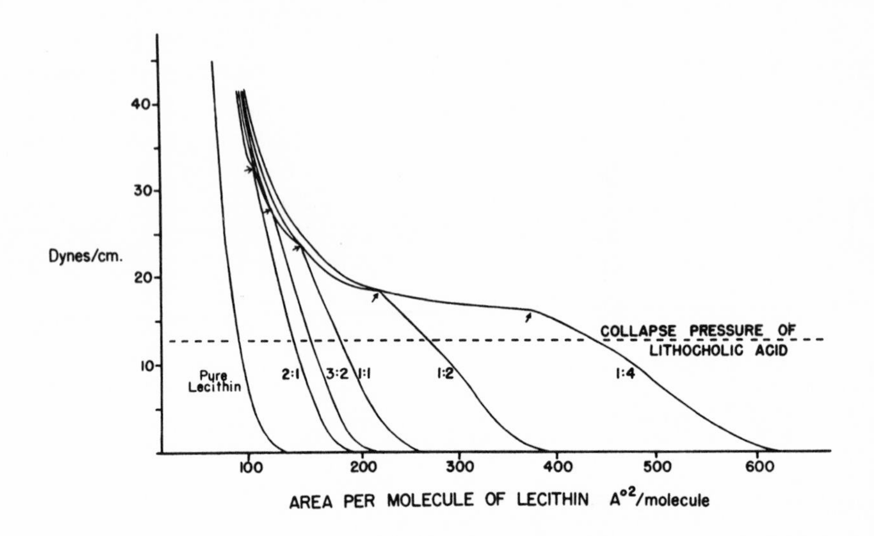

Fig. 15. Compression isotherm of lecithin–lithocholic acid mixtures. The numbers next to the curve refer to the lecithin–lithocholic acid ratio. The arrows represent the pressure at which lithocholic acid is forced out of the film as a second phase. For further explanation see text.

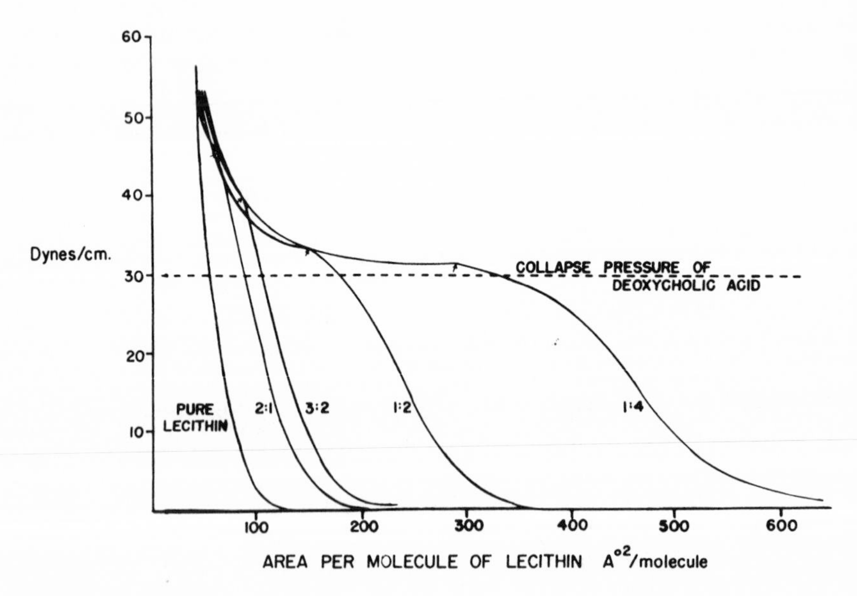

Fig. 16. Compression isotherm of lecithin–deoxycholic acid mixtures. Explanation as in Fig. 15.

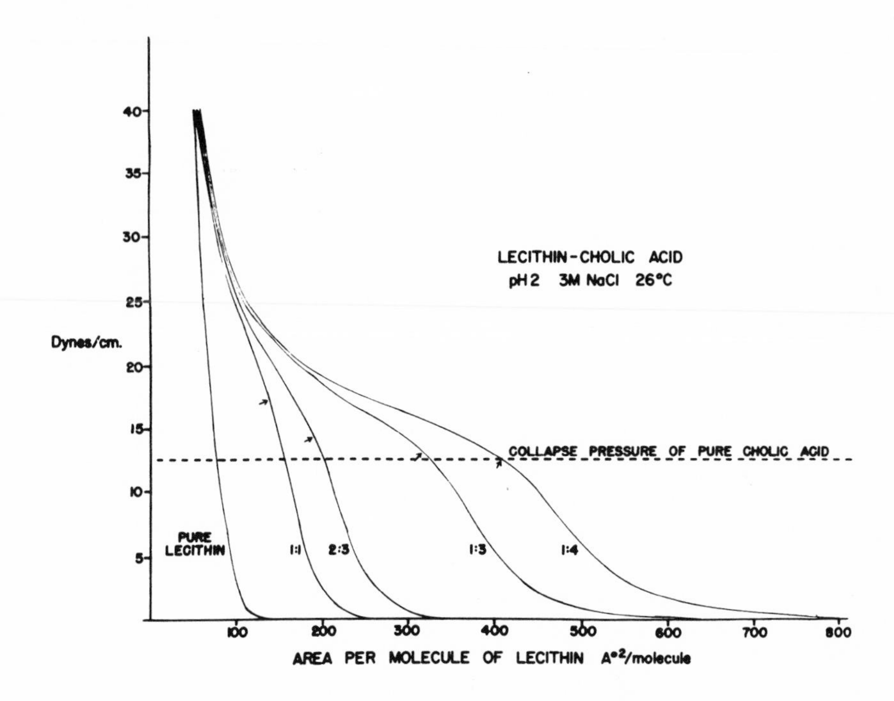

Fig. 17. Compression isotherm of lecithin–cholic acid mixtures.

on water at *p*H 2 are shown in Fig. 15. The collapse pressure of lithocholic acid on this substrate is considerably less than that on 3 *M* NaCl (see Fig. 12). The collapse pressure of lithocholic acid on water at *p*H 2 is about 13 dynes/cm compared to 20 dynes/cm on 3 *M* NaCl. When small quantities of lecithin (right side of Fig. 15) are added to lithocholic acid, the collapse pressure of the lithocholic acid is increased. For instance, when the molecular ratio is one lecithin:four lithocholic acids, the collapse pressure is increased to 16 dynes/cm from 13 dynes/cm. As the molecular ratio of lecithin: lithocholic acid increases, the collapse pressure of lithocholic acid increases. Crisp (119) has called this increasing collapse pressure with increasing molecular ratio an "envelope curve." At each sudden change in the slope of the curve, lithocholic acid is squeezed out of the surface to form a bulk phase. The fact that it is not squeezed out at every molecular ratio at the original collapse pressure of lithocholic acid (13 dynes/cm) suggests that there is interaction or *intimate miscibility* in the mixed monolayer. In other words, the components are not acting as separate islands from which the lithocholic acid would continue to collapse out of the film 13 dynes/cm no matter what the molecular ratio of lecithin. Thus the lecithin holds the lithocholic acid in

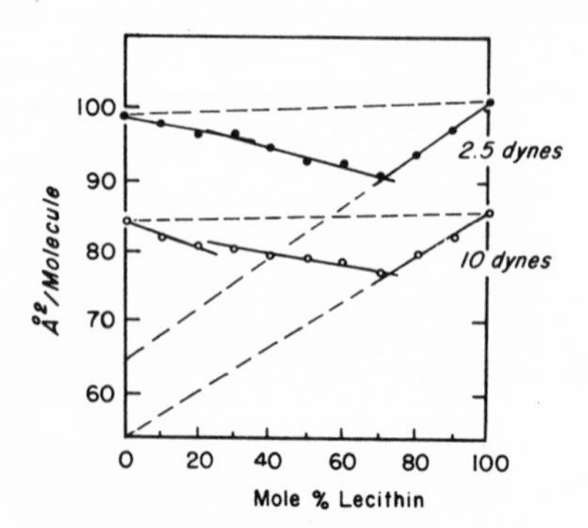

Fig. 18. Average area per molecule of the mixture in lecithin–lithocholic acid films at pressures of 2.5 dynes/cm and 10 dynes/cm, expressed as mole % lecithin. For further explanation see text.

the monolayer at higher pressures than would be expected. It is also of interest to notice that even at high pressures some lithocholic acid remains in the film. That is, with each mixture there seems to be a small amount of lithocholic acid that is not squeezed out of film, even at pressures of about 40 dynes/cm.

Further information may be gained from making a plot of the average area per molecule in Å²/molecule versus the mole % of lecithin. Theoretically, at any given pressure, if there were no intimate miscibility or interaction between the two components and each acted as a separate island of molecules, the area of any mixture would fall on a straight line connecting the areas of the pure substances. This line is called the "additivity line" and implies no interaction between the two substances. A plot for lithocholic acid-lecithin mixtures at two different pressures (2.5 dynes/cm and 10 dynes/cm) is shown in Fig. 18. The areas do not follow the typical additivity line, but the area for every mixture is somewhat less than one would expect if the components were acting as individual islands of molecules. By extrapolating the four points falling on a straight line on the right-hand side of the curve to zero lecithin concentrations one can get an idea of the area occupied by lithocholic acid in the monolayer at the pressure noted. It is significant that the area of lithocholic acid corresponds to 64 and 55 Å²/molecule, which is much too small an area to find if the molecule were lying with both the OH group and the carboxyl group in the aqueous phase. One can suggest that the molecules, even at these low pressures, are maintained in an erect position by the lecithin molecules in the monolayer.

Figure 16 shows representative curves for mixtures of deoxycholic acid and lecithin spread on 3 *M* NaCl at *p*H 2. The collapse pressure of deoxycholic acid (see Fig. 11) is about 30 dynes/cm. When small amounts of lecithin are added, this collapse pressure is increased. Like lithocholic–lecithin mixture an envelope curve is formed by the collapse points of deoxycholic acid with increasing quantities of lecithin. However, unlike lithocholc acid, most of the deoxycholic acid appears to be excluded from the monolayer at high

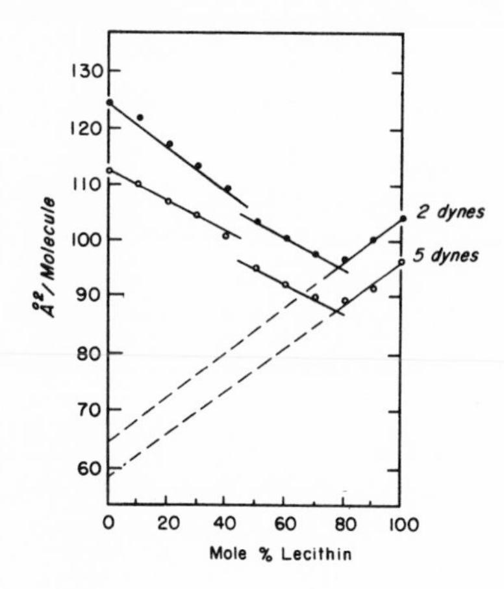

Fig. 19. Average area per molecule of mixtures of lecithin–deoxycholic acid at 2 and 5 dynes/cm pressure represented as mole % lecithin. For further explanation see text.

surface pressures. The plots of the average area in Å^2/molecule versus the mole % lecithin in each mixture (Fig. 19) show that the areas do not follow the additivity rule. Even at 2 dynes/cm pressure the extrapolated area per deoxycholate molecule is 65 Å^2/molecule, which is again too small for a molecule lying flat on the surface with both OH groups and its COOH group in the aqueous interface. This area decreases to about 55 Å^2/molecule at 5 dynes/cm pressure. The isotherms and the plots of the average molecular area versus mole % lecithin suggests that when large molecular ratios of lecithin area present the deoxycholic acid mixes with lecithin and is maintained in an erect position with the OH groups out of the water. It is interesting to note that the minimum area for the two molecular species occur at a ratio of about three lecithins to one bile salt. This is the maximum amount of bile salt that can be incorporated into a lamellar liquid crystalline lattice of lecithin (back to back monolayers of lecithin) in the bulk phase (4) containing large quantities of water. Therefore, deoxycholic acid on a substrate, which tends to salt out the OH groups, behaves somewhat like lithocholic acid, except for the fact that at high pressures all of the deoxycholic acid is excluded from the film. The isotherms of mixtures of cholic acid and lecithin on 3 *M* NaCl at *p*H 2 are given in Fig. 17. When molecular ratio of cholic acid to lecithin is high (1:4) the cholic acid collapses out of the monolayer at the same pressure as pure cholic acid (12.5 dynes/cm). However, when the molar ratio approaches 1:1 the cholic acid appears to be pushed out of the film at about 17 dynes/cm or somewhat higher than the collapse pressure of pure cholic acid on this substrate. Although small amounts of cholic acid appear to stay in the film up to 40 dynes/cm, no clear envelope curve is formed and above that pressure no cholic acid is left in the film. If we examine a plot of the area per molecule versus the mole % lecithin at several different

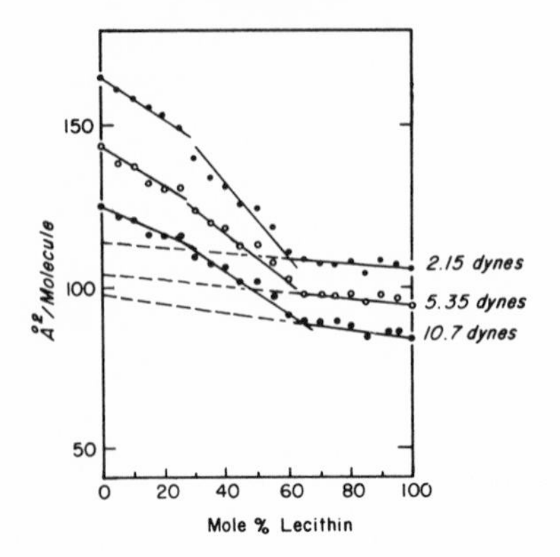

Fig. 20. Average molecular areas of mixtures of lecithin–cholic acid at pressures of 2.15 dynes/cm, 5.35 dynes/cm, and 10.7 dynes/cm, expressed as mole % lecithin. For further explanation see text.

pressures (Fig. 20), we can see that at low lecithin mole % the curve is nearly additive (left side, Fig. 20), showing that up to about 30% lecithin very little miscibility occurs. However, beyond that, with greater amounts of lecithin, the curve deviates from the additivity curve. A clear break in this curve occurs at about 60% lecithin–40% cholic acid, and extrapolating the right-hand side of this curve to 0% lecithin shows that the cholic acid molecule occupies an area in these mixtures of about 98–112 $Å^2$/molecule, depending on the surface pressure. In contrast to lithocholic acid and deoxycholic acid, these areas are similar to those found for the bile salts lying flat on the surface with the three OH groups and the polar group in the aqueous phase. This suggests that at all of these pressures the cholic acid molecules are lying flat on the surface as they do in the monolayer even though there is some slight interaction with the lecithin. The fact that at high cholic acid concentrations the additivity rule is followed (Fig. 20) and that the collapse pressure of the cholic acid is the same as pure cholic acid suggests that there is no interaction of these compounds in molar ratios greater than three cholic acids to one lecithin.

Thus these studies show that the trihydroxy bile acid, cholic acid, and lecithin have limited miscibility in monolayers, and when there is miscibility the cholic acid appears to be lying flat on the surface. Deoxycholic acid has a larger range of interaction at the interface and small proportions of deoxycholic acid appear to be held in the upright position in the monolayer of lecithin. Lithocholic acid, on the other hand, is freely miscible in all proportions with lecithin and, even at high pressures, a certain amount of lithocholic acid is held tightly by the lecithin and cannot be expelled from the interface as a bulk phase. Although the interpretation that deoxycholic acid and lithocholic acid are standing erect in the surface seems most reasonable, it is possible that trilayers of bile acids may be interacting with lecithin molecules to give the smaller areas noted in the average area versus mole % lecithin plots.

C. Surface Tension of Soluble Bile Salts

Alkaline metal salts of di- and trihydroxy bile acids are very soluble in water and can be studied by the technique of surface tension. Surface tension studies have been carried out in a number of laboratories to determine critical micellar concentration. If a plot of surface tension in dynes/cm versus the log concentration is made, a sharp break in the curve will occur when the surface becomes saturated with molecules of soluble amphiphile. This saturation of the surface with amphiphile corresponds to the saturation of the bulk phase and thus approximately to the critical micellar concentration. As will be summarized in Table VII of Section VIII, surface tension has been utilized mainly for the study of critical micellar concentration. However, using the simplified form of the Gibbs adsorption isotherm equation $\Gamma = -(1/RT)(d\gamma/d\ln C)$, where R is the gas constant, T the absolute temperature, $d\gamma/d\ln C$ the slope of the experimental curve of surface tension (γ) versus ln concentration, the surface excess (Γ) can be calculated. The area per molecule (Å^2/molecule) at surface saturation can be estimated by $A = -1/\Gamma N$, where N is Avogadro's number and A is Å^2/molecule. Surface areas have been calculated for sodium glycodeoxycholate and sodium taurodeoxycholate by Kratohvil and Dellicolli (120), who found areas of about 90 Å^2/molecule for both salts. These areas are similar to those found for insoluble monolayers of deoxycholic acid (Fig. 11) at its collapse pressure of 30 dynes/cm. It should be noted here that, in the technique for measuring surface tension of bile salts, the surface tension versus time effects are important and should not be ignored. The best method is probably the use of a Wilhemy plate, rather than some of the other more commonly used methods such as the Denouy ring or the drop method, both of which have problems inherent in the technique. The study of surface tension of mixtures of bile salts and other amphiphilic compounds is complicated but can yield most interesting results.

VI. SOLUBILITY IN ORGANIC SOLVENTS

The solubility of the common bile acids and two common sodium salts in water and organic solvents are given in Table IV. Most of the values in Table IV have been extracted from the common references (27, 74). However, the values for the solubility of deoxycholic acid and cholic acid in water have been taken from the careful work of Ekwall *et al.* (121). In some cases

TABLE IV. Solubility of Bile Acids

	Solubility g/100 ml solvent						
	Water	Alcohol	Acetone	Ether	Chloroform	Benzene	Acetic acid
Cholanic acid	insol. RT[d]	0.6[b]	0.3[b]	0.9[b]	6.6[b]	1.7[b]	sol. RT
Lithocholic acid	insol. RT	1.1[b]	0.1	0.2[b]	0.7[b]	0.1[b]	sol. RT
Chenodeoxycholic acid	insol. RT	3.4[b]	2.2[b]	0.3[b]	0.3[b]	<0.04[b]	6.2
Deoxycholic acid	0.00431[a]	22.07	1.05	0.12	0.29	0.01	0.81
Cholic acid	0.00776[a]	3.06	2.82	0.12	0.59	0.04	15.2
3,7,12-Triketo cholanic acid	0.018	0.33	0.78	0.046	0.9	0.1	0.74
Sodium deoxycholate	33.4[e]	3.21	0.16	0.01	0.01	0.15	2.99
Sodium cholate	56[ce]	1.81	0.03	0.01	0.14	0.01	9.52

[a] From Ekwall *et al.* (121), 37°C.
[b] Crude estimations of solubility carried out by the author at 25°±3°C.
[c] From Small *et al.* (2), 20°C.
[d] RT (room temperature).
[e] g/100 ml solution

solubilities were crudely determined in our laboratory at ambient temperatures to get an idea of the order of magnitude of solubility of various bile acids in various organic solvents. As would be expected, cholanic acid tends to be more soluble in nonpolar solvents than polar solvents, whereas the opposite is true for the di- and trihydroxylated bile acids. The explanation for the extraordinarily high solubility of deoxycholic acid in alcohol and cholic acid in acetic acid is not clear. In general, as the number of hydroxyl groups on bile salt is increased, its solubility in nonpolar organic solvents decreases.

Recently Bennet *et al.* (122) showed that the methyl ester of cholic acid forms small aggregates in relatively nonpolar organic solvents. These aggregates are made up of about four molecules of the methyl ester hydrogen bonded to one another through the hydroxyl groups and therefore might be considered as a form of "reversed micelle." These derivatives do not form micelles in a more polar solvent, such as alcohol, but are present in true solution (122).

VII. SOLUBILITY IN WATER

A. Bile Acids and Their Salts

While the sodium and potassium salts of the free di- and trihydroxy bile acids (Na^+A^-, K^+A^-) are very water-soluble, protonated free bile acids (HA) are very insoluble (Table IV). Ekwall *et al.* (121) estimated the solubility of cholic acid gravimetrically at about 9.18 mg per 100 ml of water at 20°C. The author has repeated these measurements on cholic acid and deoxycholic acid using microbalance techniques and found the solubility at 37°C to be 12.0 mg/100 ml for cholic acid and 4.5 mg/100 ml for deoxycholic acid. The above values are considerably lower than those in the older literature (27, 74) but are probably closer to correct values. Due to partial hydrolysis of the acid, values measured in this way include a small amount of dissociated acid (A^-). Ekwall *et al.* (123), using the methods of Back and Steenberg (124), calculated the solubility of two bile acids in dilute solutions of their sodium salts. As long as the total bile salt concentration is below the CMC the solubility of the acid was constant at 7.76 mg/100 ml for cholic acid and 4.31 mg/100 ml for deoxycholic acid. At concentrations above the CMC, bile acid is solubilized in the micelle and thus its solubility is increased. Lithocholic acid and cholanic acid are even more insoluble. The homologous di- and trihydroxy glycine-conjugated acids are also quite insoluble in the protonated form (HA). The fact that monolayers of these acids (Section V) are stable for hours at high surface pressures attests to their insolubility in the bulk. Cholic acid and glycocholic acid are the exceptions in that their monolayers do dissolve gradually in the substrate (water at $p\mathrm{H} = 2$) over a period of hours.

The taurine conjugates of di- and trihydroxy bile acids are very strong acids and are very soluble down to $p\mathrm{H} = 1$ (125–127).

B. The Effect of *p*H on the Solubility of Bile Acids

The effect of *p*H on the physical state of bile acids and their sodium salts is perhaps best explained by examining a titration curve of typical bile salt obtained by titrating an alkaline bile salt solution with hydrochloric acid.

In the case of a free or glycine-conjugated bile salt, the initial solution contains the soluble sodium salt of the bile acid which is fully ionized at high *p*H (10.5). Titration with HCl is carried out until all the bile acid anion is in the form of an insoluble acid. The chemical reaction in simplified form is

$$Na^{+}A^{-} + HCl \longrightarrow HA + Na^{+}\,Cl^{-} \qquad (1)$$

where $Na^{+}A^{-}$ is the fully ionized bile salt and HA is the insoluble acid. Although the solubility of bile acid alone (HA) is very small (see above), the bile salt anion (A^{-}) is a colloidal electrolyte and, above a certain concentration (see Section VIII. D), forms micelles and will solubilize a certain amount of bile acid (HA). The solubility of the acid, as well as the $p\mathrm{K}'_a$ of the acid, can be calculated from the titration curve (124).

An ideal titration curve of a free bile salt at a concentration above its CMC is given in Fig. 21. The initial part of the curve represents the titration of the small excess NaOH. At the inflection point *W*, the convex slope of the curve suddenly becomes concave, indicating the first equivalence point (i.e., the point at which the reaction given in [Eq. (1)] starts). The curve then flattens slowly until at point *X* precipitation suddenly occurs and, due to the trapping of the H^{+} ions in the crystalline precipitate, the *p*H rises sharply to *X′* without the further addition of acid. This precipitation is often heralded by the appearance of a Tyndall effect at about *T*.

With each successive addition of HCl, an equivalent amount of bile acid is precipitated, which appears to act as a buffer producing a plateau in the curve, in which there is very little change in the *p*H. Toward the end of the bile salt titration, this plateau portion changes to a convex slope, and finally when the reaction shown in [Eq. (1)] is complete, the curve shows a second inflection point *Z*, which is the final equivalence point. Extrapolation of the plateau section of the curve horizontally from point *X′* to point *Y* indicates the point at which precipitation of bile acid crystals would have occurred, had there been no supersaturation. Between points *Y* and *X*, therefore, the solution is supersaturated and is not in physical equilibrium.

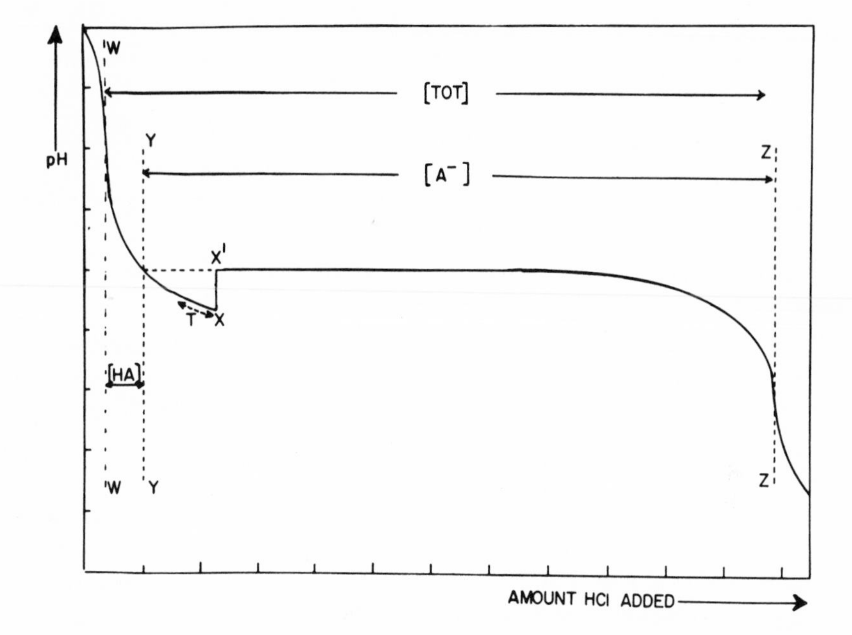

Fig. 21. Hypothetical titration curve for solutions of free bile salts or for glycine conjugates. W = first equivalence point where titration of bile salt with hydrochloric acid commences. Y = last point where bile salt solution is in thermodynamic equilibrium as a single aqueous phase. T = Tyndall effect noted in this region of titration curve. X = point where precipitation of bile acid crystals commences. X' = equilibrium pH at point of bile acid precipitation. Z = second equivalence point where titration of bile salt with hydrochloric acid is complete. TOT = total amount of acid required to complete the titration. HA = the amount of acid added from the first equivalence point (W), to point Y, which represents the maximum solubility of the bile acid (HA) in the bile salt solution (A^-). For further explanation of the symbols, see text.

From the start of precipitation (X) onward, there are two phases present (liquid plus crystal), and since solutions between Y and X are unstable, the only portion of the curve that is in thermodynamic equilibrium for the single liquid phase is portion WY. Point Y, therefore, represents the maximum solubility of bile acid (HA) in the bile salt solution (A^-) before supersaturation occurs.

From such titration curves it is possible to calculate (123, 124, 128) a number of characteristics for each bile salt, such as

a. the total number of moles of hydrochloric acid (TOT) required to titrate a given amount of bile salt (point W to point Z).
b. the pH at which bile acid precipitates from a supersaturated solution (pH at X), and the pH at which the first bile acid molecules

would have precipitated had there been no supersaturation (pH at X'), in other words, the equilibrium pH at the point of precipitation.

c. the maximum solubility of the acid in the bile salt solution. Above the CMC the number of molecules of bile salt necessary to "carry" one molecule of bile acid in a micellar solution (A^-/HA) may be obtained from

$$\frac{A^-}{HA} = \frac{\text{moles of acid from } Y \text{ to } Z}{\text{moles of acid from } W \text{ to } Y} \quad (2)$$

d. a simplification* of the formula proposed by Back and Steenbeerg (124) can be used to calculate the pK'_a† of the bile acid from the portion WY of the curve in the following manner:

$$pK'_a = pH_y + \log\frac{HA}{TOT - HA} + \frac{0.5\sqrt{\mu}}{1+\sqrt{\mu}} \quad (3)$$

where pH_y is the pH at point Y, TOT is the total amount of acid needed to carry the reaction [Eq. (1)] to completion (from W–Z), and HA is the amount of acid added from the first equivalence point (W) to the point Y. μ is the total ionic strength at point Y.

Sodium cholate and sodium deoxycholate have been studied in concentrations below and above their CMC's by Ekwall *et al.* (123). The variation in precipitation pH and pK'_a are summarized in Figs. 22 and 23. The precipitation pH for both increase rapidly at very low concentrations, gradually leveling out as they approach the CMC (123). The pK_a is constant until micelles begin to form. Over a concentration range where some added bile salt is being bound to micelles and some is present as monomer the pK'_a increases. When all added bile salt becomes bound to micelles, the pK'_a remains fairly constant. Chenodeoxycholic acid (Table V) behaves very much like DCA except that its pK'_a at any given concentration is slightly higher. Unfortunately, the pK'_a of the conjugated bile acids have not been so exhaustingly studied. However, the data in Table V suggest that the marked

*The full formula is

$$pK'_a = pH_y + \log\frac{HA + [OH^-] - [H_3O^+]}{TOT - HA - [OH^-] + [H_3O^+]} + \frac{0.5\sqrt{\mu}}{1+\sqrt{\mu}}$$

where $[OH^-]$ and $[H_3O^+]$ are concentrations of the OH^- and H_3O^+ ions at point Y. In the simplified formula used in this paper, the ionic product of water has not been included since the calculation showed that by inclusion of this correction factor, the resultant difference in pK'_a was small and was within the experimental error of the method.

†pK'_a refers to the fact that the value has been calculated from the concentration and not from the activities of the species involved.

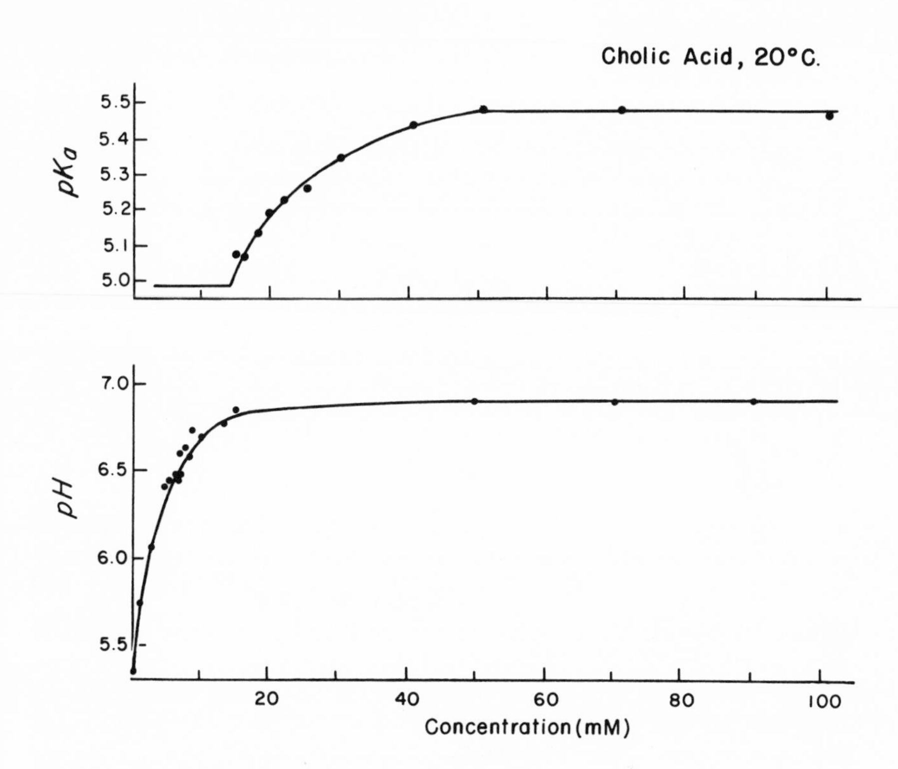

Fig. 22. The apparent pK_a and the pH of precipitation of cholic acid as a function of concentration in water 20°C [from Ekwall *et al.* (123)].

dependence of pK'_a of the free acids on concentration is not nearly as evident with either glycine or taurine conjugates. Thus, concentration seems to have little effect on the pK'_a's of the conjugates.

Table V also gives the pH of precipitation and the number of bile acid anions needed to solubilize one molecule of bile acid (A^-/HA).* The values in Table V were determined from the titration curves of the bile salts in water at 37°C (128).

Examples of the titration curves for a pure bile salt (NaC) and pure conjugated bile salt solutions (NaGC and NaTC) are shown in Fig. 24. The curves for sodium cholate and sodium glycocholate are similar to the ideal curve (Fig. 21), but since sodium taurocholate is the salt of a strong acid and remains completely ionized, the solution remains clear throughout the titration and therefore neither the supersaturation nor the precipitation characteristic of the free and glycine-conjugated bile salts takes place.

*This value (A^-/HA) is actually the sum of the small amount of acid soluble as individual molecules plus the acid solubilized in the bile salt micelle.

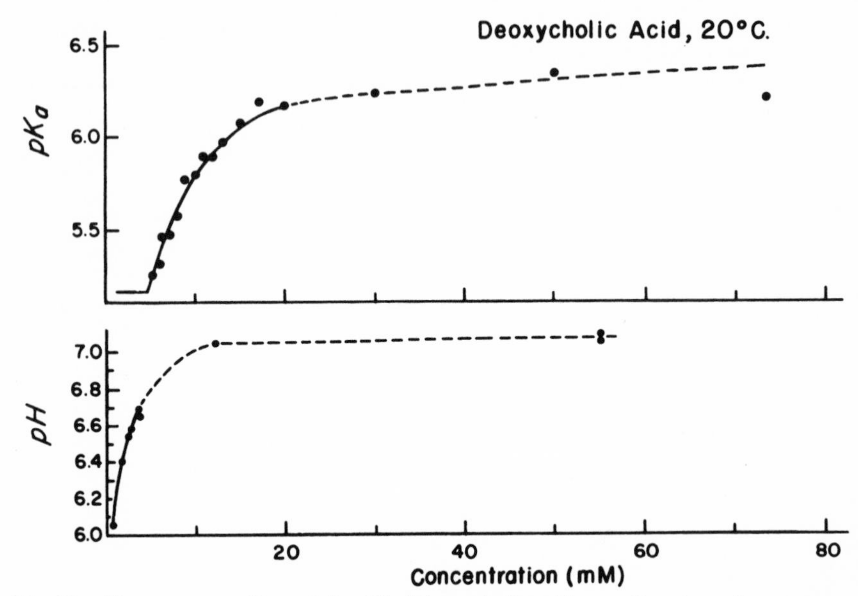

Fig. 23. The apparent pK_a and the pH of deoxycholic acid as a function of concentration of the acid in water, 20°C [from Ekwall *et al.* (123)].

There are several obvious differences in the titration curves for the free NaC solution and the corresponding glycine conjugate (NaGC). First, the free bile acid precipitates at a near neutral pH (6.5)—a level frequently found in the normal jejunum—whereas glycocholic acid does not precipitate until relatively acid levels are reached (pH 4.3)(Fig. 24). Secondly, the amount of acid required to produce precipitation is much less for the free bile salt than for the conjugate. Although this is in part due to a longer supersaturation effect for NaGC (*Y–X*), calculation of the A^-:HA ratio for the two curves shows that there is a real difference between the two solutions. The ratio (A^-/HA), that is, of bile salt molecules required to "carry" one bile acid molecule in solution, is 15:1 for NaC, whereas for NaGC it is 4:1 (see Table V).

C. The Effects of *p*H on Binary Mixtures of Free and Conjugated Bile Acids (128)

1. *NaC and NaTC*

The individual titration curves for various different mixtures of NaC and NaTC are shown in Fig. 25 and the measurements of precipitation pH's and the A^-/HA ratios obtained from these curves are plotted in Fig. 26.

In general, as the percentage of conjugated bile salts in the mixture increases, the pH at which the free bile acid precipitates falls in a linear fashion. Initially, however, the addition of increasing amounts of NaTC up to 30%

TABLE V. pK'_a, pH of Precipitation, and A^-/HA of Free and Conjugated Bile Acids Calculated from the Titration Curves of Sodium Salts with HCl

Bile acid	Bile salt concentration, moles/liter	pK'_a	pH of precipitation	A^-/HA
Free bile acids				
CA	2	4.98		
	22	5.21	6.5	15
	80	5.5		
DCA	2	5.3		
	23	6.21	6.92	6
	80	6.3		
CDCA	2.2	5.88		
	10.1	6.07		
	23	6.18	7.0	6
	101.0	6.53		
Glycine conjugates				
GCA	1.8	3.95		
	4.3	3.85		
	8.6	3.81		
	20	3.80	4.32	4
	86	4.09		
GDCA	2.1	4.69		
	10.5	4.84		
	20.0	4.77	4.96	3
GCDCA	1.5	4.23		
	7.6	4.34	4.77	3
	76.0	4.20		
Taurine conjugates				
TCA	18.5	1.85		
	37.0	1.85		
TDCA	19.0	1.93		
	38.0	1.95		

does not influence the pH at which cholic acid precipitates. At the other end of the scale, when the mixture contains 80% more NaTC (molar ratio conjugated to free—4.8:1) free bile acids did not precipitate. Therefore, the free acids in molar ratios greater than 4.8:1 are solubilized by the taurine conjugates. At 75% NaTC and 25% NaC (molar ratio conjugated:free, 3.6:1), precipitation was seen (Fig. 25, curve 9).

The number of moles of bile salt necessary to solubilize one mole of bile acid in solution is also shown in Fig. 26. As the amount of conjugated bile salt increases in the mixture, the ratio of bile salt needed to "carry" one mole of bile acid in a clear micellar solution decreases steadily until the mixture contains 60% or more NaTC, when the ratio becomes fixed between 5 and 5.5:1. When the bile salt in the mixture is predominantly NaC, three times the amount of bile salt is necessary to "carry" free bile acids in solution than when the predominant bile salt is the taurine conjugate.

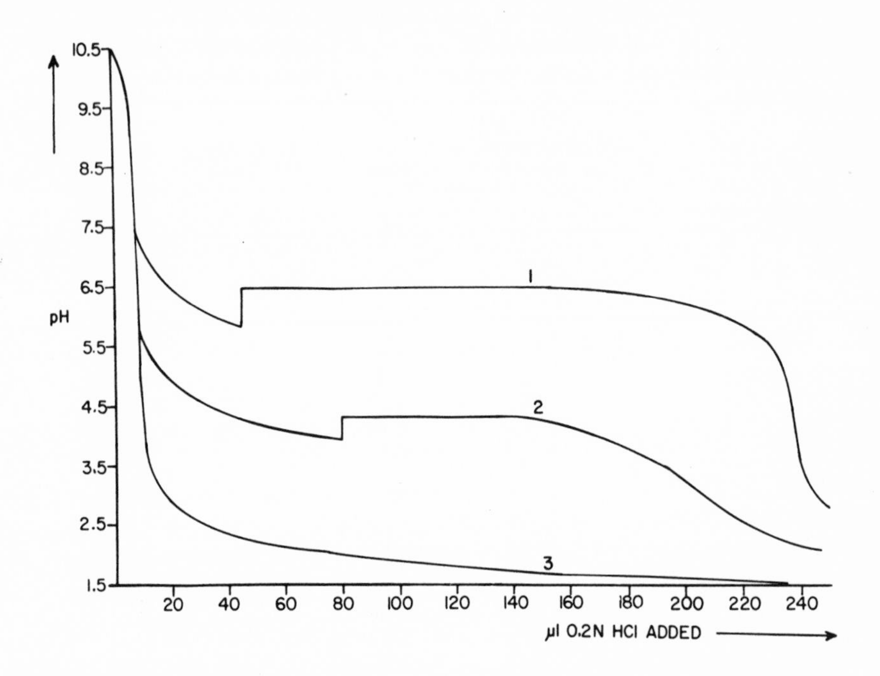

Fig. 24. Titration curves of sodium cholate, sodium glycocholate, and sodium taurodeoxycholate as 1% solutions in water at 37 °C. Sodium cholate (1), sodium glycocholate (2), and sodium taurocholate (3).

2. NaC and NaGC

Unlike the mixtures of NaC and NaTC in which the conjugate always remains in solution, both cholic and glycocholic acid form crystalline precipitates (Fig. 24) and therefore solutions containing mixtures of sodium salts of these bile acids exhibit a different type of titration curve (Fig. 27). The titration curves for pure NaC and NaGC are shown in broken lines together with the curve for a mixture containing 50% each of free and conjugated bile salt.

Initially the curve W–X–X'–Z follows a similar pattern to the ideal curve shown on Fig. 21. The precipitation at point X–X' is due to cholic acid crystals (confirmed by polarizing microscopy and determination of melting points). Point Z probably represents the completion of the titration of NaC since, in practice, the molar amount of acid added from W to Z, closely corresponds with the theoretical calculation of the number of moles of acid required to titrate the known amount of NaC. However, at point Q, further precipitation occurs, with a corresponding jump in pH (Q–Q'). This precipitate has an appearance grossly different from the milky opalescence of the cholic acid crystals in solution. Microscopy and melting point

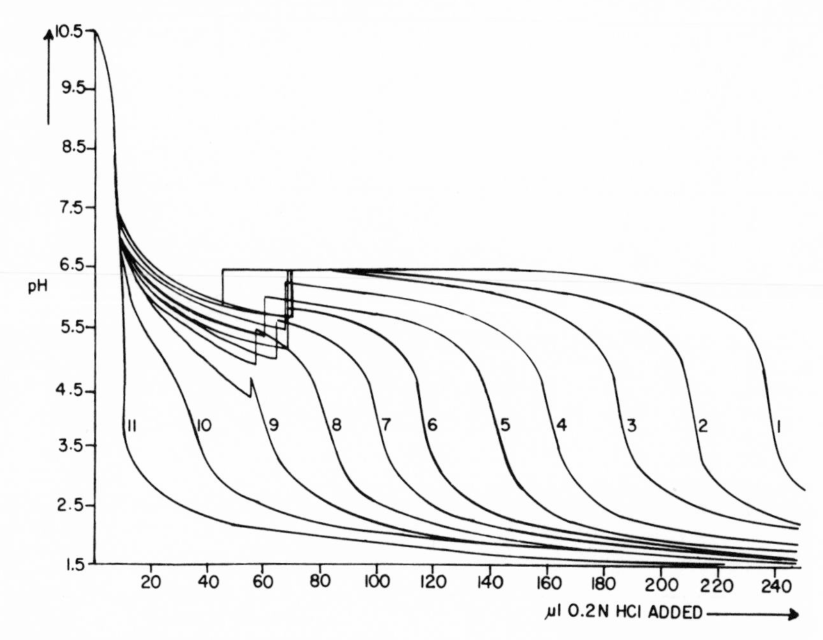

Fig. 25. Representative titration curves for mixtures of 1% solutions of NaC and NaTC. (1) NaC and NaTC. (2) 90% NaC, 10% NaTC. (3) 80% NaC, 20% NaTC. (4) 70% NaC, 30% NaTC. (5) 60% NaC, 40% NaTC. (6) 50% NaC, 50% NaTC. (7) 40% NaC, 60% NaTC. (8) 30% NaC, 70% NaTC. (9) 20% NaC, 80% NaTC. (10) 10% NaC, 90% NaTC. (11) 100% NaTC.

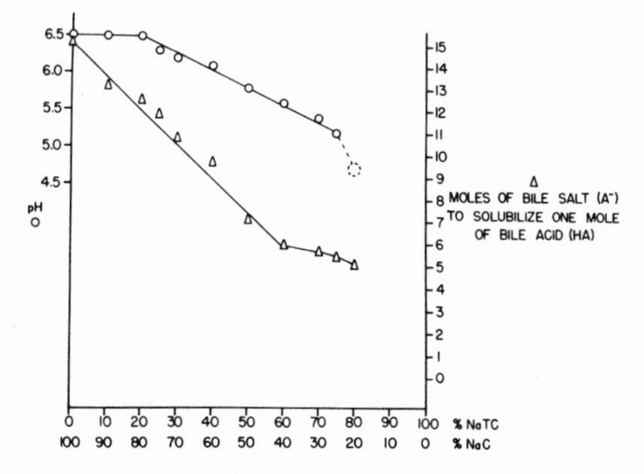

Fig. 26. Equilibrium pH levels at point of precipitation of cholic acid (O) and ratios of the number of molecules of bile salt necessary to solubilize one molecule of bile acid (Δ) from titration of varying mixtures of 1% solutions of NaC and NaTC. The broken circle was taken from curve 9 in Fig. 25 and does not represent a true equilibrium pH.

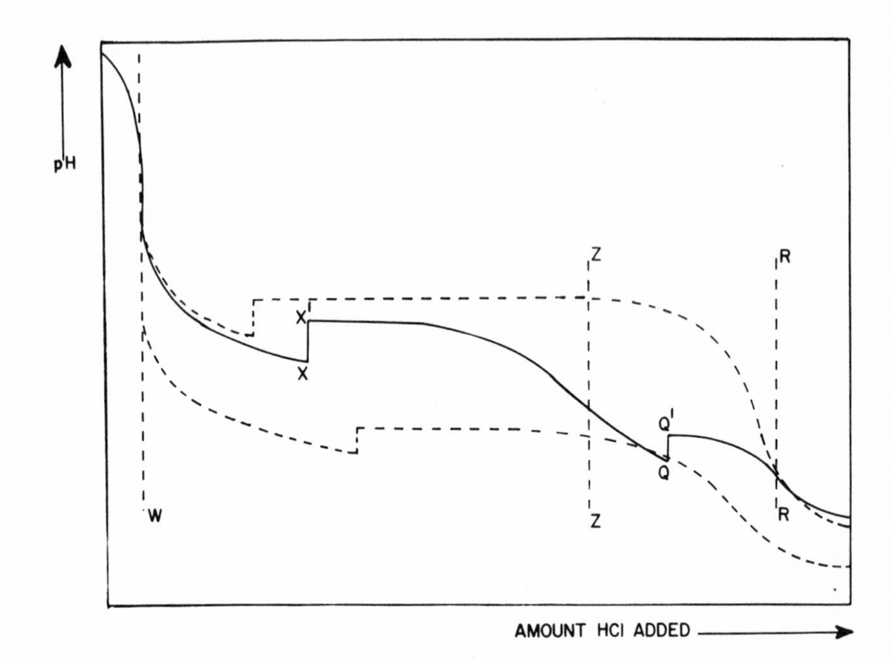

Fig. 27. Hypothetical titration curve for a mixture containing two bile acids (cholic and glycocholic acid), both of which precipitate from solutions of their sodium salts (solid line). The broken lines represent the titration curve for the free bile salt alone, NaC (above) and the pure glycine conjugate alone, NaGC (below). W—Z = amount of hydrochloric acid required to titrate the NaC in the mixture. Z—R = amount of hydrochloric acid required to titrate the NaGC in the mixture. X' = equilibrium *p*H at the point of precipitation of cholic acid. Q' = equilibrium *p*H at the point of precipitation of glycocholic acid.

determinations confirm that the second precipitation is due to crystals of glycocholic acid. The final equivalence point, R marks the completion of the NaGC titration. From this "bimodal" titration curve it is therefore possible to make the same calculations for the individual components in the mixture as were made from the curve for a single bile salt.

The compilation of individual titration curves for mixtures containing varying percentages of NaC and NaGC, together with measurements made from these curves, are shown in Figs. 28 and 29. With increasing amounts of NaGC up to 70%, the *p*H at which the free cholic acid crystals precipitate again falls. With higher concentrations of NaGC than 70% supersaturation again occurs and it is, therefore, not possible to calculate precipitation *p*H's for cholic acid beyond this point. The broken circles in Fig. 29 represent points where Tyndall effects were seen due to cholic acid microcrystals, but

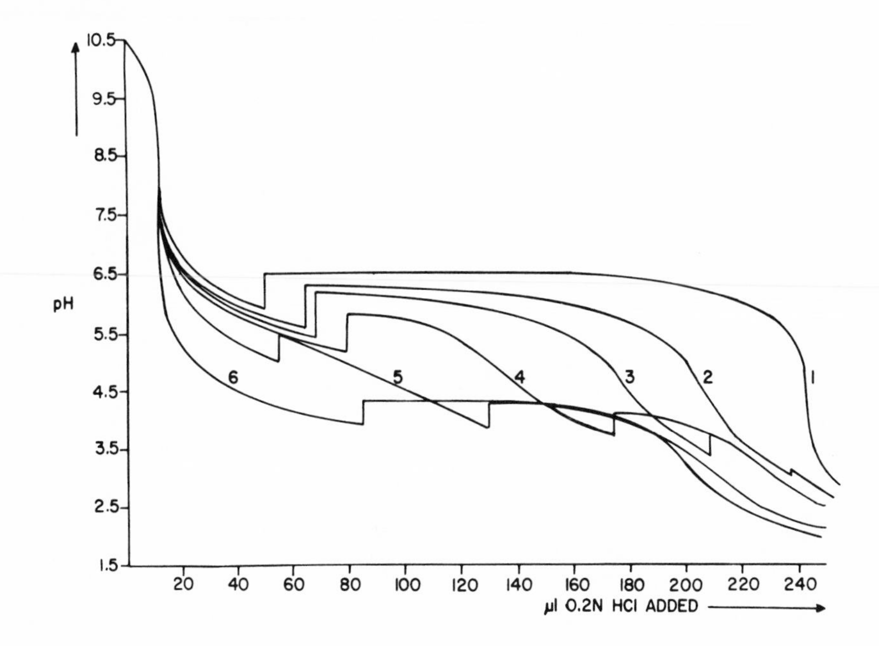

Fig. 28. Representative titration curves for mixtures containing 1% solutions of NaC and NaGC. (1) 100% NaC. (2) 80% NaC, 20% NaGC. (3) 70% NaC, 30% NaGC. (4) 50% NaC, 50% NaGC. (5) 30% NaC, 70% NaGC. (6) 100% NaGC.

without frank precipitation. Since they do not represent true equilibrium *p*H's, the dip in the line joining the precipitation *p*H's is drawn as a broken line and does not represent an equilibrium state. The calculations of the ratios of moles of bile salt A^- required to solubilize one mole of bile acid HA [Eq. 2] when the mixtures contain different percentages of NaGC are similar to those for NaC and NaTC.

3. *Titration Curves for Mixtures Containing Sodium Deoxycholate and Its Glycine and Taurine Conjugates*

The titration curves for combinations of NaDC and NaTDC are shown with their attendant calculations in Figs. 30 and 31 and for NaDC and NaGDC in Figs. 32 and 33.

In general, the deoxycholate mixtures showed the same pattern as was found for the free and conjugated mixtures of cholic acid salts.

Pure DCA precipitates from a 1% solution of NaDC at *p*H 6.92 and GDCA from a 1% solution of NaGDC at 4.96. As with NaTC, NaTDC

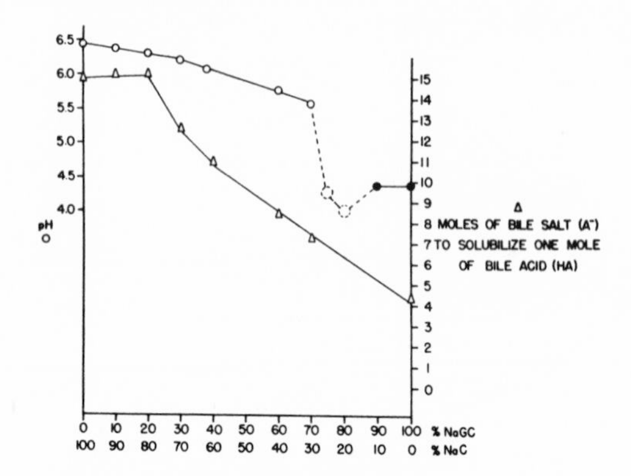

Fig. 29. Equilibrium *p*H at point of precipitation of cholic acid (○) and of glycocholic acid (●), and the ratios of the number of molecules of bile salt necessary to solubilize one molecule of bile acid (Δ) from titrations of varying mixtures of 1% solutions of NaC and NaGC. The broken circles represent the points where Tyndall effects were seen (due to cholic acid microcrystals) but without a jump in the titration curve of frank precipitation. These broken circles do not represent equilibrium *p*H's.

remains ionized throughout the titration and therefore does not precipitate. again, with increasing concentrations of either glycine or taurine conjugates in the mixture, there is a progressive fall in the *p*H level at which the reciprocal amount of DCA precipitates.

Unlike the cholate titrations, however, with increasing concentrations of the conjugates in the mixture there is little change in the ratio of the number of moles of anion required to solubilize one mole of bile acid. When the solution contains pure NaDC, the ratio of moles of bile salt needed to "carry" one molecule of bile acid in solution (A^-/H) is 6:1 and with increasing amount of conjugated deoxycholate in the solution, gradually falls to 4:1 for NaGDC and 3.9:1 for NaTDC.

Bile salts titrated in 0.1 *N* NaCl instead of water give titration curves virtually identical to those done in water. Further, the A^-/HA and the pK'_a are the same in water and 0.1 *N* NaCl.

The pK'_a of the free bile acids calculated from mixtures was surprisingly constant despite the fact that the concentration of the conjugated species varied markedly. These values are: pK'_a cholic acid $= 5.23 \pm 0.01$ and pK'_a deoxycholic acid $= 6.12 \pm 0.05$. This fact, when compared to the change in pK'_a of the free acids with decreasing concentration in the absence of conjugated bile salts, suggests that if adequate conjugated bile salt is present to

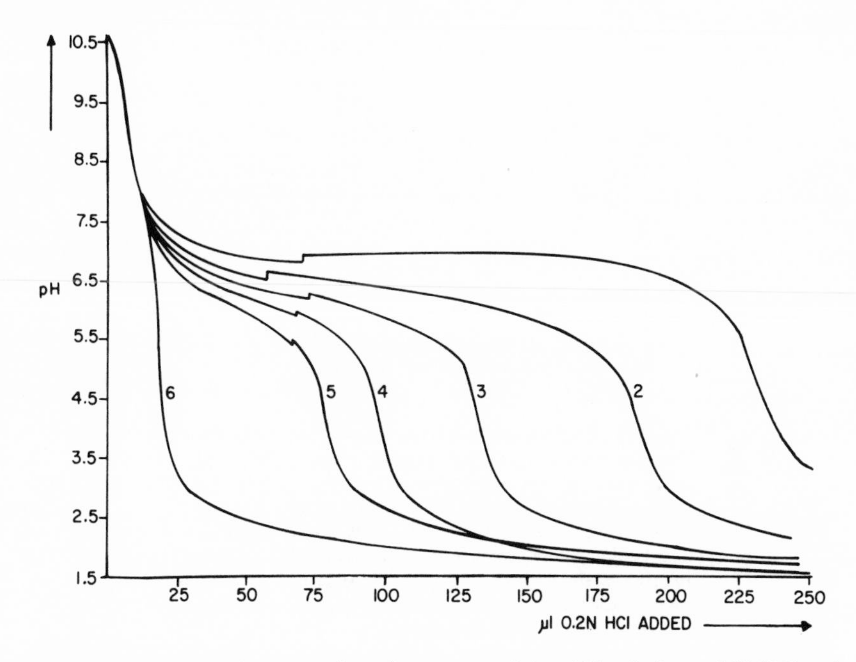

Fig. 30. Representative curves for mixtures containing 1% solutions of NaDC and NaTDC. (1) 100% NaDC. (2) 75% NaDC, 25% NaTDC. (3) 50% NaDC, 50% NaTDC. (4) 30% NaDC, 70% NaTDC. (5) 25% NaDC, 75% NaTDC. (6) 100% NaTDC.

form micelles the free bile salt present will also be part of the micelle and thus will have a pH'_a of "micellar" bile acid.

D. Alkali Metal Salts of Bile Salts

Since the pK'_a of the bile acids is fairly low, one might expect that at physiological *p*H's they would be partly ionized. Thus, the interaction with univalent ions is important and will be considered in detail.

1. *Salts of Cholanic Acid*

All alkali metal salts of cholanic acid are extremely insoluble in water from temperatures of 0°C to 100°C. The sodium and potassium salts of its glycine and taurine conjugates are also very insoluble.

2. *Salts of Lithocholic Acid*

Lithocholic acid is a monohydroxy bile acid which is found in trace amounts in the bile of man and other animals. In contrast to the predominant

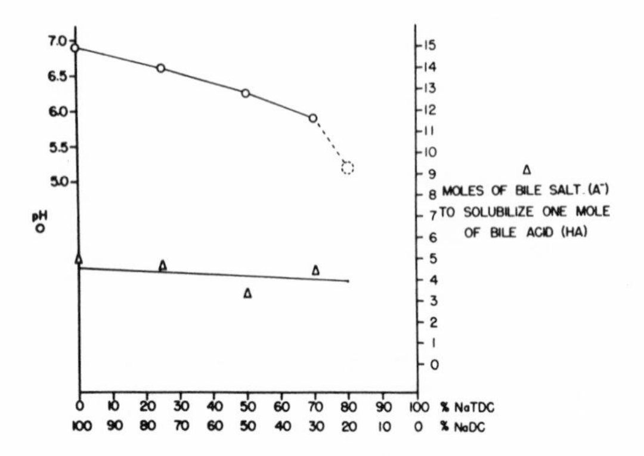

Fig. 31. Equilibrium *p*H levels at point of precipitation of NaDC (○) and ratios of the number of molecules of bile salt necessary to solubilize one molecule of bile acid (Δ) from titration of 1% solutions of varying mixtures of NaDC and NaTDC. The broken circle represents a Tyndall effect due to microcrystals of deoxycholic acid but without frank precipitation or a jump in the titration curves. It does not represent an equilibrium *p*H.

bile salts in man (cholate, deoxycholate, and chenodeoxycholate), lithocholate is highly toxic.* Also unlike the common di- and trihydroxy bile salts, salts of lithocholic acid are rather insoluble at body temperatures, but, unlike salts of cholanic acid, salts of lithocholic acid show a sharp increase in solubility with temperature. This behavior is reminiscent of detergents and soaps which are amphiphilic compounds. The term amphiphile indicates that those molecules possess hydrophilic and hydrophobic properties. A characteristic of the soluble amphiphile is the ability to form multimolecular aggregates (micelles) in water (42, 132). Micelle formation in relatively dilute solutions can occur only above a certain solute concentration, the critical micellar concentration (CMC), and at solution temperature above the critical micellar temperature (CMT) (1, 10, 44b).

The CMC and CMT can best be explained by considering the effects of temperature and concentration on the physical state of mixtures of soluble

*It has been shown to produce severe inflammation and fever when injected subcutaneously in man (129), cholestasis when infused into rats and hamsters (25, 26), liver damage and gallstones when fed to rats (14, 15, 16), and cirrhosis when fed to a number of different species of laboratory animals (17–24). Further, lithocholate has recently been implicated in the etiology of human liver disease (130, 131). The common denominator between all of these pathologic processes may possibly be due to accumulation, especially in the liver, of lithocholate in an insoluble form (45).

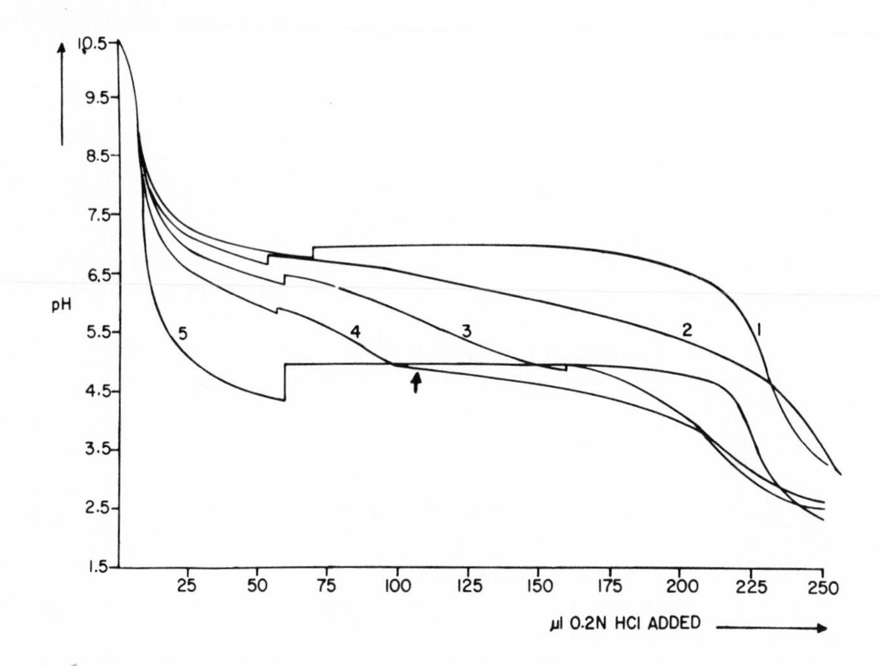

Fig. 32. Representative titration curves for mixtures containing 1% solutions of NaDC and NaGDC. (1) 100% NaDC. (2) 75% NaDC, 25% NaGDC. (3) 50% NaDC, 50% NaGDC. (4) 20% NaDC, 80% NaGDC. (5) 100% NaGDC. The arrow on curve 4 indicates a Tyndall effect due to microcrystals of NaGDC.

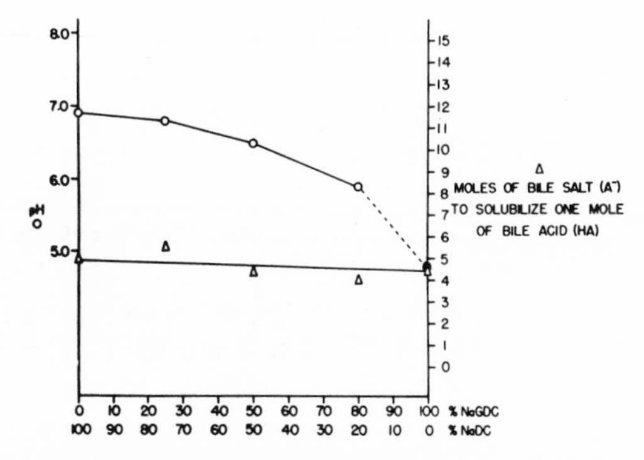

Fig. 33. Equilibrium pH at point of precipitation of deoxycholic acid (○) and of glycodeoxycholic acid (●) and ratios of the number of molecules of bile salt necessary to carry one molecule of bile acid in solution (Δ) in titrations of varying mixtures of 1% solutions of NaDC and NaGDC.

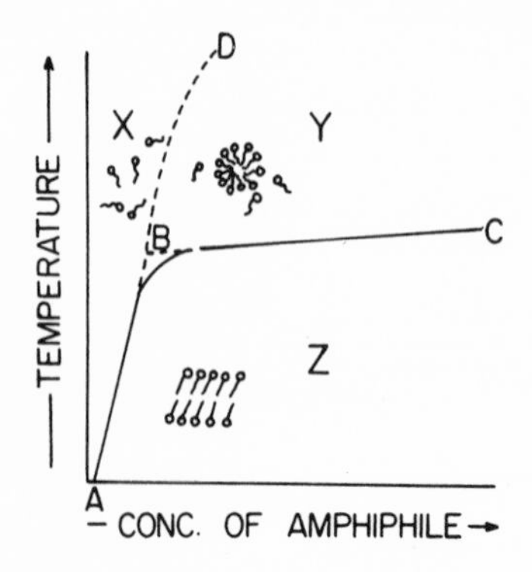

Fig. 34. Hypothetical representation of the binary phase diagram at regions of very dilute concentrations of amphiphilic compound. Small pollywog-like creations represent molecules of an amphiphilic compound, such as a detergent. The orderly array in zone *Z* represent crystals. The spherical array in zone *Y* represent micelles. In zone *X* only individual molecules are present. For explanation of *A*, *B*, *C*, and *D* see text.

amphiphiles in water as is shown in Fig. 34. At 0 °C little amphiphile is soluble (point *A*). However, as the temperature is increased the amount of amphiphile in solution (as individual molecules) increases slightly in a nearly linear fashion along *AB*. After point *B* there is a rather sharp increase in solubility over a narrow range of temperature (*BC*).

Why does this occur? Consider a solution in zone *X* at a temperature well above *B*. Under these conditions the individual molecules of amphiphile are freely distributed throughout the water forming a true solution. As the concentration of amphiphile is increased (moving toward zone *Y*) a point of maximum molecular solubility is reached at line *BD*. Any further addition of amphiphile results in the formation of micelles (zone *Y*). Therefore, concentration at which micelle formation first occurs (*BD*) is the CMC. It is influenced slightly by temperature.

Consider now a mixture falling in zone *Z* with a concentration of amphiphile greater than *B*. This mixture is a turbid suspension of two phases—a liquid phase having a composition lying along line *AB* and a solid phase composed of the crystalline amphiphile. As the temperature is increased a point is reached on *BC* where the solution clears. This clearing is due to melting of the crystals of amphiphile and subsequent micelle formation resulting in a clear solution. Thus, *BC* gives the relation between temperature and the maximum solubility of the amphiphile in a micellar solution while *AB* gives the relation between temperature and the maximum solubility of the amphiphile as individual molecules.

The temperature of clearing (*BC*) for solutions of salts of long-chain fatty acids increases only slightly with increases in the concentration of amphiphile above the CMC (*BD*) (133, 134).

This has led previous investigators to ignore the effect of detergent concentration on solubility above the CMC. In fact, the Krafft point, often used synonymously with CMT, originally referred to the slight temperature varia-

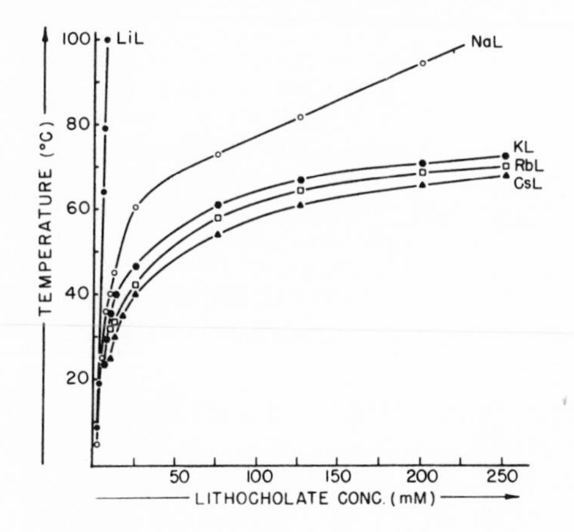

Fig. 35. The solubility of the alkali metal salts of lithocholic acid in water as a function of temperature. LiL, lithium lithocholate; NaL, sodium lithocholate; KL, potassium lithocholate; RbL, rubidium lithocholate; CsL, cesium lithocholate (45).

tion in *BC* for soaps of long-chain fatty acids (133). In contrast, the CMT of lithocholates increase significantly with increases in the concentration above the CMC (as will be seen in Fig. 35). For this reason, I have chosen not to use the terms CMT and Krafft point synonymously, as is common practice. Rather I have restricted the use of the latter term to the temperature at which clearing occurs in solutions in which the concentration of amphiphile is approximately at the CMC. Therefore, the Krafft point represents a single point (*B*) on the curve shown in Fig. 34 and the CMT represents a range of temperature which varies with the concentration of amphiphile (and counterion) present (*BC*). Point *B* is analogous to a triple point since both temperature and concentration are fixed.

A major and striking physicochemical difference between lithocholic acid and the dihydroxy and trihydroxy bile acids is the insolubility of the sodium salts of the former (10, 45). Sodium salts of the common bile acids (taurine and glycine conjugates of cholic acid, deoxycholic acid, and chenodeoxycholic acid) are very soluble in water and physiological saline, even at 0 °C. The solubility of the ammonium, lithium, sodium, potassium, rubidium, and cesium salts of lithocholic acid (NH_4L, LiL, NaL, RbL, CsL) have been studied in water as a function of temperature (45).

The effects of different counterions on the solubility of lithocholate is shown in Fig. 35. The lithium salt is quite insoluble. The solubility increases in a linear fashion with temperature and reaches a maximum of approximately 0.6 mmole/liter at 100 °C. Although the solubility of sodium lithocholate is slight at temperatures between 0 and 60 °C, there is a sharp increase in the solubility above 60 °C. This break in the solubility curve, although less sharp than analogous curves obtained with soaps of long-chain fatty acids (135), gives an estimate of the Krafft point. Similar curves for the potassium,

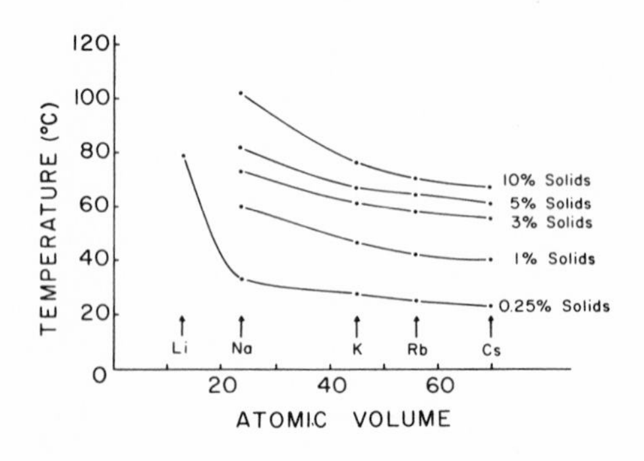

Fig. 36. Critical micellar temperature (CMT) of alkaline metal salts of lithocholic acid as a function of the atomic volume of the alkaline metal. CMT vertical axis, atomic volume horizontal axis. Percent solids refers to the total amount of bile salt in g/100 ml water. Note that as the atomic volume of the alkaline metal decreases, the Krafft point for any given concentration of bile salt increases. There is a striking rise with lithium, which has the smallest atomic volume.

rubidium, and cesium lithocholates show that the Krafft point decreases in that order. Thus, the solubility is a function of not only the temperature, but the type and concentration of counterion. The solubility of the alkaline metal salts, at any given temperature, increases according to the ionic radius or atomic volume of the metal (Fig. 36). Thus, the order of solubility is: LiL $<$ NaL $<$ KL $<$ RbL $<$ CsL. Similar results have been obtained with alkali metal soaps of long-chain fatty acids (135). Also, as the concentration of the salt is increased, its counterion concentration (Na^+, K^+, Rb^+, etc.) also increases. Thus the increase in counterion concentration decreases the solubility, especially in the case of Na^+.

The effect of conjugation on the solubility of lithocholic acid salts is shown in Fig. 37. Conjugation with either glycine or taurine decreases the solubility for any given temperature. Thus, the physiological mechanism of conjugation of lithocholic acid by the liver actually decreases the solubility of the conjugates compared to the nonconjugated form. Palmer has found that the glycine and taurine conjugates of lithocholic acid undergo a second

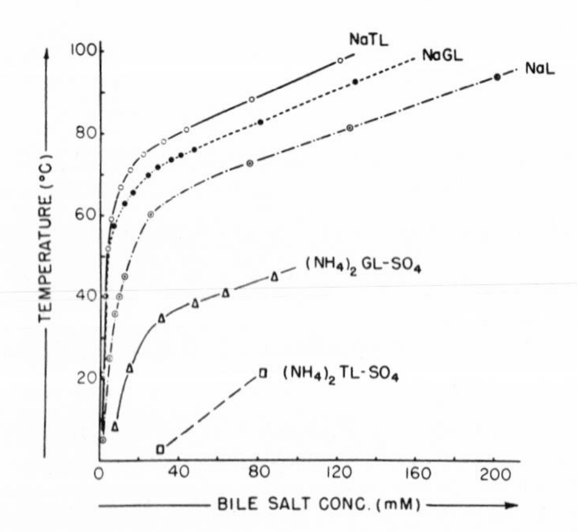

Fig. 37. The solubility of lithocholate conjugates in water as a function of temperature. NaTL, sodium taurolithocholate; NaGL, sodium glycolithocholate; NaL, sodium lithocholate; $(NH_4)_2$TL-SO_4, diammonium glycolithocholate–sulfate; $(NH_4)_2$TL-SO_4, diammonium taurolithocholate–sulfate (the sulfates were a gift of Dr. R. Palmer) (45).

conjugation at the 3α hydroxyl group with sulfate (126). The solubility of the diammonium salts of glycolithocholate–sulfate and taurolithocholate–sulfate is quite high compared with the simple lithocholate conjugates. Thus, by adding another strong polar group to the opposite end of the molecule, the solubility has been increased.

E. Solubilization of Lithocholate by Other Bile Salts

The change in the solubility of sodium lithocholate brought about by the addition of other common di- or trihydroxy bile salts (45) is shown in Fig. 38. When large amounts of the common bile salts are present (left-hand side of the diagram) the lithocholate can be easily solubilized. Although these results show that free di- and trihydroxy bile salts solubilize either free or conjugated lithocholates, similar studies using taurine or glycine conjugates of the di- and trihydroxy salts show that they also effectively solubilize lithocholate. This solubilization takes place in mixed micelles. As the mole % lithocholate is increased in the mixture there is a decrease in its solubility. These curves are similar to those reported for binary mixtures of detergents (137) or soaps (138). At each molar ratio the taurine and glycine conjugates of lithocholate are less soluble than the free bile salt. The effect of increased counterion concentration is to decrease the solubility at every level, whereas the effect of changing the counterion to potassium (the major intracellular cation) is to increase the solubility at every level.

Concerning the pathological conditions attributable to either infusion or increased dietary intake of lithocholate (14–22), it has been speculated (45) that increased quantities of lithocholate present in serum are probably bound by protein (32) and therefore do not precipitate. On entering the liver cell

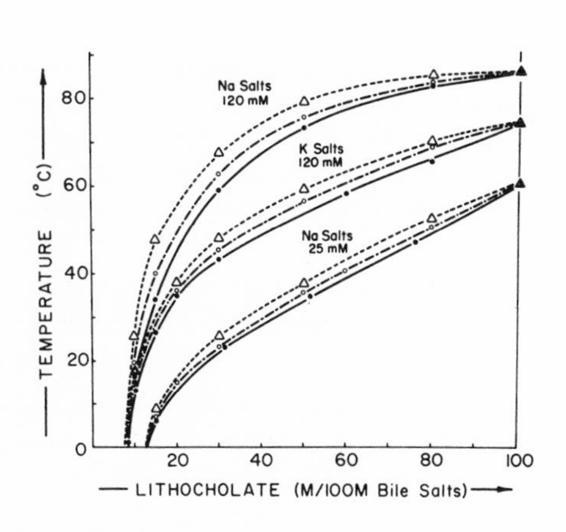

Fig. 38. The effect of dihydroxy and trihydroxy bile salts on the solubility of lithocholate. The upper and lower sets of three curves (Na salts, 120 m*M*; Na salts, 25 m*M*) represent the solubility of NaL as a function of temperature in varying amounts of sodium chenodeoxycholate (*Δ*), sodium deoxycholate (○), and sodium cholate (●). Total bile salt concentrations are 120 m*M* in upper three curves and 25 m*M* in lower three curves. The middle set of three curves (K salts, 120 m*M*) represent solubility of KL in potassium chenodeoxycholate (*Δ*), potassium deoxycholate (○), and potassium cholate (●), at a total bile salt concentration of 120 m*M* (45).

they may be maintained in soluble form as potassium salts since potassium is the major intracellular cation, but on being excreted into the bile canaliculi where sodium is the predominant cation, their solubility may be exceeded unless adequate amounts of other normal conjugated bile acids are present to render them soluble. Precipitation of these bile salts in canaliculi (14, 15, 16) could lead to inflammation, intrahepatic cholestasis, liver damage, cirrhosis, and stone formation.

F. Salts of Polyvalent Cations

Little is known about the effects of other cations on bile salts (1). Calcium ions precipitate free bile salts readily. Calcium and magnesium also precipitate dihydroxy glycine conjugates, but not trihydroxy glycine conjugates or taurine conjugates. Since certain types of gallstones are primarily the calcium salts of abnormal bile acids [see (9) for summary of mechanism], it is important to establish the solubilities of bile salts of polyvalent cations alone and in mixtures with other ions.

VIII. MICELLE FORMATION AND CRITICAL MICELLAR CONCENTRATION (CMC) OF BILE SALTS

The critical micellar concentration of any detergent may be determined by a number of different methods, including the solubilization of insoluble dye, osmotic pressure, conductivity, surface tension, light scattering, nuclear magnetic resonance, refractive index, freezing point determination, vapor pressure, sound velocity, etc. (141). Each method may give a somewhat different value for CMC.

A. Summary of Previous Work

The CMC's of the various bile salts have been studied in a number of different laboratories by different techniques (12, 44b, 118, 120, 142–153). These studies are summarized in Tables VI, VII, and VIII. Each bile salt is listed individually and the method of determining the CMC, temperautre, *p*H, and the type of medium in which the studies were carried out are listed. In a number of cases the temperature, *p*H, or medium were not given. Further, the purity of the bile salts was not always stated. Despite the variability of results it appears that (1) the trihydroxy bile salts tend to have CMC's somewhat higher than the dihydroxy bile salts and that (2) the CMC decreases with increasing concentrations of counterion (e.g., Na^+) present in the

TABLE VI. Critical Micellar Concentration of Cholates

Bile salt	Reference		Method	T, °C	pH	Medium	mmoles/liter
NaC	Ekwall	(142)	Solubilization	20°C	NS[a]	Water	13
	Norman	(143)	Solubilization (methylcholanthrene)		NS	Water	12
	DeMoerloose	(144)	Light scattering	20°C	6.6–7.6	0.1 *N* NaCl	20
	Furusawa	(145)	Surface tension	10°C	NS	Physiological saline	3.3
	Miyake	(146)	Surface tension	20°C	NS	NS	5
	Small	(152)	Surface tension	22°C	9.0	Water	12
	Small	(152)	Surface tension	22°C	7.4	*M*/15 Na phosphate	4.9
	Small	(152)	Surface tension	22°C	7.4	*M*/5 Na phosphate	3.25
NaTC	Ekwall	(142)	Solubilization	20°C	NS	Water	10
	Hofmann	(12)	Solubilization (azobenzene)	37°C	6.3	0.15 *M* Na+	10.0
	Hofmann	(12)	Solubilization (1-mono-olein)	37°C	6.3	0.15 *M* Na+	4.2
	Furusawa	(145)	Surface tension	10°C	NS	Physiological saline	3.4
	Miyake	(146)	Surface tension	20°C		NS	1.5
	Norman	(143)	Solubilization (20-methylcholanthrene)	RT[b]	NS	Water	12.0
	Bates	(147,148)	Solubilization (Griseofulvin)	37	5.4–6.6	Water	8.0
	Bates	(147,148)	Solubilization (Hexestrol)	37	5.4–6.6	Water	14.0
	Carey *et al.*	(149)	Electron spin resonance (ESR) (Nitroxide probe)	33	6.8	Water	6.0
	Carey *et al.*	(150)	Shift in maximum wavelength of emission spectra (Rhodamine 6G)	25	6.8	Water	4.5
	Woodford	(153)	Diffusion	25	NS	Water	6.7

TABLE VI (continued)

Bile salt	Reference		Method	T, °C	*p*H	Medium	mmoles/liter
NaGC	Ekwall *et al.*	(142)	Solubilization	20 °C	NS	Water	10
	Hofmann	(12)	Solubilization (azobenzene)	37 °C	6.3	0.15 *M* Na^+	8.0
	Hofmann	(12)	Solubilization (1-mono-olein)	37	6.3	0.15 *M* Na^+	4.2
	Furusawa	(145)	Surface tension	10	NS	Physiological saline	3.1
	Miyake	(146)	Surface tension	20	NS	NS	1.25

[a] NS (not stated).
[b] RT (room temperature).

TABLE VII. Critical Micellar Concentration of Deoxycholates

Reference	Method	T, °C	*p*H	Medium	mmoles/liter
NaDC					
Johnston and McBain (151)	Freezing point depression	0	NS	Water	70
Ekwall (142)	Solubilization	20	NS	Water	5
Norman (143)	Solubilization (methylcholanthracine)			Water	5
Furusawa (145)	Surface tension	10	NS	Physiol. saline	1.7
DeMoerloose (144)	Surface tension	20	6.6–7.6	Water	2
DeMoerloose (144)	Surface tension	20	6.6–7.6	0.1 *N* NaCl	1
Miyake (146)	Surface tension	20	NS	NS	5
NaTDC					
Ekwall (142)	Solubilization and other methods	20	NS	Water	4
Norman (143)	Solubilization (20-methylcholanthrene)	RT	NS	Water	5.0
Hofmann (12)	Solubilization (Azobenzene)	37	6.3	0.15 *M* NaCl	1.9
Small and Carey (43, 44)	Light scattering	25	8.0	Water	1.5
Small and Carey (43, 44)	Light scattering	25	8.0	0.1 *M* NaCl	1.4
Small and Carey (43, 44)	Light scattering	25	8.0	0.3 *M* NaCl	1.1
Small and Carey (43, 44)	Light scattering	25	8.0	0.5 *M* NaCl	1.0
Kratohvil and Dellicolli (120)	Light scattering	25	6.0	Water	2.9
Kratohvil and Dellicolli (120)	Light scattering	25	6.0	0.15 *M* NaCl	1.8
Kratohvil and Dellicolli (120)	Light scattering	25	6.0	0.50 *M* NaCl	1.3
Kratohvil and Dellicolli (120)	Surface tension	25	6.0	Water	3.1
Kratohvil and Dellicolli (120)	Surface tension	25	6.0	0.015 *M*	2.4
Kratohvil and Dellicolli (120)	Surface tension	25	6.0	0.050 *M* NaCl	1.9
Kratohvil and Dellicolli (120)	Surface tension	25	6.0	0.15 *M* NaCl	1.7
Kratohvil and Dellicolli (120)	Surface tension	25	6.0	0.30 *M* NaCl	1.2
Kratohvil and Dellicolli (120)	Surface tension	25	6.0	0.50 *M* NaCl	1.2
Carey *et al.* and Small (149)	Electron spin resonance (ESR) (nitroxide probe)	33	6.8	Water	3.5

TABLE VII (continued)

Reference	Method	T, °C	*p*H	Medium	mmoles/liter
Carey *et al.* and Small (150)	Shift in maximum wavelength of emission spectra (Rhodamine 6G)	25	6.8	Water	1.5
NaGDC					
Ekwall (142)	Solubilization and other methods	20	NS	Water	4
Hofmann (12)	Solubilization (azobenzene)	37	6.3	0.15 *M* Na+	1.9
Hofmann (12)	Solubilization (1-mono-olein)	37	6.3	0.15 *M* Na+	0.6
Furusawa (145)	Surface tension	10	NS	Physiological saline	2.6
Miyake (146)	Surface tension	20	NS	NS	1.25
Kratohvil and Dellicolli (120)	Surface tension	20	6	Water	2.12
Kratohvil and Dellicolli (120)	Surface tension	20	6	0.015 *M*	1.7
Kratohvil and Dellicolli (120)	Surface tension	20	6	0.05 *M* NaCl	1.3
Kratohvil and Dellicolli (120)	Surface tension	20	6	0.15 *M* NaCl	1.1
Kratohvil and Dellicolli (120)	Surface tension	20	6	0.30 *M* NaCl	0.9
Kratohvil and Dellicolli (120)	Surface tension	20	6	0.50 *M* NaCl	0.73
Kratohvil and Dellicolli (120)	Light scattering	20	6	0.15 *M* NaCl	1.1
Kratohvil and Dellicolli (120)	Light scattering	20	6	0.50 *M* NaCl	0.74

medium. Because the results on the same bile salts studied by different techniques in different laboratories are often in conflict, we have carried out a systematic study of the effects of temperature and salt concentration on the CMC of a trihydroxy bile salt (NaTC), a dihydroxy bile salt (NaTDC), and an equimolar mixture of the two bile salts (44). The taurine conjugates were chosen to eliminate the effect of *p*H. The technique of spectral shift of a water-soluble dye (Rhodamine 6G) was employed (141) because it can be used in high salt concentrations and at different temperatures. The results are given in Tables IX, X, and XI.

B. Effects of Temperature

In Fig. 39 the CMC's for NaTC, NaTDC, and NaTC/NaTDC mixtures

TABLE VIII. Critical Micellar Concentration of Chenodeoxycholate and Other Miscellaneous Bile Salts

Bile salt	Reference	Method	T, °C	*p*H	Medium	mmoles/liter
NaCDC	Norman (143)	Solubilization (methylcholanthrene)		NS	Water	6
NaCDC	Miyake (146)	Surface tension	20°C	NS	NS	30
NaTCDC	Hofmann (12)	Solubilization (azobenzene)	37°C	6.3	0.15 *M* Na^+	2.5
NaTCDC	Hofmann (12)	Solubilization (1-mono-olein)	37°C	6.3	0.15 *M* Na^+	0.8
NaTCDC	Miyake (146)	Surface tension	20°C	NS	NS	1.5
NaGCDC	Hofmann (12)	Solubilization (azobenzene)	37°C	6.3	0.15 *M* Na^+	2.4
NaGCDC	Hofmann (12)	Solubilization (1-mono-olein)	37°C	6.3	0.15 *M* Na^+	0.8
NaGCDC	Miyake (146)	Surface tension	20°C	NS	NS	8
Na lithocholate	Small (118)	Spectral shift (44)	60°C	8	Water	0.2
Na glycolithocholate	Miyake (146)	Surface tension	20°C	NS	NS	1
Na taurourocholate	Miyake (146)	Surface tension	20°C	NS	NS	1.5
Na glycourocholate	Miyake (146)	Surface tension	20°C	NS	NS	1.5
Na hyodesoxycholate	Miyake (146)	Surface tension	20°C	NS	NS	30
Na dehydrocholate	Ekwall (142)	Solubilization and other methods	20°C	NS	Water	140
mixture of conjugated bile salts	Hofmann (12)	Solubilization (azobenzene)	37°C	6.3	0.15 *M* Na^+	3
mixture of conjugated bile salts	Hofmann (12)	Solubilization (1-mono-olein)	37°C	6.3	0.15 *M* Na^+	1.4
mixture of conjugated bile salts	Small (152)	Surface tension	22°C	5.3	*M*/15 Na phosphate	0.7
mixture of conjugated bile salts	Small (152)	Surface tension	22°C	7.5	*M*/15 Na phosphate	0.7

TABLE IX. CMC's of NaTC at Eight Temperatures (10–80 °C) in Water and Seven Added Counterion Concentrations of 0.02–3.0 *M* NaCl (44)

	NaTC, mmoles/liter								
NaCl (*M*)	10	20	30	40	50	60	70	80	Rounded off Ag♯ 20 °C[a]
0	3.2	2.8	3.1	3.0	3.3	3.4	4.5	5.2	4
0.02	3.9	3.3	3.5	3.4	4.0	3.7	4.6	6.0	4
0.04	3.9	2.9	3.5	3.7	3.5	4.4	4.5	5.4	4
0.08	3.7	3.1	3.5	3.9	4.2	4.2	4.6	5.4	5
0.15	4.1	2.7	4.0	4.0	4.1	4.7	5.0	6.0	5
0.3	3.3	3.3	3.1	3.2	3.3	4.4	4.4	5.0	6
1.0	2.0	2.1	2.4	2.5	2.8	3.2	3.2	4.2	9
3.0	0.6	0.7	0.6	0.7	1.1	1.1	1.3	2.0	16

[a] Aggregation numbers (Ag♯) from (43) for each counterion concentration at 20 °C.

TABLE X. CMC's of NaTDC at Eight Temperatures (10–80 °C) in Water and Six Added NaCl Concentrations (0.2–0.6 *M*) (44)

	NaTDC, mmoles/liter								
NaCl (*M*)	10	20	30	40	50	60	70	80	Rounded off Ag♯ 20 °C[a]
0	1.8	1.5	1.8	1.8	2.1	2.4	3.0	3.2	6
0.02	1.6	1.7	1.6	1.7	1.9	2.1	2.7	3.3	7
0.04	1.6	1.7	1.6	1.5	1.9	2.4	2.7	3.3	10
0.08	1.1	1.1	1.1	1.4	1.7	2.1	2.7	3.1	13
0.15	0.9	0.8	1.0	1.2	1.6	1.8	1.8	2.7	22
0.3	0.9	0.9	0.8	1.1	1.6	1.7	1.9	2.0	32
0.6	0.8	0.8	0.8	0.8	0.9	1.0	1.3	1.5	55

[a] Aggregation numbers from (43) at 20 °C.

TABLE XI. CMC's of an Equimolar Mixture of NaTC/NaTDC. Temperature Range 10–80 °C and Added Counterion Concentration 0.02 to 0.8 *M* NaCl (44)

	NaTC:NaTDC mixture, mmoles/liter								
NaCl (*M*)	10	20	30	40	50	60	70	80	Rounded off Ag♯ for equimolar mixture of conjugated di- and trihydroxy bile salts, 20C °[a]
0	2.3	2.3	2.1	1.6	2.9	4.2	4.5	4.2	—
0.02	1.9	1.7	2.1	2.2	2.6	2.9	3.6	4.2	—
0.04	1.6	1.9	1.8	2.1	1.9	3.0	3.8	5.6	9
0.08	1.8	1.8	1.8	1.8	2.1	2.1	3.5	4.0	—
0.15	1.5	1.5	1.6	1.3	1.7	2.7	2.7	3.1	15
0.3	1.1	1.2	1.4	1.2	1.9	2.9	3.1	3.2	19
0.8	0.9	0.8	0.8	0.8	1.1	1.2	1.6	1.8	—

[a] Aggregation numbers at 20°C in 0.04, 0.15, and 0.3 *M* NaCl for an equimolar mixture of conjugated di- and trihydroxy bile salts (44a, 44b).

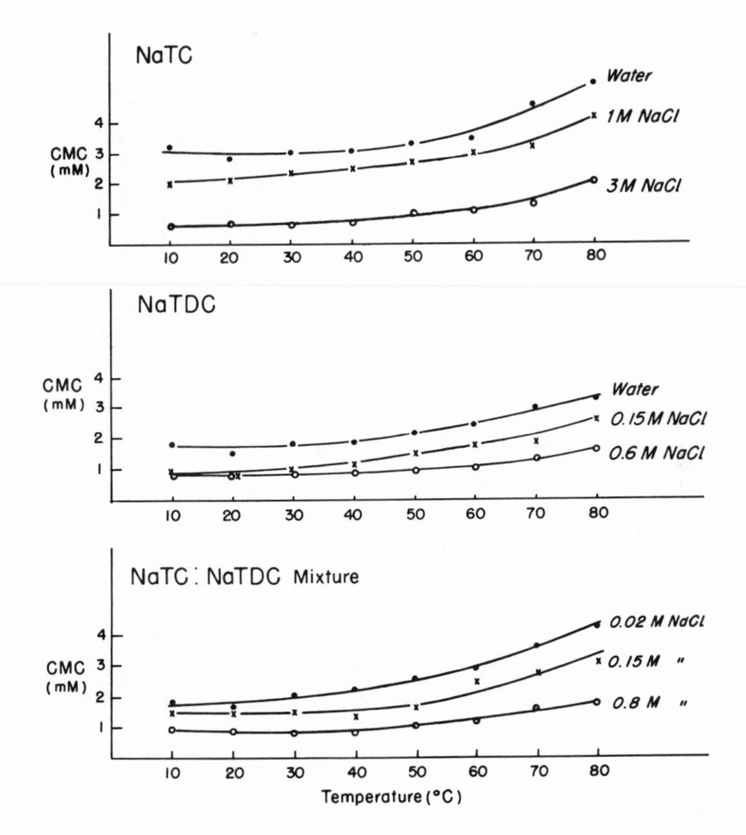

Fig. 39. Example of CMC values (m*M*) plotted against temperature (10–80 °C) for NaTC in water, 1 and 3 *M* NaCl (top), NaTDC in water, 0.15 and 0.6 *M* NaCl (middle), and NaTC/NaTDC mixtures in 0.02, 0.15, and 0.8 *M* NaCl (bottom) (44).

are plotted against temperatures at three NaCl concentrations. The following effects of temperature on the bile salts are noted.

1. NaTC (Table IX and Fig. 39, Top)

A fall in the CMC occurs as the temperature is increased from 10 to 20 °C; this phenomenon is seen in the aqueous and salt solutions up to 0.3 *M*. In the presence of NaCl concentrations of 1.0 and 3.0 *M*, no minimum occurs at 20 °C. Above 60 °C the CMC increases steadily with temperature in all the NaCl solutions.

2. NaTDC (Table X and Fig. 39, Middle)

In water a minimum value for CMC was noted at 20 °C. For each NaCl concentration the CMC remains nearly constant up to 40 °C, after which it

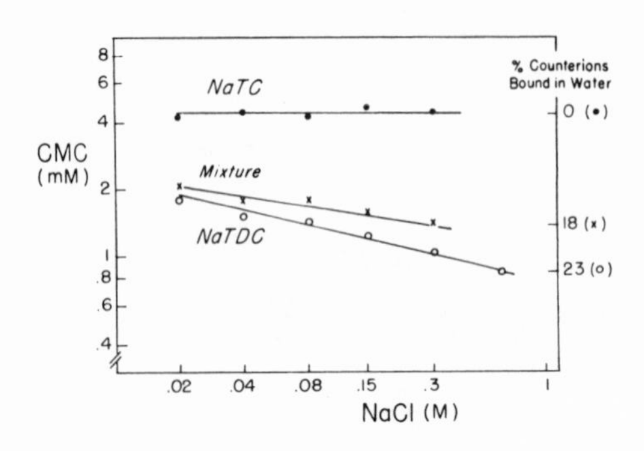

Fig. 40. Examples of log critical micellar concentration *vs* log counterion (NaCl) concentration for NaTC, an equimolar mixture of NaTC and NaTDC (mixture), and NaTDC. The percent of counterions bound to the micelle in water is indicated on the right-hand side of the diagram: 0% for NaTC, 18% for the mixture, and 23% for NaTDC. For further explanation see text.

increases gradually with increasing temperature, though the increment with temperature was less than with NaTC.

3. *NaTC/NaTDC Mixtures (Table XI and Fig. 39, Bottom)*

The values obtained for the CMC were closer to NaTDC than NaTC. The variations of CMC with temperature were also more like that of the dihydroxy bile salt. At each NaCl concentration a slight fall or plateau occurs in the CMC at lower temperatures. The CMC does not begin to rise much until 50 °C and then does so gradually in water and all salt concentrations.

C. Effects of Added NaCl (Counterion)

Examples of log-log plots of NaCl concentrations with corresponding CMC values at 20 °C of NaTC, NaTDC, and NaTC/NaTDC mixtures are shown in Fig. 40. The effects of increasing counterion concentration can be summarized as follows:

1. *NaTC*

Below an extrapolated NaCl concentration of 0.7 *M* the slope of the log CMC/log NaCl plots are practically horizontal. Above this salt concentration (not shown in Fig. 40, but see Table IX) a steep fall in the log CMC occurs with a slope of 45°.

2. *NaTDC*

This bile salt was studied only up to 0.6 *M* NaCl due to insolubility in higher NaCl concentrations. The log–log plots lie nearly on straight lines with negative slopes of 23°.

3. NaTC/NaTDC Mixtures

Except for the CMC in 0.8 *M* NaCl (see Table XI) the points lie on a line with a slope of 18°. The CMC at 0.8 *M* NaCl is well below the curve extrapolated from lower NaCl concentrations.

D. Explanation of Temperature Effects

Since the transfer of hydrocarbons from an aqueous to a nonpolar environment is endothermic we anticipate that "hydrophobic bonds" should become more stable as the temperature is increased (154). Micelle formation can be considered as an example of "hydrophobic bonding" (155); thus, not considering other factors micelles should become more stable with rising temperature, i.e., the CMC should decrease. This is true of several typical anionic detergents within a certain temperature range (e.g., sodium *n*-decyl sulfate, sodium 2-decyl sulfate, sodium *n*-dodecyl sulfate, and sodium 4-tetradecyl sulfate) (141, 156, 157). This is the case with the CMC of NaTC (Table IX) as the temperature is increased above 10°C. This decrease is of the same order as that for ionic detergents (156, 157). NaTC micelles are most stable in the middle temperature range (20–60°C). This phenomenon is explained by assuming that monomer solubility and other factors (dielectric effects, etc.) balance hydrophobic interactions which tend to lower the CMC, thus the CMC shows little variation. At higher temperatures (60°C and greater) the dielectric of the medium is significantly lowered, so one might expect the repulsive force between the ionic heads of the molecules to increase and lessen the tendency for monomers to associate to form micelles. Thus the CMC increases. NaTDC (Table X, Fig. 39, middle) has lower CMC's than NaTC, but behaves in a similar fashion. At temperatures between 10 and 40°C hydrophobic forces are balanced by other forces (ionic repulsion, etc.) and the CMC remains fairly constant.

E. Micellar Charge

Major differences in response to added NaCl are related to the fact that NaTDC has one less hydroxyl group than NaTC, the rest of the molecules being identical. Added counterion up to about 0.7 *M* seems to have little or no effect on the CMC of NaTC. Above 0.7 *M* NaCl a steep fall in CMC (Table IX) occurs. The NaCl depresses the CMC of NaTDC and the NaTC/NaTDC mixture at all salt concentrations. Counterions lower the CMC's of both ionic and nonionic detergents by different mechanisms (141, 158). Although the major effects of added electrolyte on the CMC of ionic detergents are due to charge neutralization, Mukerjee (155) has suggested that salting-out effects, generally neglected, can be quite substantial. The effect of electrolytes on nonionic detergents has also been attributed to the salting-out mechanism (158). The magnitude of these effects depends on the electrolyte concentration and the lyotropic number of the added counterions (159). A linear dependence of log CMC on excess salt concentration is the rule with most systems of nonionic detergents (160). When the log CMC of NaTC is plotted against NaCl concentrations of 0.3–3 *M* a straight line is approximated indicating essentially nonionic behavior of NaTC in high salt concentrations.

Thermodynamically a linear dependence of log CMC on log NaCl concentrations can be predicted for ionic detergents (155). Bile salt solutions above their CMC's can be considered to consist of monodisperse micelles and monomers in equilibrium. With this assumption the law of mass action can be applied (161). Let A^- represent a bile salt, n the number of bile salt anions present in the micelle, Na^+ its univalent counterion and m the number of counterions bound to a micelle (M). Then at equilibrium between A^-, Na^+, and M

$$nA^- + m\,Na^+ \rightleftarrows M^{(n-m)-} \tag{4}$$

where $n - m$ is the number of free charges on the micelle as m is less than n. From Eq. (4) the equilibrium constant Km for the system is

$$Km = \frac{[M]}{[A^-]\,n[Na^+]\,m} \tag{5}$$

It follows that the free energy change per monomer (ΔF) is

$$\Delta F = \frac{RT\ \ln Km}{n} \tag{6}$$

where R is the molar gas constant and T the absolute temperature. Combining Eqs. (5) and (6) and replacing natural logarithms with log 10, we get

$$\frac{\Delta F}{2.303RT} = \frac{-\log [M]}{n} + \log [A^-] + \frac{m}{n} \log [Na^+] \tag{7}$$

As $[A^-]$ is the concentration of bile salt anions at equilibrium, this concentration is equivalent to the CMC. If, in the case of ionic detergents, the free energy change at micellization is assumed to remain fairly constant with added counterion the whole of the left side of Eq. (7) becomes constant. Since at the CMC, [M], which is the concentration of micelles, i.e., the overall detergent concentration minus the CMC, is very small, then the micellar term $-\log [M]/n$ is negligible irrespective of the magnitude of n, the aggregation number (Ag #). Thus, Eq. (7) becomes

$$\frac{\Delta F}{2.303RT} = \log \text{CMC} + \frac{m}{n} \log [Na^+] \tag{8}$$

which is the standard formula for a straight line relating log CMC to log NaCl with a slope of m/n. The slope m/n should be a fair approximation of the fraction of counterions bound to micelles in water (no added NaCl). The percent of counterions bound in water is given on the right side of Fig. 40. The number of bound counterions on NaTC micelles in water is negligible. That is, the micelle is completely charged in water. As counterion is increased to 0.7 *M* NaCl, more and more counterions are bound. One can estimate that above 0.7 *M* NaCl, the charges on the micelle are bound.

NaTDC has a certain calculable but small percentage of counterions bound in water. The mean slope of the curve is 23°, giving a m/n slope of -0.23. Thus, the number of counterions bound to NaTDC micelles made up of six monomers is between 1 and 2 at zero excess counterion concentration.

The mean slope of the curves for the NaTC/NaTDC mixture is 18°, giving a slope of -0.18. The percentage of counterions bound to these mixed micelles is quite small, being about 1 Na^+ per 10 bile salt ions. The slope of log CMC–log NaCl plot for a typical anionic detergent (e.g., sodium dodecyl sulfate) is about -0.50 (141); thus, the surfactant binds more counterions than either bile salt.

In a light-scattering study of dihydroxy bile salts, Kratohvil and Dellicolli (120) estimated the fractional micellar charge, p/n, of NaTDC in 0.15 *M* NaCl from Debye plots. Here p represents the number of fundamental charges per micelle and n the number of monomer ions per micelle. Since m is the number of counterions bound to a micelle, then p/n is equal to $1 - m/n$. The fractional micellar charge p/n was 0.29 in 0.15 *M* NaCl. From the present data p/n for *zero* excess counterion (i.e., in water) concentration is 0.77.

The CMC values of Kratohvil and Dellicolli (120) plotted as log NaTDC *vs* log NaCl concentration were compared with the data of Carey and Small (44) and with CMC values determined by light scattering (44). The slopes of these curves showed excellent agreement. All predicted the percentage of counterions bound to NaTDC micelles as 23% in water. In other words, in water the micelle is highly charged with less than one quarter its ionic groups bound to counterion. With increasing added NaCl, more counterions are bound to the micelle and it becomes less charged. It has been suggested that NaGDC micelles are uncharged in 0.5 *M* NaCl (120).

A comparison of three CMC values for NaTC, NaTDC, and 1:1 mixture of the two is shown in Fig. 41. The CMC's of the NaTC/NaTDC mixture lie much nearer to NaTDC than NaTC. The most likely explanation is that in each micelle of the mixture there are more NaTDC than NaTC molecules, whereas NaTC makes up a greater proportion of the monomer. This is a result of the lower monomeric solubility of NaTDC.

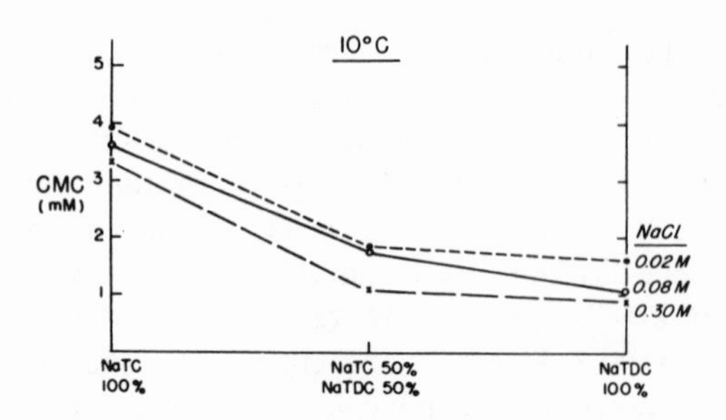

Fig. 41. The CMC values (mM) of an equimolar mixture of both bile salts compared to the CMC values of NaTC and NaTDC at 0.02, 0.08, and 0.3 M NaCl, respectively. The CMC's of the mixture lie closer to NaTDC (44).

F. Thermodynamics of Bile Salt Micelle Formation

In order to make a rough approximation of the thermodynamic constants associated with bile salt micellization, the mass action approach from which Eq. (8) is derived has been employed (44). In water m/n, i.e., the slope of the log CMC–log counterion concentration is zero in the case of NaTC micelles; therefore, Eq. (8) reduces the simple form of

$$\Delta F_m = 2.303 RT \log \text{CMC} \tag{9}$$

for change in free energy calculations. It can also be used with relative accuracy for NaTDC micellization as $m/n \log [\text{Na}^+]$ is very small with no added counterion and can be ignored. Using the same model, the Clausius–Clapeyron equation can be applied to estimate the enthalpy change (ΔH_m) from the temperature dependence of the CMC (162)

$$\Delta H_m = \frac{-RT^2 \, d \ln (\text{CMC})}{dt} \tag{10}$$

where ΔH_m is the change in heat content when 1 mole of bile salt ions in monomeric form aggregate to the micellar form. The change in entropy ($T\Delta S_m$) associated with micellization follows from the Gibbs relationship

$$T\Delta S_m = \Delta H_m - \Delta F_m \tag{11}$$

The assumptions implied in this treatment are (a) that the system is monodisperse, (b) that an abrupt change in the system occurs at the CMC, and (c) that there is a constant concentration of the monomer above the CMC. More refined calculation should take into account the micelle charge, counterion activity, and electrical free energy at the micelle surface (163). With these limitations in mind one can examine the values of ΔF_m, ΔH_m, and $T\Delta S_m$ for NaTC and NaTDC in water (Table XII).

TABLE XII. Standard Thermodynamic Functions Expressed in kcal/mole for Micellization of NaTC and NaTDC in Water at Three Temperatures (44)

	ΔF_m		ΔH_m		$T\Delta S_m$	
T, °C	NaTC	NaTDC	NaTC	NaTDC	NaTC	NaTDC
20	−3.43	−3.79	−0.31	−1.03	+3.12	+2.76
50	−3.70	−3.97	−1.31	−3.94	+2.39	+0.03
80	−3.70	−4.04	−5.21	−3.92	−1.51	+0.12

The magnitude of the negative ΔF values obtained indicates that bile salt micellization is always a spontaneous process, i.e., no external work is needed on the system and the detergent ions in the micelle have a lower energy state than in molecular solution. ΔF_m for NaTDC is less than NaTC, and thus NaTDC can be considered (163) to form more stable micelles. As would be expected from its lower CMC, NaTDC forms larger micelles (see Tables IX and X and Section VIII.G.)

At 50 °C the entropy changes at the CMC of NaTC and NaTDC are positive. The most likely explanations for the entropy changes are that when a hydrocarbon is solubilized in water it induces increased structuring of the water molecules about it (164), bringing about a negative entropy effect (165, 166). In bile salt solutions below the CMC's clathrates of structural water form around bile salt molecules. At the concentration of maximum monomeric solubility of the bile salts, the molecules escape from their water cages to form micelles. The loss of water structure when monomers enter micelles and the disorder produced by the more bulky micelles give rise to positive entropy changes.

At temperature higher than 50 °C, entropy changes for NaTC are negative at the CMC and for NaTDC may be still slightly positive. At these temperatures the driving force for micelle formation would appear to be primarily the heat component or the magnitude of the negative ΔH_m and not the entropy factor.

G. Bile Salt Micelle Size and Shape — Effects of Structure, Concentration, Counterion, *p*H, Temperature, Urea, and Other Solvents

A number of studies have been carried out to determine the aggregation number (number of molecules per micelle) of bile salt micelles. In common with results of previous studies on the critical micellar concentration (CMC) these studies have given variable results which give aggregation numbers for bile salts from 1 (that is, no micelle formation) to well over 1000 associated molecules (73, 144, 145, 120, 167–169).

A summary of selected studies on the micellar weights and corresponding aggregation numbers of bile salt micelles carried out in various other laboratories is given in Table XIII. Where they are known, the temperature, counterion, concentration, and *p*H are given. Certain studies suffer from not having controlled the *p*H (145). Failure to control the *p*H in these studies led to extraordinarily high aggregation numbers for sodium deoxycholate of over 1000 molecules (1, 43) (see below concerning the effect of *p*H). Even though conditions and methods varied it is clear from Table XIII that trihydroxy bile salts form smaller micelles than the dihydroxy bile salts. Because it seemed obvious that experimental conditions would affect the aggregation numbers of each bile salt, a systematic study of the effects of bile salt concentration, counterion concentration, *p*H, temperature, and urea concentration on the aggregation number of all of the common bile salts and their glycine and taurine conjugates was carried out (43). This study showed that the type of bile salt, the *p*H, the temperature, and the counterion concentration can all affect appreciably the size and most probably the structure of bile salt micelles.

The aggregation numbers reported in Table XIV were obtained by equilibrium centrifugation (43). The values were checked by light scattering for sodium taurodeoxycholate and found to be substantially the same (43).

1. Effects of Bile Salt Concentration

The effect of concentration of bile salt was studied in the case of sodium glycocholate, sodium glycodeoxycholate, and sodium taurodeoxycholate in very dilute salt concentrations or in water. Under these conditions, aggregation numbers did not vary with concentration. The following results are given in apparent aggregation numbers (Ag #) and not for aggregation numbers extrapolated to zero micelle concentration. Kratohvil (120) has pointed out in the case of sodium glycodeoxycholate in low salt concentrations that the apparent aggregation number may be slightly low. However, for general comparison of all of the major bile salts the apparent aggregation numbers serve adequately.

2. Effect of Counterion Concentration: The Difference Between Dihydroxy and Trihydroxy Bile Salts

The aggregation number of the micelle (number of molecules of bile salt per micelle) as a function of the concentration of NaCl is given for free bile salts (Fig. 42), taurine conjugates (Fig. 43), and glycine conjugates (Fig. 44), respectively, at 20 °C and *p*H values from 8 to 9.

In water at 20 °C, the bile salts, free or conjugated, both dihydroxy and trihydroxy, from very small aggregates, usually dimers. Unlike the dimers

TABLE XIII. Anhydrous Particle Weights of Bile Salt Micelles

Type of bile salt	Method	T, °C	Counterion conc. (*M*) and type	*p*H	Anhydrous micellar weight	Aggregation number	Reference
Trihydroxy bile salts							
cholate	LS[a]	20	(H_2O)	6.6–7.6	430	1	(144)
	LS	10	0.15 Na^+	NS	3,700	8	(145)
	LS	20	0.15 Na^+	NS	1,600	4	(167)
	LS	20	0.50 Na^+	6.6–7.6	2,500	6	(144)
	LS	20	1.00 Na^+	6.6–7.6	3,230	8	(144)
taurocholate	LS	20	0.15 Na^+	NS	3,700	7	(145)
glycocholate	LS	20	0.15 Na^+	NS	3,000	6	(167)
	LS	10	0.15 Na^+	NS	4,000	8	(145)
	LS	20	0.04 Na^+	7.4	1,900	4	(168)
	LS	20	0.08 Na^+	7.4	2,200	5	(168)
	LS	20	0.16 Na^+	7.4	3,200	7	(168)
Dihydroxy bile salts							
deoxycholate	LS	20	0.10 Na^+	6.6–7.6	5,320	13	(144)
	LS	NS	0.15 Na^+	NS	5,400	13	(167)
taurodeoxycholate	LS	NS	0.15 Na^+	NS	11,000	22	(167)
	UC[b]	20	0.15 Na^+	5.8	11,900	24	(73)
	GF[c]	20	0.15 Na^+	NS	11,000	22	(169)
	LS	20	0.15 Na^+	NS	12,000	24	(120)
	LS	20	0.5 Na^+	NS	25,000	50	(120)
glycodeoxycholate	LS	20	0.15 Na^+	NS	11,500	26	(120)
	LS	20	0.5 Na^+	NS	25,000	56	(120)

[a] Light scattering.
[b] Ultracentrifugation.
[c] Gel filtration.

TABLE XIV. Apparent Anhydrous Particle Weights and Aggregation Numbers of Bile Salt Micelles by Equilibrium Ultracentrifugation (43)

Bile salt	T, °C	Solvent	*p*H	*M*	Ag ‡
NaC	20	H_2O	8–9	853	2.0
		0.01 *M* NaCl	8–9	1,227	2.8
		0.05 *M* NaCl	8–9	1,650	3.8
		0.15 *M* NaCl	8–9	2,061	4.8
	36	0.15 *M* NaCl	8–9	2.062	4.8
	20	0.3 *M* NaCl	8–9	2,519	5.8
		1.0 *M* NaCl	8–9	3,108	7.2
	36	1.0 *M* NaCl	8–9	3,137	7.3
NaGC	20	H_2O	8–9	925	1.9
		0.05 *M* NaCl	8–9	2,455	5.0
		0.15 *M* NaCl	8–9	2,816	5.8
	36	0.15 *M* NaCl	8–9	2,750	5.6
	20	0.3 *M* NaCl	8–9	3,435	7.0
NaTC	20	0.05 *M* NaCl	8–9	2,243	4.2
		0.15 *M* NaCl	8–9	2,447	4.6
	36	0.15 *M* NaCl	8–9	2,395	4.5
	20	0.3 *M* NaCl	8–9	3,215	6.0
		0.01 *M* NaCl	8–9	1,916	3.6
	4	0.15 *M* NaCl	8–9	2,154	4.0
		0.3 *M* NaCl	8–9	2,788	5.2
	20	1.0 *M* NaCl	8–9	4,844	9.0
		4 *M* urea + 0.15 *M* NaCl	8–9	22,82	4.2
		6 *M* urea + 0.15 *M* NaCl	8–9	943	1.8
		3 *M* NaCl	8–9	8,270	16.0
NaDC	20	0.01 *M* NaCl	8–9	1,781	4.3
	36	0.01 *M* NaCl	8–9	1,582	3.8
	20	0.05 *M* NaCl	8–9	3,678	8.9
		0.15 *M* NaCl	8–9	6,290	15.2
	36	0.15 *M* NaCl	8–9	5,496	13.3
	20	0.3 *M* NaCl	8–9	12,107	29.2
		6 *M* urea + 0.15 *M* NaCl	8–9	2,688	6.5
		0.15 *M* NaCl	7.3	228,809	551.9
	36	0.15 *M* NaCl	7.3	5,587	13.5
	20	0.15 *M* NaCl	7.85	6,975	16.8
	36	0.15 *M* NaCl	7.85	5,521	13.3
NaGDC	20	H_2O	8–9	919	2.0
		0.05 *M* NaCl	8–9	5,520	11.7
		0.15 *M* NaCl	8–9	9,139	19.4
	36	0.15 *M* NaCl	8–9	7,618	16.2
	20	0.3 *M* NaCl	8–9	17,836	36.9
		0.5 *M* NaCl	8–9	30,143	63.9
	36	0.5 *M* NaCl	8–9	21,975	46.6
	20	2 *M* urea + 0.5 *M* NaCl	8–9	14,059	29.8
	36	2 *M* urea + 0.5 *M* NaCl	8–9	10,693	22.7
	20	4 *M* urea + 0.5 *M* NaCl	8–9	10,018	21.2
		6 *M* urea + 0. 5 *M* NaCl	8–9	3,293	7.0
		0.15 *M* NaCl	4.9	1,030,000	2184.0
		0.15 *M* NaCl	6.2	8,644	21.4
		0.15 *M* NaCl	7.2	10,076	18.3
NaTDC	20	0.01 *M* NaCl	8–9	3,259	6.2
		0.05 *M* NaCl	8–9	5,885	11.3

TABLE XIV (continued)

Bile salt	T, °C	Solvent	*p*H	*M*	Ag ‡
		0.15 *M* NaCl	8–9	11,367	21.8
	36	0.15 *M* NaCl	8–9	9,367	18.0
	20	0.3 *M* NaCl	8–9	15,154	29.1
		0.5 *M* NaCl	8–9	24,402	46.8
		2 *M* urea + 0.5 *M* NaCl	8–9	16,332	31.3
		4 *M* urea + 0.15 *M* NaCl	8–9	5,798	11.1
	36	4 *M* urea + 0.15 *M* NaCl	8–9	5,293	10.1
	20	4 *M* urea + 0.5 *M* NaCl	8–9	10,991	21.1
		6 *M* urea + 0.15 *M* NaCl	8.0	4.732	9.1
		6 *M* urea + 0.5 *M* NaCl	8–9	8,321	16.0
		0.15 *M* NaCl	1.6	13,446	25.8
		0.15 *M* NaCl	3.6	11,431	21.9
		0.15 *M* NaCl	6.5	11,068	21.2
NaCDC	20	0.05 *M* NaCl	8–9	2,709	6.5
		0.15 *M* NaCl	8–9	4,705	11.3
	36	0.15 *M* NaCl	8–9	4,502	10.9
	20	0.3 *M* NaCl	8–9	8,385	20.2
		0.15 *M* NaCl	7.00	7,812	18.8
		0.15 *M* NaCl	7.05	7,254	17.5
NaGCDC	20	0.05 *M* NaCl	8.0	4,983	10.6
		0.15 *M* NaCl	8.0	9,706	20.6
	36	0.15 *M* NaCl	8–9	8,603	18.2
	20	0.3 *M* NaCl	8–9	13,417	28.4
NaTCDC	20	0.15 *M* NaCl	8–9	4,956	9.5
		0.15 *M* NaCl	8–9	10,155	19.5
	36	0.15 *M* NaCl	8–9	7,945	15.2
	4	0.15 *M* NaCl	8–9	13,313	25.5
	20	0.3 *M* NaCl	8–9	15,141	29.0
NaDHC[a]	20	0.15 *M* NaCl	8–9	485	1.1
		0.3 *M* NaCl	8–9	497	1.2
		1.0 *M* NaCl	8–9	1,199	2.8
TLCS[b]	20	H_2O	11	1,300	2.0

[a] Sodium dehydrocholate.
[b] Taurolithocholate sulfate, diammonium salt (from R. Palmer).

formed by some aliphatic detergent molecules that form at low concentrations below the CMC (155), bile salts form dimers over a wide range of concentrations well above the CMC. As the counterion concentration is increased, a marked difference manifests itself between the trihydroxy and dihydroxy bile salts. The trihydroxy bile salt micelles increase from dimers in water up to a maximum of about seven to nine molecules per micelle at 1.0 *N* NaCl. On the other hand, the counterion concentration has a marked effect on the size of the dihydroxy bile salt micelles, even at relatively low salt concentrations—for instance, Na glycodeoxycholate, which forms dimers in water, forms large micelles with an aggregation number of about 63 in 0.5 *N* NaCl (Fig. 44). Since both deoxycholate and chenodeoxycholate show this marked increase in micelle size with an increase in counterion concentration, it does

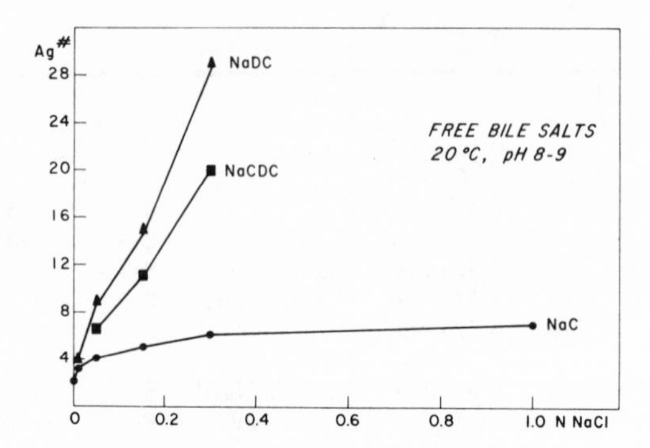

Fig. 42. Apparent aggregation number *vs* counterion concentration of free bile salts at 20°C, *p*H 8–9 (43). ● Sodium cholate. ■ Sodium chenodeoxycholate. ▲ Sodium deoxycholate.

not appear to be the specific position of the OH group (7α or 12α) on the molecule but rather the number of hydroxyl groups that is important. Further, higher counterion concentrations (above about 0.8 *N* NaCl) cause precipitation of the dihydroxy bile salts at 20°C, while trihydroxy bile salts remain in solution above 3 *N* NaCl. It, therefore, appears that the number of hydroxyl groups are of importance, not only in the type of micelles formed, but also in the solubility of bile salts. The dihydroxy bile salts "salt out" at a much lower NaCl concentration than trihydroxy bile salts.

As noted in Sections II and V the bile salts have a peculiar molecular structure compared with ordinary detergents, possessing both an ionic polar group and nonionic hydrophilic groups. Since both the charged portion and

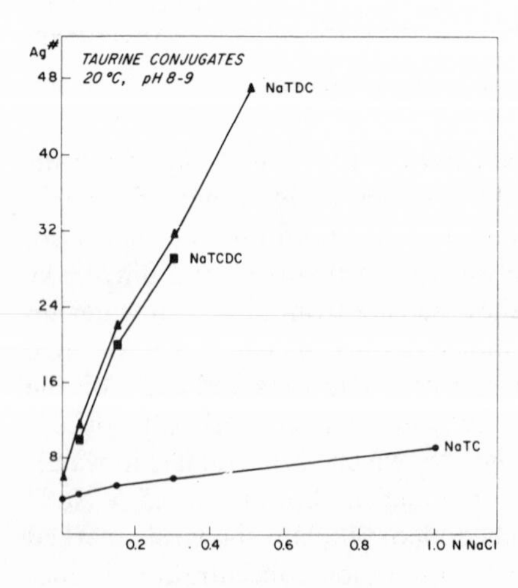

Fig. 43. Apparent aggregation numbers *vs* counterion concentration of taurine conjugates at 20°C, *p*H 8–9 (43). ● Sodium taurocholate. ■ Sodium taurochenodeoxycholate. ▲ Sodium taurodeoxycholate.

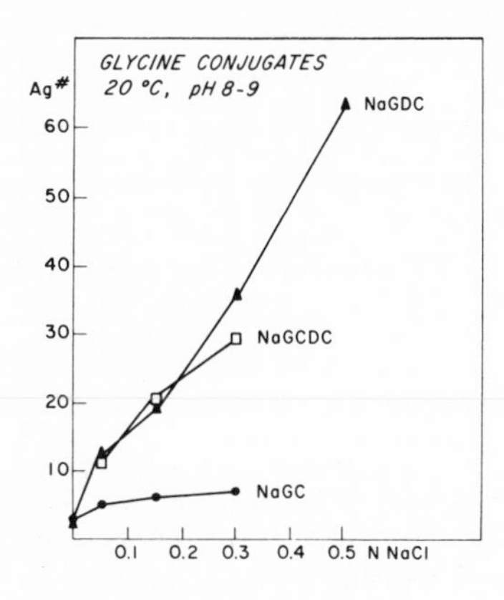

Fig. 44. Apparent aggregation number *vs* counterion concentration of glycocine conjugates at 20°C, *p*H 8–9 (43). ● Sodium glycocholate. ■ Sodium glycochenodeoxycholate. ▲ Sodium glycodeoxycholate.

the hydrophobic part of the molecule are identical for any given pair—e.g., the taurine conjugates NaTC and NaTDC—the major difference in response to increased counterions lies with the nonionic polar part of the molecule. Generally speaking, the increase in the aggregation number of nonionic detergents is related to the dehydration of polar groups. Since there is a marked difference between dihydroxy and trihydroxy bile salts, it seems reasonable to conclude that the dihydroxy bile salts with only two hydroxyl groups are dehydrated much more readily that those with three groups. The behavior of the CMC in response to salt (Section VIII) also supports this notion.

3. *Effect of pH–The Difference between Dihydroxy and Trihydroxy Bile Salts*

Decreasing the *p*H of trihydroxy bile salt solutions (even to the point of precipitation of NaC or NaGC) has little effect on the Ag # of these bile salts. The effect on dihydroxy bile salts is strikingly different as illustrated by the effect of *p*H on the deoxycholate series of bile salts in 0.15 *N* NaCl at 20°C and 36°C (Fig. 45). Solutions of free and glycine conjugated bile salts were studied between *p*H 9.2 and their precipitation *p*H (NaDC = 6.95, NaGDC = 4.9) (see Section VII. B). Taurine conjugates were studied between *p*H values 10 and 1.6.

Sodium deoxycholate above *p*H 8 in 0.15 *N* NaCl has an aggregation number of 15 molecules. This increases moderately between *p*H values 8 and 7.8. As the *p*H falls, a marked increase in Ag # occurs from 18 to over 500 at *p*H 7.3. This change corresponds to the formation of a thin gel. These gels have been studied previously (170–173). Sodium glycodeoxycholate behaves in similar fashion except that the *p*H at which the marked increase in aggregation number occurs is much lower (about 5.0). In both cases the solutions are clear. The amount of free acid necessary to produce the gel stage has been

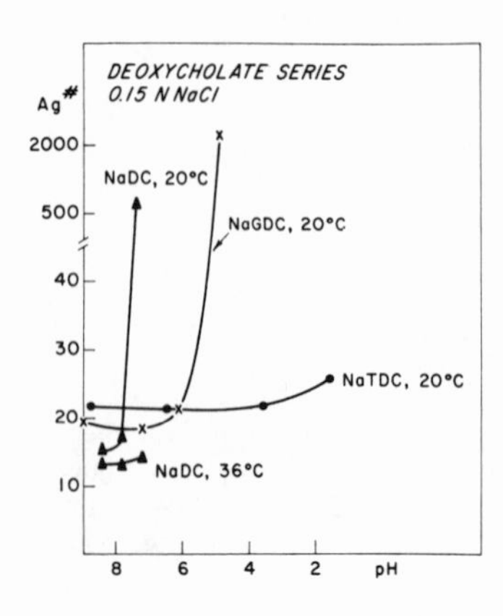

Fig. 45. Apparent aggregation number *vs* *p*H of deoxycholate series (0.15 *N* NaCl) (43). ▲Sodium deoxycholate. × Sodium glycodeoxycholate. ● Sodium taurodeoxycholate.

calculated from the titration curves of the bile salts and corresponds to less than 1 molecule of bile acid for 5 molecules of bile salt. These large aggregates are temperature-sensitive. The large aggregates of Na deoxycholate are reduced to smaller ones if temperature is raised from 20 °C to 36 °C (see Fig. 45). The temperature-dependence of the gels of Na deoxycholate has been pointed out by Sobotka (170). Since hydrophobic bonding increases (see Section VIII. D) with increasing temperature, one must consider that the mechanism of gel formation in these bile salts is through hydrogen bonding of the OH group at the 3α position to part of the carboxyl group. At somewhat lower temperatures (e.g., 5 °C) chenodeoxycholate also forms gels when the *p*H is lowered toward the point of precipitation.

4. *Effect of Temperature*

All of the bile salts in 0.15 *N* NaCl at *p*H values from 8 to 9 were studied at 20 °C, 36 °C, and in some cases at 4 °C (Fig. 46). Trihydroxy bile salts which form small micelles (Ag # 2–9) are unaffected between 4° and 36 °C. Further, low salt concentrations are not affected by temperature. However, the sizes of the micelles of dihydroxy bile salts having aggregation numbers greater than 10 are decreased as the temperature is increased.

Increasing temperature may cause a number of opposite effects (see Section VIII. D). It may favor hydrophobic bonding and thus promote an increase in micellar size and stability. It may increase the repulsive forces between the anionic polar heads and thus result in decreased micellar stability and structure breaking effects. It may cause partial dehydration of non-ionic polar groups and promote increased micelle size. It may decrease intermolecular H bonding. Since temperature shows little or no effect on the small micelles formed by trihydroxy bile salts, it can be assumed that these opposing effects are roughly balanced. The large micelles formed by dihydroxy

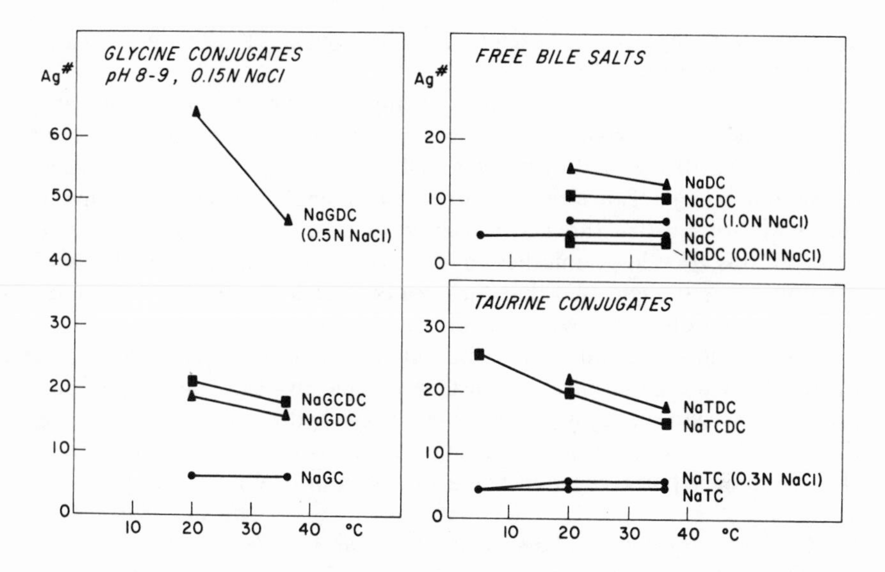

Fig. 46. Apparent aggregation number *vs* temperature; *p*H 8–9, 0.15 *N* NaCl (43). Left, glycine conjugates, *p*H 8–9, 0.15 *N* NaCl. Upper right, free bile salts. Lower right, taurine conjugates.

bile salts in NaCl solutions are markedly decreased in size by temperature. This behavior is opposite that observed for nonionic detergents, whose aggregation numbers increase with temperature (141, 173). Therefore, it would appear that the nonionic polar parts of the bile salt are hidden in these large micelles and that the anionic polar moiety predominates. Thus, the decrease in size may be partly explained by increased charge repulsion (since the ionic groups would be closer together in the large micelles), but the most likely explanation is that the salt causes some intermolecular H bonding of the dihydroxy bile salts which increases their size. An increase in the temperature breaks these hydrogen bonds and results in decreased micelle size.

5. *Effect of Added Urea*

The results of studies (43) of several bile salts under conditions giving different aggregation numbers studied in increasing concentrations of urea support these conclusions. The addition of 4 *M* urea to the small aggregates (Ag $\# < 8$) formed by trihydroxy bile salts has little or no effect. On the other hand, all of the dihydroxy aggregates under conditions giving aggregation numbers greater than 18 are markedly decreased in size by 2, 4, and 6 *M* urea. Concentrations of 6 *M* urea reduce the apparent aggregation number of NaGDC in 0.5 *N* NaCl at 20 °C from 63 to 6. Therefore, in urea, small micelles do not behave like the large micelles. Small micelles resist change until concentrations of urea reach 6 *M*, while large micelles are very sensitive to low concentrations of urea.

The effect of low concentrations of urea (2 *M*) on the large dihydroxy bile salt micelles is striking, while similar concentrations have no effect on the small trihydroxy or dihydroxy micelles. The effects of urea on micelle formation and aggregate size are undoubtedly complicated (174) and involve changes in solvent structure and thus hydrophobic bonding and hydration of polar groups. For large micelles of dihydroxy bile salt (formed in NaCl solutions) it is thought (43) that the urea competes effectively at relatively low concentrations (2 *M*) with NaCl to reestablish hydration of the nonionic hydroxyl groups and thus reduce micelle size. If this is true, some of the hydroxyls in the large micelles must be completely dehydrated. The most probable mechanism is by hydroxyl–hydroxyl hydrogen bonding between the bile salts, such as occurs with bile acid derivatives in organic solvents (122) or in amorphous melts of bile acids (Section II. C).

6. Concept of Primary and Secondary Bile Salt Micelles

These observations suggest that bile salts form two types of micelles, "primary" and "secondary."

Primary Micelles. Primary micelles are formed by the trihydroxy series of bile salts, free or conjugated. They are also formed by the dihydroxy bile salts in water or at low salt concentrations ($>$0.03 *M* NaCl) at 20°C. They are small, with aggregation numbers ranging from 2 (dimers) to 6 or 7. On the basis of the orientation of bile salts at air–water interfaces (see Section V) and a consideration of their molecular structure, studied from space-filling models (see Section II. B), the micellar structure shown in Fig. 47 has been proposed (43). In this model the hydrocarbon backs of the steroid nucleus (the only hydrophobic part of the bile salt molecule) associate. In water, even at bile salt concentrations many times greater than the CMC, dimers are present. Somewhat larger micelles are formed at low NaCl concentrations. In the case of trihydroxy bile salts, the aggregate increases to about nine monomers at high NaCl concentrations. The increase in aggregation number in the case of the small micelles with increasing counterion concentrations is probably caused by several factors, including decreased repulsion of the charged anionic polar groups, partial dehydration of the nonionic hydroxyl groups, and salting out of the hydrocarbon part of the molecule. Noting the marked decrease in CMC of NaTC in NaCl concentrations greater than 0.7 *M* attributed to salting out of OH (Section VIII. C), one might think the size of the NaTC micelle should get larger above 0.7 *M* counterion. It does. It increases to 9 in 1 *M* NaCl and to 16 in 3 *M* NaCl (Table XIV). Thus, when the CMC begins to fall, larger micelles form. The micelle at a salt concentration just below 0.7 *M* NaCl would contain about 6 monomers

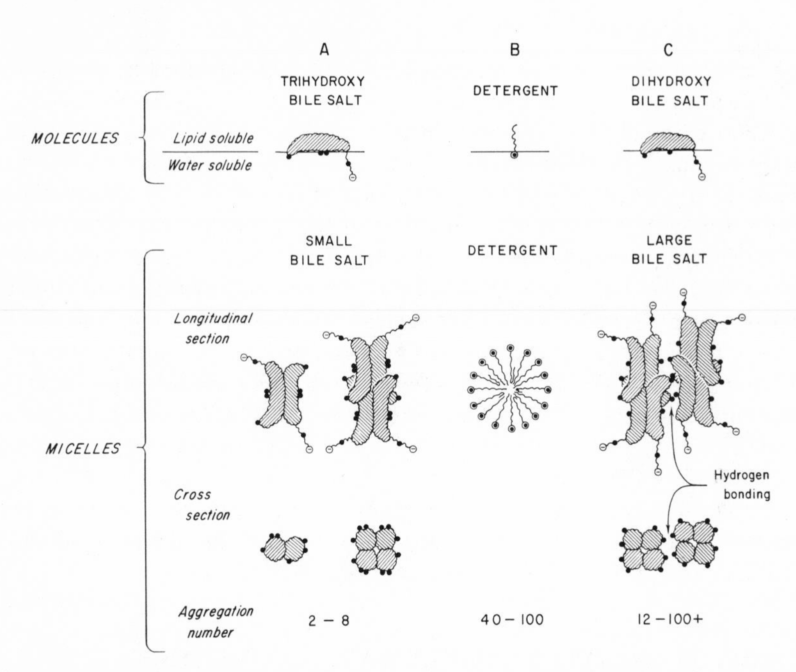

Fig. 47. Comparison of proposed structure of primary and secondary bile salt micelles and a classical detergent micelle (43). (A, C) Longitudinal and cross sections of primary and secondary bile salt micelles, respectively. (B) Ordinary detergent micelles. Top—Orientation of molecules at oil–water interface (refer to Section V). Wavy line—Hydrocarbon chains of detergent molecules. Shaded area—Lipidsoluble cyclic hydrocarbon part of bile salt molecule. ⊙ Polar head of detergent molecule. ● OH or ester groups. ⊖ Negatively charged ionic group of bile salt.

of taurocholate. Six molecules fit together comfortably so that the hydrophobic parts are in contact. Attempts to fit together hydrophobically a number of molecules greater than 10 either leaves a space in the center, thus exposing the hydrophobic parts to water, which seems an unlikely structure, or necessitates contact with other molecules through the hydrophilic parts of the molecules. Thus primary micelles, though slightly asymmetric and of unfamiliar structure when compared with a representation of an aliphatic detergent micelle (Fig. 47B), are probably analogous to detergent micelles, in that the proposed bonding is hydrophobic. The bonding of the bile salts, however, is back to back rather than chain to chain as with detergents. Strong evidence for this type of association comes from NMR studies of bile salts in deuterium oxide (67) in which the motion of protons on the back side of the bile salts was impeded above the CMC while the motion of protons on the hydrophilic side was not affected.

Secondary Micelles. Dihydroxy bile salts in the presence of moderately increased counterions and perhaps trihydroxy bile salts in 3 *M* NaCl form these larger micelles. Secondary micelles are probably formed by the aggregation of primary micelles. Since the surface available for hydrophobic interaction is expended in the formation of the primary micelles, the bonding that takes place is probably between some of the hydrophilic parts of the bile salts. It is suggested (Fig. 47C) that in increased counterion concentrations dehydration of the weaker nonionic part of the dihydroxy bile salts occurs and hydrogen bonding between the hydroxyls of the bile salts takes place. Other types of aggregates, such as helices, rods, oblate and prolate spheroids, which could explain the increase in micelle size with increasing counterion concentration, are very improbable since studies that might suggest the shape of bile salt micelles (viscometry, light scattering) appear to show that the micelles are nearly spherical (38, 118, 120, 167, 168).

7. *Effect of Other Solvents*

Sodium cholate and sodium deoxycholate were studied in alcohols and mixtures of ethanol and water. They did not form micelles in methanol, ethanol, or in 1:1 or 1:2 ethanol–water mixtures (118).

IX. MIXED MICELLES OF BILE SALTS AND OTHER LIPIDS*

A. What Are Mixed Micelles?

The definition of a mixed micelle is any micelle made up of more than one lipid-like chemical species. At least one of the chemical species must be able to form micelles alone in aqueous solutions. If we refer to micelles formed by long-chain soaps or detergents and another chemical species, three different groups of micelles are readily definable. First, the micelle may be made up of two chemically different detergents. Both of the detergents are present in the micelle and both are present as individual molecules outside the micelle. Thus, both contribute to the measured CMC. Usually the species with the lower CMC has a greater concentration in the micelle and a conversely lower concentration as monomer than its overall concentration in the total solution, i.e., the species with lower CMC partitions favorably into the micelle. The second type of micelle is one formed by a rather insoluble amphiphilic substance such as a long-chain alcohol or monoglyceride and a detergent. This might be called "solubilization by association" in that the insoluble amphiphile is interdigitated between the detergents with its polar

*Since this review was written a fairly extensive critique of mixed micelles has been published (44b). This study should be consulted for further discussion.

portion in the aqueous phase. It could be assumed that the solubility of the insoluble amphiphile in the nonmicellar aqueous phase is not increased by the presence of detergent and therefore virtually all is present in the micelle. It would follow that the measured CMC is due only to detergent molecules. Unfortunately, this assumption, implicit in many studies on mixed micelles of this type, is very difficult to test. It is possible that an appreciable increase in the molecular species of the insoluble amphiphile is brought about by the detergent. Thus part of the measured CMC may be due to the amphiphile. The decrease in CMC often seen in such mixed micelles is due to the increase in size of the micelle and the decreased ionic repulsion resulting from separation of charges induced by the interdigitating insoluble species. The third general type of mixed micelle is one in which an insoluble nonpolar substance such as hexadecane is assimilated by the detergent micelle. It has been assumed that the insoluble species is hidden in the hydrocarbon core of micelles and is not present at the micelle surface or in the surrounding fluid. This assumption is on somewhat firmer ground, although not fully tested. These micelles have a very limited capacity to solubilize large nonpolar molecules, and the CMC and size change only slightly from the pure micelle.

Throughout this review the author has stressed the difference in chemical and physical structure of bile salt and long-chain detergents. The problems of mixed micelles involving bile salt amplify these differences. It is well known that the hydrocarbon chains of detergent micelles are in a liquid-like state and capable of a large range of rotational, twisting, flexing, and bending motion (133–135, 141, 175). The rigid hydrophobic back side of the bile salt molecule can hardly be considered liquid, although the angular methyl groups and individual steroid nucleus protons can spin quite freely. Further, the several hydrophilic groups (hydroxyls and the anionic tail) give the molecule a large hydrophilic area. Thus, one might expect differences in the type of mixed micelles formed by bile salts and other lipid species (41, 44b). Differences are present and are particularly marked in the case of insoluble polar and nonpolar lipids. These differences are very important from a physiologic point of view for bile salts in high concentrations have been found in bile and small intestinal content. In the proximal small intestine and in bile the bile salts keep company with a variety of other lipid species as well as proteins, carbohydrates, and small molecules. In the bile the other lipids are principally lecithin and free cholesterol, while in jejunal content a number of lipid species may be present, including the phospholipids, glycolipids, lysophosphatides, glycerides, sterols, sterol-esters, and partially ionized fatty acids. The study of the micellar weight, aggregation number, and critical micellar concentration of pure bile salts in NaCl is fraught with experimental difficulties. Thus, addition of other lipid species and counterions makes the number of variables large and systematic study of the effects of these variables virtually impossible. Nevertheless, an attempt has been made to study mixed micelles of bile salts and other lipid types in counterion concentrations which are in the physiologic range. These results are summarized in Table XV. The micellar weights were carried out by the author either by the method of equilibrium ultracentrifugation (43) in which a number of different apparent micellar weights at different concentration differences were extrapolated to zero micelle concentration, or by the method of sedimentation and diffusion (73). The agreement between these two methods is quite good in the case of sodium taurocholate–potassium oleate mixed micelles. It must be stressed that, although these are probably relatively accurate micellar weights for the conditions stated, effects of temperature, *p*H, and inorganic salt concentration make marked differences in the size of these micelles and these effects cannot be ignored when comparing these to other values. Many of the changes observed with simple bile salt systems described previously in Section VIII are present in these systems, but compounded by the presence of another lipid species. With these forewarnings we will consider mixed micelles of bile salt with a soap of a long-chain fatty acid, with a monoglyceride, with free cholesterol, and with phospholipid (egg lecithin). Bile salt has very little effect on the solubilization of the larger glycerides, such as di- and triglycerides, or of waxes and cholesterol esters of long-chain fatty acids, although possibly some slight increase in solubility of these compounds occurs in the presence of relatively large concentrations of bile salts.

B. Mixtures of Bile Salts

In Section VIII. B–E the CMC of a mixture of NaTC and NaTDC was discussed. In respect to this type of micelle the CMC behaves in a fashion analogous to other detergents. Normal human bile and intestinal contents contain a mixture of six conjugated bile salts, the glycine and taurine conjugates of CA, DCA, and CDCA (1). In Fig. 48 the aggregation numbers of a mixture simulating the proportions of these six bile salts in bile are compared to the range of aggregation numbers of pure di- and trihydroxy conjugates as a function of NaCl concentration at 20 °C. At each NaCl concentration the aggregation number, like the CMC, falls in between those of the trihydroxy conjugates and those of the dihydroxy conjugates. The mixture in physiologic concentrations of saline (0.15 *N* NaCl) at 20 °C has a micellar weight of 6900 or an aggregation number of 13 or 14. Such a mixture of bile salts and inorganic salts might be expected to exist in the ileum of man. Although this specific mixture was not studied at 37 °C one can extrapolate from the effect of temperature on pure bile salt micelles (Section VIII. G, Fig. 46) and predict that the aggregation number might be slightly less at

TABLE XV. Apparent Anhydrous Micellar Weights of Mixed Micelles of Bile Salt with Other Lipids at 20 °C

Bile salt	Other lipid	Molar ratio bile salt/lipid	Method	Medium	pH	V_{ap},[c] cc/g	Micellar weight	Approximate ♯ BS in micelle	Approximate ♯ lipid in micelle
NaTC	K oleate	4.9:1	UC[a]	0.1 *M* KCl	10	0.764	4650	9	2
”	K oleate	4.9:1	SD[b]	0.1 *M* KCl	10	0.764	4530		
”	K oleate	2.2:1	UC	0.1 *M* KCl	10	0.792	6780	11	5
”	K oleate	2.2:1	SD	0.1 *M* KCl	10	0.792	6530		
”	K oleate	1.3:1	UC	0.1 *M* KCl	10	0.806	9750	13	10
”	K oleate	1.3:1	SD	0.1 *M* KCl	10	0.806	9960		
”	K oleate	0.54:1	UC	0.1 *M* KCl	10	0.858	15000	15	28
”	K oleate	0.54:1	SD	0.1 *M* KCl	10	0.858	14700		
NaTC	1-monoolein	10.7:1	UC	0.15 *M* NaCl	8	0.768	5500	10	1
”	1-monoolein	7.9:1	UC	0.15 *M* NaCl	8	0.778	5900	11	1–2
”	1-monoolein	3.6:1	UC	0.15 *M* NaCl	8	0.802	7150	11	3
”	1-monoolein	2.3:1	UC	0.15 *M* NaCl	8	0.823	9200	14	6
NaTC	cholesterol	100:1	UC	0.15 *M* NaCl	8	0.76	3600	-	-
”	cholesterol	100:1	UC	0.15 *M* NaCl	8	0.76	4000	-	-
NaTDC	cholesterol	37:1	UC	0.15 *M* NaCl	8	0.76	9800	-	-
NaTC	egg lecithin	13:1	UC	0.30 *M* NaCl	8	0.75	7600	12	1
NaTC	egg lecithin	13:1	SD	0.30 *M* NaCl	8	0.75	6400		
NaGC	egg lecithin	6.2:1	UC	0.15 *M* NaCl	8	0.790	6600	12	2
”	egg lecithin	2.3:1	UC	0.15 *M* NaCl	8	0.849	15700	18	8
”	egg lecithin	1.6:1	UC	0.15 *M* NaCl	8	0.875	26600	26	17
”	egg lecithin	0.66:1	UC	0.15 *M* NaCl	8	0.923	89000	52	80

[a] Equilibrium ultracentrifugation.
[b] Sedimentation and diffusion.
[c] Apparent specific volume of mixed micelle by pycnometry (20 °C).

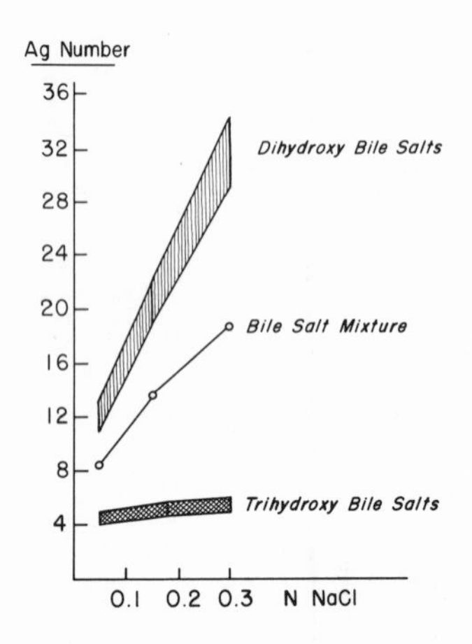

Fig. 48. Comparison of the aggregation number (Ag #) of a mixture of conjugated bile salts and the range of aggregation numbers of trihydroxy and dihydroxy bile salts. At each different counterion concentration (NaCl) the average size of the mixed micelle lies between the size of the trihydroxy and dihydroxy bile salt micelles.

37 °C—say 10–12. Therefore, provided the bile salt concentration reaching the ileum is above the CMC, the bile salt mixture presenting itself to that organ would be made up of monomers and micelles of rather small aggregation numbers.

C. Bile Salts and Soaps, Sodium Taurocholate–Potassium Oleate Mixed Micelles

Table XV shows that the two methods used (equilibrium/ultracentrifugation and sedimentation/diffusion) give comparable results. The sodium taurocholate micelle is swollen appreciably by the presence of even small amounts of potassium oleate. As the weight ratio increases the micelle increases in size.* It is not possible to say whether there are two different species of micelle present, although this seems unlikely since the schlieren sedimentation curve was symmetrical and showed no shoulders or second bumps that would suggest polydispersity. It is probable that sodium taurocholate and potassium oleate form a mixed micelle that increases in size as more oleate is added. Since both these compounds are soluble amphiphiles (42) they will be present in both the micelle and as monomers. At present it is impossible to know how the species are partitioned. If one assumes that the micelle composition is similar to that of the whole solution (a valid assumption at high micelle concentrations) then the number of molecules of each

*Note added after preparation of text. In a recent article, Benzonana (202) studied the NaDC–sodium oleate system in 0.1 M NaCl, pH 9.0, by light scattering. The micellar weight was noted to increase from 4950 (no oleate) to 7810 for solutions containing 1 mole oleate to 5 moles deoxycholate.

species in the average micelle may be calculated from the micellar weight. These values are tabulated in Table XV.

D. Bile Salts and Monoglycerides, Sodium Taurocholate–Monoolein Mixed Micelles

A phenomena very similar to that seen with potassium oleate is seen with 1-monoolein. Even small amounts of 1-monoolein, roughly 1 molecule per micelle, swell the taurocholate micelle appreciably. As the weight proportion of monoolein increases, the micellar weight also increases. Again, polydispersity was not suggested by the schlieren sedimentation traces. As the number of molecules of 1-monoolein increases from 1 to 6 in the micelle only a small increase in bile salts (from 10 to 14) occurs.

E. Bile Salt–Cholesterol Micelles

Cholesterol is poorly solubilized by bile salts. Although the solubilization maximum of cholesterol varies somewhat from individual bile salt to individual bile salt, it is a fact that it takes about 30 to 100 bile salts to solubilize one molecule of cholesterol (2, 6, 44b, 153) in solutions of bile salt well above the CMC. This is also true of a mixture of conjugated bile salts (6, 47). If each micelle held just one cholesterol molecule the aggregation number ought to be between 30 and 40. To test this hypothesis, studies were carried out to determine the aggregation number of micelles of NaTC and NaTDC nearly saturated with cholesterol by the equilibrium centrifugation technique (43). Apparent micellar weights were obtained at several different concentration differences and extrapolated to zero micelle concentration to obtain the micellar weights.

In 0.15 *M* NaCl the aggregation numbers (calculated from the micellar weights) of cholesterol-saturated micelles were: NaTC at 20°, 8; NaTC at 36°, 8. These are slightly higher than the aggregation numbers of NaTC without cholesterol. The aggregation number of NaTDC with cholesterol was unchanged from that of pure NaTDC. These results probably mean that while some of the micelles might contain a cholesterol molecule, each micelle *cannot* contain a molecule of cholesterol at the same time. Therefore the principle of at least "one molecule of solubilizate per micelle" (44b) is not followed and the real mechanism of cholesterol solubilization by bile salts is not known. While it is possible that there are two kinds of micelles present —ordinary-sized micelles with aggregation numbers of about 5, and a few very large micelles containing 1 cholesterol molecule with an aggregation number of about 25 to 35—as suggested by Woodford from free diffusion data (153), this situation is unlikely from a thermodynamic point of view.

One must ask the question—why don't the rest of the bile salts form large micelles with one cholesterol molecule per micelle? Unless this is a thermodynamically unstable situation the existence of two markedly different kinds of bile salt micelles in the same solution at cholesterol saturation seems unlikely since one must postulate that some bile salt micelles have a specific affinity for cholesterol while other chemically identical molecules do not. These observations may also apply to other compounds solubilized only slightly by bile salts [e.g., large fat-soluble dyes, methyl cholanthracine, griseofulvin, and glutethemide (12, 44b, 143, 147–148, 176)].

F. Bile Salt–Lecithin Micelles

The CMC of the lecithin–bile salt system has been investigated using the spectral-shift technique (177). The CMC of the mixtures, expressed as the concentration of bile salts needed to form the mixed micelle, is given as a function of the mole ratio of bile salt-to-lecithin in Table XVI. Three different counterion concentrations were studied. Interestingly, the CMC of the mixed micelles was lower in low salt concentrations. The opposite is found for ionic aliphatic detergents (132, 141) and for dihydroxy bile salts (see Section VIII. C, Fig. 40). When counterion is increased to 0.15 or 0.3 *M* NaCl the CMC increases. This unusual effect might be explained by "salting in" of the lecithin molecules, thus increasing the micellar charge and repulsive forces between the molecules. Except for the values obtained in low salt concentrations (0.05 *M* NaCl), high bile salt-to-lecithin ratios (27:1) result in very little CMC change. As the ratio increases the CMC falls abruptly and reaches a low and nearly constant value at about 2 bile salts to 1 lecithin. Further, the size of the lecithin–bile salt micelle increases markedly as the molar ratio of bile salt to lecithin increases (Table XV).

Using phase equilibrium techniques (discussed in Section X) it was shown that bile salts are extremely efficient solubilizers of lecithin (2, 4).

TABLE XVI. Critical Micellar Concentration (CMC) of NaTC–Lecithin Micelles in Different NaCl Concentrations, *p*H 6.9, 20 °C

Molar ratio NaTC:lecithin	CMC, mmoles bile salt per liter		
	0.05 NaCl	0.15 NaCl	0.3 NaCl
NaTC (no lecithin)	3.1	3.1	3.1
27:1	1.2	2.6	2.8
13:1	1.0	1.3	1.7
6:1	0.5	0.8	1.1
3:1	0.4	0.5	0.5
1.4:1	0.2	0.3	0.3
0.6:1	0.1	0.2	0.2

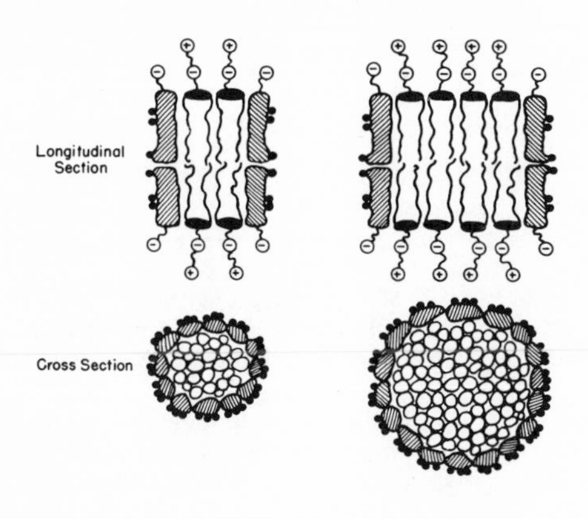

Fig. 49. Diagrammatic representation of the lecithin–sodium cholate micelle. On the left are represented small micelles containing a larger proportion of bile salt to lecithin. Above are longitudinal sections and below are cross sections of these micelles. On the right are shown larger micelles which have less bile salt and more phospholipid. The micelles are disc-shaped bimolecular leaflets of lecithin surrounded on their hydrophobic parts by a perimeter of bile salt molecules. Wavy lines or hollow circles, alkyl chains of lecithin; ⊕—⊖, phosphoryl choline of lecithin (67).

One mole of bile salt can solubilize a maximum of 2 moles of lecithin. This has since been confirmed by others using several different chemical species of lecithin (178, 179). Aliphatic detergents are far less efficient. For instance, it takes about 20 moles of lysolecithin to solubilize 1 mole of lecithin (42). How can we account for this phenomenon? Since lecithin is insoluble in water we will assume that virtually all the lecithin solubilized by bile salt is in the micelle and that only a negligible amount is in the aqueous phase outside of the micelle. Therefore, knowing the CMC, we may calculate from the micellar weight and bile salt:lecithin molar ratio how many bile salt molecules and how many lecithin molecules would be in each micelle at different molar ratios. These calculations show that as the number of lecithin molecules in the micelle increases from 2 to 80 the number of bile salts in the micelle increases from 12 to only 52. In other words, as the mole % lecithin increases, relatively less bile salt is needed to solubilize each lecithin molecule. This phenomenon is best explained by assuming a disc-shaped micellar model (41) shown in Fig. 49 (67). A disc-shaped bimolecular leaflet of lecithin molecules forms the core of the micelle. The ends of the disc are covered with the water-soluble phosphoryl choline groups but the sides of the disc are composed of the hydrocarbon alkyl chains. The hydrocarbon side of the bile salt molecule associates hydrophobically with the alkyl chain and part of the lecithin disc and thus hides these hydrocarbon chains from the surrounding aqueous phase. The hydrophilic parts of the bile salts (hydroxyls and anionic groups) protrude into the aqueous phase. With each increase in the mole % lecithin the number of lecithin molecules in the core disc increases. This increase causes an increase in the volume of the core and a corresponding increase in the cross-sectional area (Fig. 50). Note, however, that the new hydrocarbon surface produced is not a function of the increase in area ($\pi D_L^2/4$) but of the increase in perimeter of the disc πD_L. Thus, the

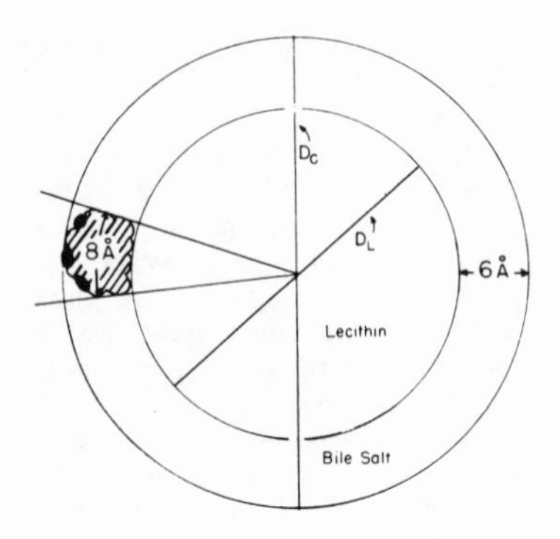

Fig. 50. Schematic cross section of the lecithin–bile salt micelle. D_L = diameter of lecithin disc. D_c = diameter of lecithin–bile salt micelle. Since the bile salt molecule is approximately 6 Å across, $D_c = D_L + 12$ Å.

increase in the number of bile salts needed to solubilize a given increase in the number of lecithin molecules is a function of the increase in perimeter, D_L, produced by the increase in the diameter of the disc. With certain assumptions, using the disc-shaped model, one may calculate the theoretical micellar weight of the bile salt–lecithin micelle formed by any bile salt–lecithin molar ratio. The assumptions are: (1) the bile salt molecule has a cross-sectional dimension of 6 × 8 Å. This value is a very realistic assumption (see Sections II. B and V. A). (2) The cross-sectional dimension of each lecithin molecule in the core is 70 Å^2. This value is taken from the maximum area of the lecithin molecule in a bilayer structure in equilibrium with excess water (7). (3) The height of the disc is 2 molecules thick, i.e., it is a bilayer.

Increasing numbers of lecithin molecules give a disc of larger area having an increasing diameter (D_L). The perimeter of the disc will increase by $\Delta\pi D_L$, exposing an increased surface to be covered by bile salts. As the perimeter is increased by approximately 8 Å two more bile salts (one each for upper and lower part of the bilayer) will be needed. The sum of the number of lecithin

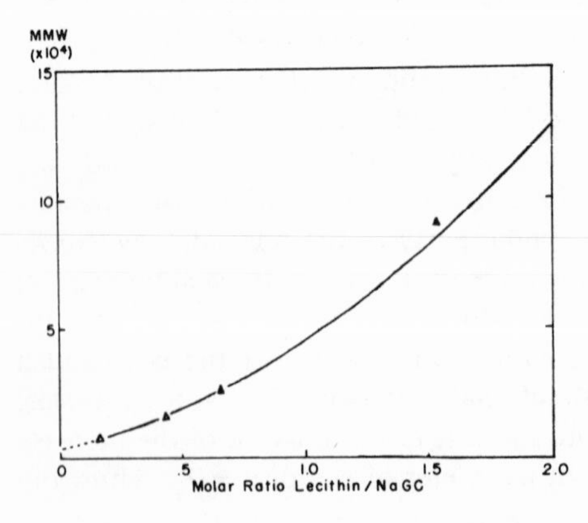

Fig. 51. Comparison of the calculated micellar weights of lecithin–bile salt micelle (solid line) with the values of lecithin–bile salt micellar weights obtained by equilibrium ultracentrifugation (triangles). The calculated and experimental values show excellent agreement.

molecules multiplied by its molecular weight plus the number of bile salts multiplied by its molecular weight will give the micellar weights for a disc of lecithin containing any number of lecithin molecules. A curve plotting the lecithin-to-bile salt ratio versus the calculated micellar weight appears in Fig. 51. The experimental values presented in Table XV are plotted as triangles and give excellent agreement with the calculated values for the same lecithin: bile salt ratio.

Further support for the model comes from X-ray diffraction studies on the hexagonal liquid crystalline phase (Fig. 52) formed by bile salts and lecithin in water (3). The hexagonal liquid crystalline phase of this system is composed of indefinitely long cylinders packed in a compact hexagonal lattice. The X-ray data give the distance from the center of one cylinder to another (D_t). This, of course, includes the total diameter of the cylinder (D_c) and the thickness of the water layer between the cylinders (D_w). The hexagonal phase breaks up into micelles when the excess water is added. The distance between cylinders (D_t) at the hexagonal phase–micellar phase transition, calculated from X-ray diffraction data (3), plotted against the molar ratio of lecithin: bile salt (NaC), shows that D_t increases in a linear fashion from about 48.5 Å to 79 Å as the molar ratio increases from 0.7 to 1.75. Using the same disc model proposed for the micelle (Fig. 50) one can calculate what the diameter of the cylinder of lecithin and bile salts (D_c) would be at different lecithin-to-bile salt ratios. Figure 52 shows that D_c is less than D_t but has the same increase as D_t with increasing lecithin: bile salt ratios. Thus, the differences between D_t and D_c represent the thickness of the

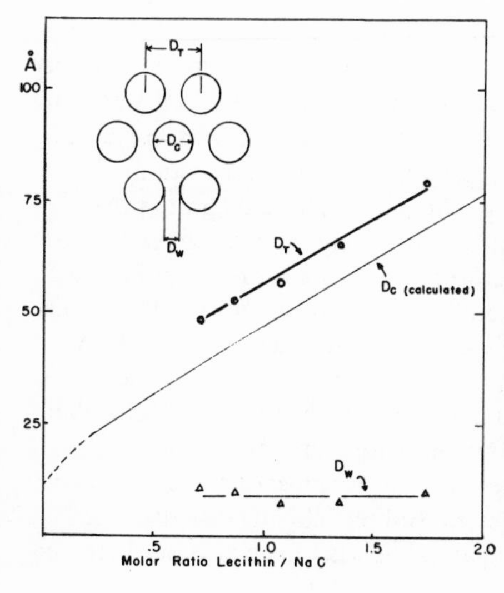

Fig. 52. X-ray diffraction data concerning the hexagonal liquid crystalline phase formed by NaC and lecithin in water at maximum swelling of the liquid crystalline phase. Data taken from (3). Vertical axis in Å; horizontal axis lecithin–bile salt (NaC) molar ratio. D_t = distance from the center of one cylinder to another (X-ray diffraction). D_c = diameter of the lecithin–bile salt cylinder. D_w = thickness of water layer between the cylinders. Insert on the left gives two-dimensional packing of lecithin–bile salt cylinders in a compact hexagonal array.

water layer D_w at a composition where the hexagonal liquid crystal phase breaks up into micelles. D_w is a constant 9 Å or about 3 water molecules thick, at this composition. Thus, both studies of micellar weight and X-ray studies of the parent phase (hexagonal–liquid crystal phase) of the micelles supports the proposed model.

X. PHASE EQUILIBRIA DIAGRAMS—THE BULK INTERACTION OF BILE SALTS AND ACIDS WITH WATER AND OTHER LIPIDS

The most complete method of presenting the interaction of one chemical species with another is to study the physical state of all possible mixtures of the two compounds under varying conditions of temperature and pressure. The construction of phase equilibria diagrams (180, 181, 182) presents such a description. Further, more complicated phase diagrams delineate the physical state (solid, liquid crystal, liquid) of any mixture of 3 or 4 components at a given temperature and pressure.

A. Binary Systems Involving Bile Salts

Binary systems (two components, A and B) at constant temperature and pressure would be represented by a line AB.

100% A	75% A	50% A	25% A	0% A
0% B	25% B	50% B	75% B	100% B

One end of the line would be 100% A and 0% B and the other end 100% B and 0% A. Any point along the line would represent mixtures of A and B. The closer the point is to 100% A the more A is in the mixture. Often only one of the components is labeled, but the other can be deduced since any point must be equal to 100% (A+B). Binary phase diagrams are usually given as a function of temperature, but could be given as a function of pressure, magnetic field, electric field, etc. In the binary phase diagram presented below A and B are bile salt and water, respectively, and the physical state will be plotted as a function of temperature at constant pressure. Figure 53 represents sodium cholate–water systems studied from 0°C to 100°C (42). A similar binary phase diagram for Na deoxycholate was published by Vold and McBain over 25 years ago (183). The striking difference between these molecules and the classical alkyl chain detergents such as soaps of long-chain fatty acids is that no liquid crystalline phases are formed (2, 4, 42, 183) and the critical micellar temperature (Section VII. D) for most of these substances is below 0°C (1, 10, 45). The micelles of sodium cholate are discussed in Section VIII. Phase diagrams of conjugated bile salts have also been studied

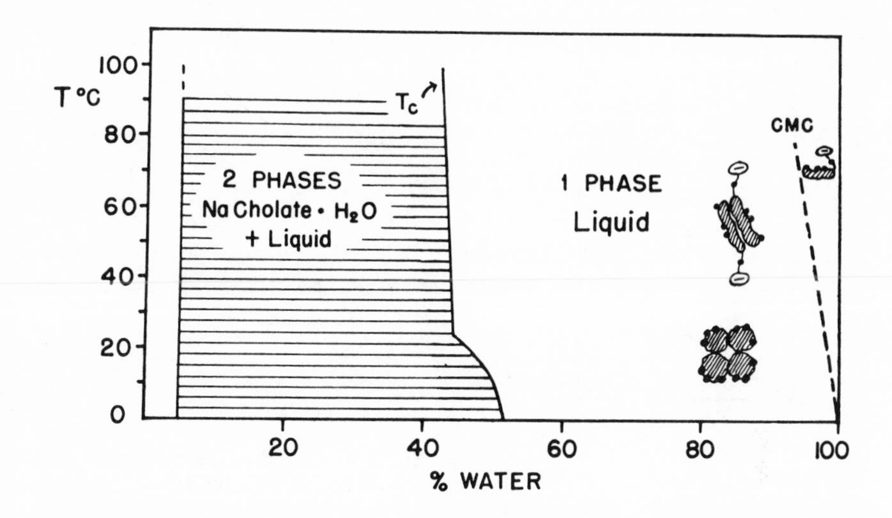

Fig. 53. Sodium cholate–water binary phase diagram. Expressed as wt %. Vertical axis, temperature degrees C; horizontal axis, percentage of water. The line at about 5% water represents a monohydrate of sodium cholate. This monohydrate is in equilibrium with a liquid phase. The dotted line marked T_c represents the solubility of sodium cholate in water. The solubility increases slightly with increasing temperature. The liquid phase in the dilute region is made up of small micelles. CMC, critical micellar concentration (this line is very approximate; see Section VIII. B) (42).

(4). Both the taurine and glycine conjugates are very soluble and seem to form one liquid phase with water in concentrations up to 92% bile salts. As a greater proportion of bile salt is present, the aqueous phase becomes more and more viscous so the 90% bile salt–10% water mixture is a glassy semi-solid. No liquid crystals have been identified in any of these systems.

B. Ternary Systems Involving Bile Salts

The phase equilibria relationships of mixtures of three components (A, B, and C) at constant temperature and pressure are best represented by an equilateral triangle ABC. Each side of the triangle represents a binary system (AB, AC, or BC). Any point within the triangle represents a given mixture dictated by the geometry of the triangle. If one wishes to study the ternary system ABC as a function of another variable, for instance temperature, one can represent the system as a regular prism as shown in Fig. 54. One then may represent the system as a series of ternary diagrams (ABC) at different temperatures. This method has been used to study the phase equilibria of triglycerides, cholesteryl esters, and cholesterol as a function of temperature (184).

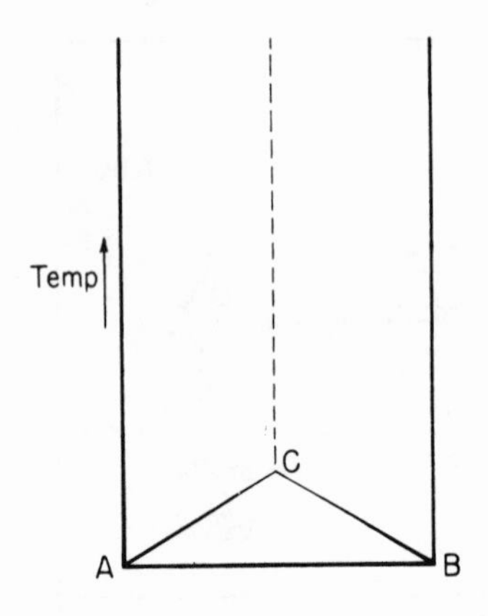

Fig. 54. Standard method for representation of a three-component system as a function of temperature.

1. Bile Salt–Decanol–Water Systems (38)

Fontell studied the solubilization of n-decanol by NaDC (38). His phase diagram is redrawn in Fig. 55 and demonstrates that bile salt in adequate quantities produces a single liquid phase extending from pure water to pure decanol. Thus, bile salt not only solubilizes decanol in an aqueous solution (left side) but also solubilizes water in a decanol solution (right side).

2. Bile Salt–Cholesterol–Water Systems (2, 6, 47)

The interactions of another alcohol, cholesterol, with water and bile salt have been studied at 20 and 37 °C (2, 6, 47). Figure 56 shows that unlike decanol bile salt does not solubilize cholesterol well. Both free and conjugated bile salts are ineffective (6, 47). Perhaps the bulky steroid nucleus,

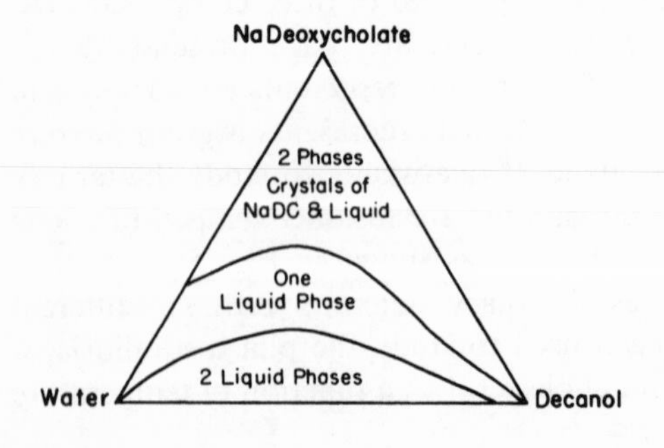

Fig. 55. Phase equilibria diagram sodium deoxycholate–decanol–water. Expressed as wt % (185).

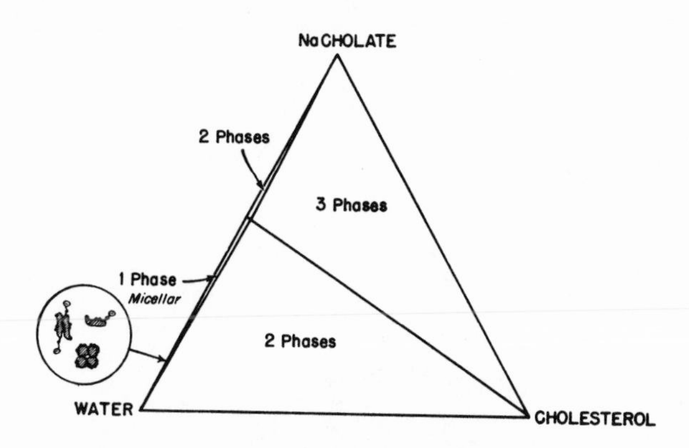

Fig. 56. Cholesterol–sodium cholate–water ternary phase diagram. The structure of the bile salt micelles is indicated in the inset. These micelles remain small in the presence of cholesterol (Section IX. E). It will be noted that the micellar zone is small and that no liquid crystalline phases are formed in this system (2, 6, 47).

the higher molecular weight, and the high melting point of cholesterol account for the marked differences in solubilization when compared to decanol. A discussion of the bile salt–cholesterol mixed micelle appears in Section IX. E. No liquid crystal phases have been noted.

3. Bile Salt–Monoglyceride–Water Systems (10)

Monoglyceride, bile salt, and water have also been studied (10). Bile salt was shown to be an effective agent to solubilize monoglyceride. Liquid crystals were noted but not examined for type or structure.

4. Bile Salt–Phospholipid–Water Systems (2, 3, 4, 8, 67, 177)

The interaction of the insoluble swelling amphiphile (lecithin) with NaC in water is shown in simplified form in Fig. 57. This system, lecithin–bile salt–water, is an important biological system, because the primary constituents, lecithin and bile salt, are the solubilizing system of bile for insoluble nonswelling amphiphiles, such as free cholesterol, which occurs in bile in high concentrations. The phases found in this diagram have been studied by a number of techniques including X-ray diffraction (3), light scattering (2), equilibrium ultracentrifugation (Section IX. F), nuclear magnetic resonance (67), viscosity, surface balance, surface tension, and picnometry. As noted in Section VIII, NaC forms very small micelles in water (Fig. 47) whereas lecithin swells in water to form a lamellar liquid crystalline phase (7). Small quantities

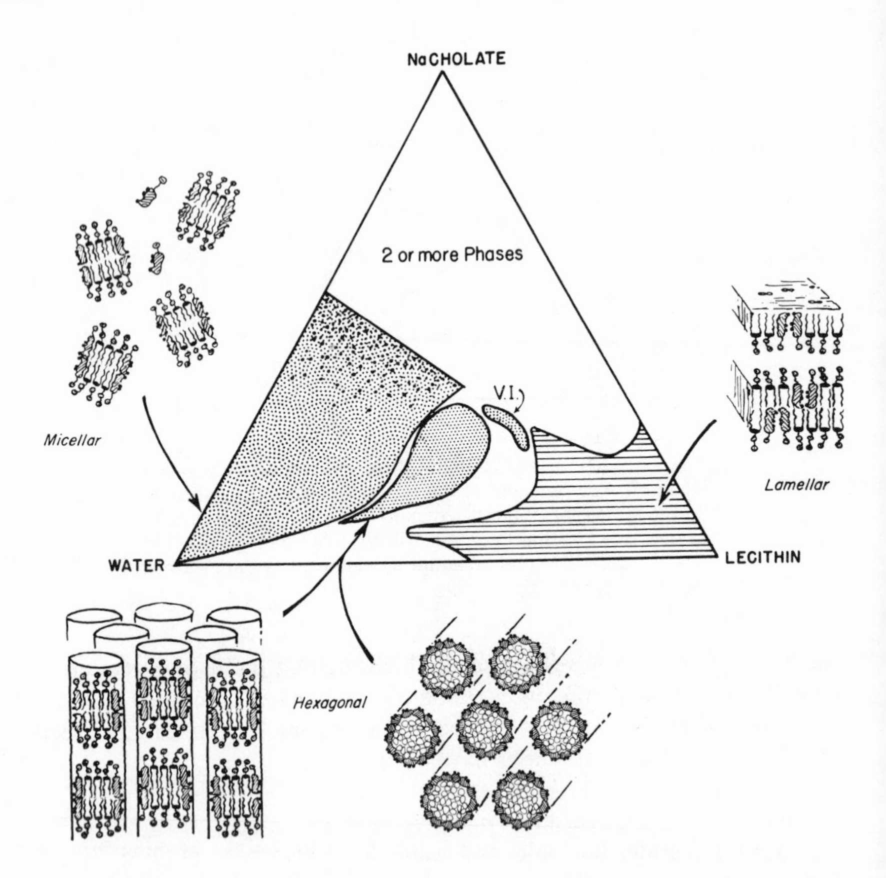

Fig. 57. Lecithin–sodium cholate–water ternary phase equilibria diagram. Expressed as wt%. The zones of separation of two or more phases have been left out to simplify the diagram. They are detailed in (4). The hexagonal and micellar zones are large. Bile salt in contrast to aliphatic detergents is an efficient solubilizer of lecithin. Only one molecule of bile salt is necessary to solubilize two molecules of lecithin. This may be because of the arrangement of the molecules in these micelles (inset on the left and Section IX. F). The probable arrangement of the molecules in the lamellar phase, the hexagonal phase, and micellar phase is shown. Both the cross section of the cylinders and a longitudinal view of the cylinders of the hexagonal phase are represented in the inset at the bottom of the figure. In the lamellar phase the molecules of bile salt form pairs or perhaps tetramers, which are probably hydrogen-bonded through their hydroxyl groups, thus exposing only their hydrophobic backs to the hydrophobic aliphatic chains of the lecithin molecules. The structure of the cubic phase (VI) has not been shown (42).

of bile salt can be incorporated in the lamellar liquid crystalline phase formed by lecithin. The bile salts are probably present as ion pairs, trimers, or tetramers, hiding their hydroxyl group—by intermolecular hydrogen bonding between the OH groups—from the surrounding lipophilic parts of the lecithin molecules in the manner noted on right side of Fig. 57. This is strongly suggested by the fact that the X-ray diffraction analysis of lecithin–bile salt mixtures at 25% water shows a decrease in the thickness of the bilayers as bile salt is added, suggesting that the bile salt is, in fact, collapsing the lecithin bilayer slightly (3). That hydrogen bonding can occur between molecules of this type in hydrophobic environment has been shown by Bennet, Elington, and Kovac (122). In the area of the diagram containing fairly large amounts of bile salt at low hydration, large amounts of bile salt can be incorporated into the lamellar liquid crystalline phase without separation as another phase. This is probably due to the fact that there is not enough water present to penetrate between the OH-bonded bile salts.

Two other interesting liquid crystalline phases are formed by these systems. (1) Cubic phase. The structure of this cubic phase is not known, but it is probably made up of mixed micelles packed in a face-centered cubic lattice (3). (2) Hexagonal phase. This phase, which covers a wide range of compositions, gives the typical "middle soap" (186) birefringent texture by polarizing microscope (4) and the spacings of a two-dimensional compact hexagonal lattice (3). X-ray examinations of 80 different mixtures in this phase have been carried out (3) and these data, coupled with studies with molecular models and observations of mixed monolayers of lecithin and bile acid (Section V. B), suggest that the structure of this hexagonal phase differs from that of soaps and other detergents (187, 188). The probable structure is that of cylinders of lecithin coated with bile salt molecules separated by water and packed in a compact hexagonal array (see Fig. 57 for both longitudinal and cross-sectional views). This type of cylinder differs in structure from that of soaps in that the cylinders are made up of stacks of short discs and contain some water within the cylinder itself. As one adds further water to these mixtures, the cylinders break up into disc-like micelles (as discussed in Section IX. E). Further confirmation of this structure comes from NMR examination of the lamellar and hexagonal phases in D_2O. Lawson and Flautt (189, 190) and Chapman *et al.* (191–196) have shown that the conventional lamellar liquid crystalline and hexagonal phases formed by soaps and phospholipids have a fairly broad NMR spectrum (about 0.1–0.4 gauss). The micellar phase, on the other hand, has a very high resolution spectrum. The spectrum of a micellar solution of NaC and lecithin (molar ratio 1.5:1) is illustrated in Fig. 58 (67). Note the very sharp signals from the $—CH_2—$, $—N(CH_3)$ (choline of lecithin), $—CH{=}CH—$, and terminal alkyl chain $—CH_3$ groups. The usually prominent agular methyl protons of the bile salt (see Section II. D) are hindered in their motion and do not appear. In Fig. 59 the high-resolution spectrum of lamellar liquid crystalline phase of 10% NaC–90% lecithin–40% D_2O is shown. It is broad and no clear-cut peaks can be seen. When viewed by the broad-line technique the spectrum is

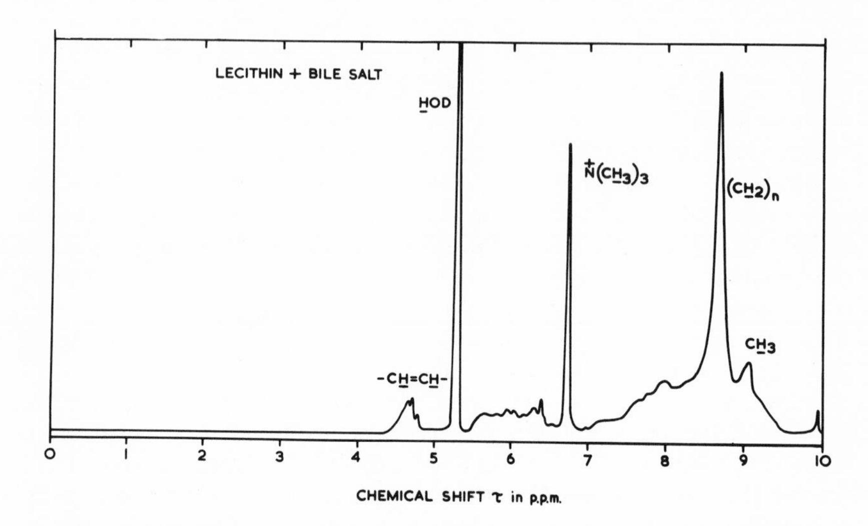

Fig. 58. High-resolution 60 Mc NMR spectra of 10% lecithin–bile salt solution in D_2O. Molar ratio of lecithin to sodium cholate, 3:2. Note the sharp peaks due to choline and the alkyl chains of the lecithin. The angular methyl groups of the bile salt are not evident. Temperature of 33.4°C (67).

about 0.2 gauss wide at mid-peak height. Above is the spectrum of the hexagonal liquid crystal phase formed by 35% NaC, 30% lecithin, and 35% D_2O. Unlike the broad spectrum seen for the lamellar phase (above) and the hexagonal phases formed by soaps and detergents (189, 190), this phase gives high-resolution spectra similar to those of the micellar phase. These data confirm the impression suggested by other techniques that the hexagonal phase formed by bile salts and lecithin consists of cylindrical stacks of disc-shaped micelles packed in a hexagonal lattice. The high-resolution spectra suggest that these micellar units have a high degree of motion within the hexagonal phase.

The NaC–lecithin–water system described above is very similar to the conjugated bile salt–lecithin–water system. Minor differences are discussed in (4).

Bile salts also solubilize other phospholipids extensively, although no phase diagrams have been published.

5. *Bile Salt–Soap–Water Systems(42)*

Figure 60 gives an incomplete phase diagram of the two soluble amphiphiles, NaC and sodium oleate in water (42). Sodium oleate, with increasing hydration, forms a small lamellar liquid crystalline phase (Fig. 60, zone

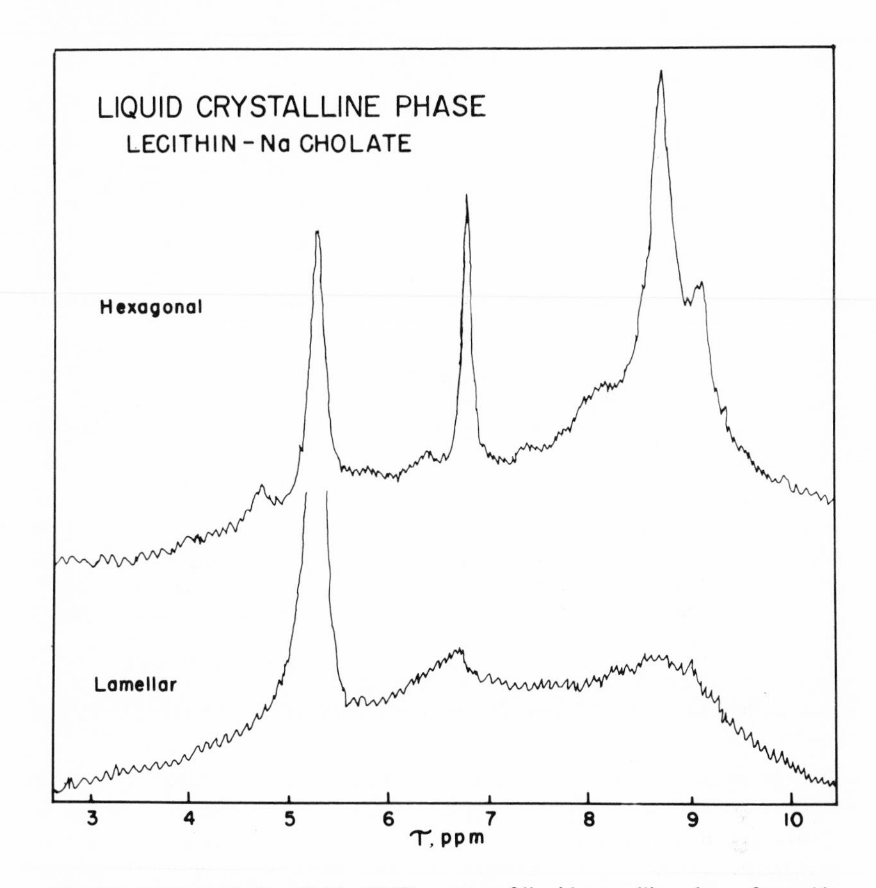

Fig. 59. High-resolution 60 Mc NMR spectra of liquid crystalline phases formed by NaC and lecithin in D_2O. Top—hexagonal liquid crystalline phase formed by 35% NaC–30% lecithin–35% D_2O. Bottom—lamellar liquid crystalline phase formed by the mixture 10% NaC–90% lecithin–40% D_2O. The peak at 5.4 ppm is due to HDO and H_2O. Note high resolution peaks due to choline at 6.7 ppm and the methyl groups about 8.7 ppm. 33.4 °C.

I), then a hexagonal phase (zone III), and finally a micellar phase (zone IV). As NaC is added to sodium oleate, the lamellar phase (zone I) disappears abruptly, showing that very little bile salt can be incorporated into the lamellar liquid crystalline phase formed by sodium oleate. At low concentrations of water, a viscous isotropic phase having a cubic lattice by X ray is present (zone II). There is a large zone of hexagonal liquid crystalline phase (zone III) which can contain up to 65% NaC by weight. This phase has been studied by X-ray diffraction at a constant 40% water, and the distance between

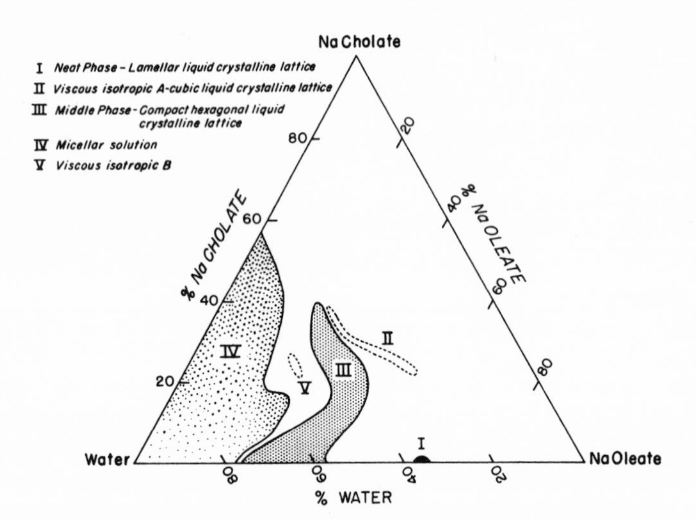

Fig. 60. Sodium oleate–sodium cholate–water ternary-phase diagram. Expressed as wt%. The various phases have been numbered I–V. The lamellar liquid crystalline phase can incorporate little bile salt into the lattice. The micellar phase is large and includes all possible combinations of sodium oleate and sodium cholate, provided the amount of water is sufficient. The structure of these micelles is not yet known, but see Section IX.C (42). 37 °C, *p*H 9.0.

cylinders increases from 35 Å at 65% NaC–35% sodium oleate to 50.7 Å at 25% NaC–75%sodium oleate. Thus, the distance between cylinders increases as the proportion of oleate increases. While the detailed molecular arrangement in the hexagonal lattice has not yet been established it may be similar to that of the NaC–lecithin–water systems. Another small zone of viscous isotropic liquid crystal (zone V) is present between the middle and micellar zones. When large proportions of water are present, a micellar phase (zone IV) is formed which is made up of mixed micelles of sodium cholate and sodium oleate similar to those described in Section IX. C.

C. Quaternary Systems Involving Bile Salts

Four-component systems (ABCD) at constant temperature and pressure may be represented as a regular tetrahedron (Fig. 61). Again, each side represents a binary system (AB, AC, AD, BC, BD, CD). Each face (an equilateral triangle) is a ternary system (ABC, ABD, ACD, BCD) and points within the volume of the tetrahedron represent a specific mixture of all 4 components. Obviously, it is difficult to use this representation directly. However, one can make planar sections through the tetrahedron parallel to one face, e.g., ABD. Then each section will represent the phase equilibria relations

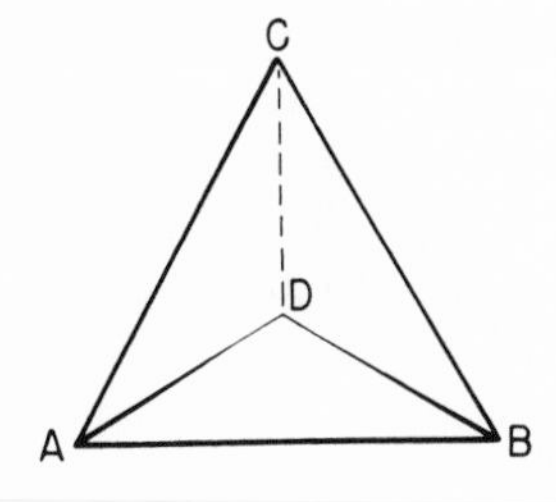

Fig. 61. Conventional representation of 4-component system ABCD as a regular tetrahedron.

of ABD in a fixed amount of C. Figure 62 represents the quaternary system cholesterol–lecithin–sodium cholate–water (2, 3, 6, 8). The ternary phase diagrams cholesterol–lecithin–water (5), cholesterol–sodium cholate–water (Section X, Fig. 56) and lecithin–sodium cholate–water (Section X, Fig. 57) represents three of the faces of the tetrahedron. The fourth face cholesterol–lecithin–sodium cholate consists of only solids and has not been studied completely. Because it is difficult to represent a three-dimensional figure in two dimensions, mixtures containing all four components have been represented

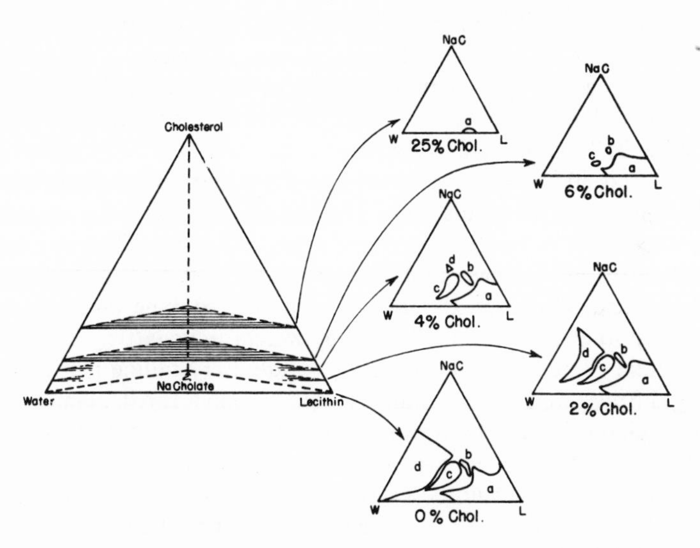

Fig. 62. The quaternary system cholesterol–lecithin–sodium cholate–water. The tetrahedron on the left is a representation of this four-component system. The five triangles of decreasing size at the right of the diagram schematically represent cuts taken parallel to the base of the tetrahedron (lecithin–sodium cholate–water), to which a given amount of cholesterol at different overall cholesterol concentrations (2%, 4%, 6%, 25%) has been added. Chol, cholesterol; L, lecithin; NaC, sodium cholate; W, water. Also: a, lamellar liquid crystalline phase; b, cubic liquid crystalline phase; c, hexagonal liquid crystalline phase; d, micellar phase (42). Expressed as wt%.

as sections through the tetrahedron, parallel to the base lecithin–sodium cholate–water. Each section represents the interactions of lecithin–NaC–water to which a fixed amount of cholesterol has been added. The resultant triangular sections are illustrated by representative cuts on the right side of Fig. 62. As the amount of cholesterol is increased, each phase becomes progressively saturated with cholesterol. Thus, the micellar phase becomes saturated at 4% cholesterol and the hexagonal and cubic phases at 6 and 7% cholesterol, respectively. The lamellar phase, however, can contain up to 30% cholesterol. It would appear that cholesterol and bile salt compete for places in the lamellar liquid crystalline lattice. All spaces between the zones of one homogeneous phase are mixtures of two or more phases. Cuts may be taken in a plane parallel to the three solid components NaC–lecithin–cholesterol. This method has been used to give the solubility of cholesterol in different micellar mixtures of bile salt and lecithin (8, 9a,b, 197). These observations have led to studies of cholesterol solubility in bile—a fluid made up primarily of these four components (8, 9, 197).

Figure 63 (left side) is a regular tetrahedron representing the quaternary system cholic acid–lecithin–sodium cholate–water. Representative cuts have been taken through the tetrahedron and are illustrated by the sections on the right side of Fig. 63. The only difference in each of the cuts is the proportion of free cholic acid to sodium cholate. Thus, these cuts are (from bottom to top): the base diagram lecithin–sodium cholate–water (Section X, Fig. 57); a cut having 25% cholic acid and 75% sodium cholate; a cut containing 50% cholic acid and 50% sodium cholate; and the ternary phase diagram cholic acid–lecithin–water. These diagrams demonstrate that as cholic acid replaces sodium cholate, the micellar phase (d), the hexagonal phase (c), and the cubic phase (b) all decrease and disappear at about 50% sodium cholate, 50% cholic acid. However, the lamellar liquid crystalline phase (a) can incorporate up to about 15% cholic acid by weight in its lattice. The structure and molecular dimensions of this lamellar liquid crystalline phase, as given by X-ray diffraction, are very similar to those shown for the lamellar liquid crystalline phase formed by lecithin and bile salt (Fig. 57). Thus, dimers, trimers, and tetramers of cholic acid are interdigitated in between the lecithin molecules in the lamellar liquid crystalline phase. In high water concentrations the effects shown would represent changes in *p*H between 10 and 2. At a *p*H of about 5.5 to 6 the micellar phase is lost.

D. Complex Multicomponent Systems Involving Bile Salts

Representation of more than four components is difficult. Often chemically similar amphiphilic substances may act as a single component over a large range of concentrations and temperatures. Thus, McBain, Vold, and

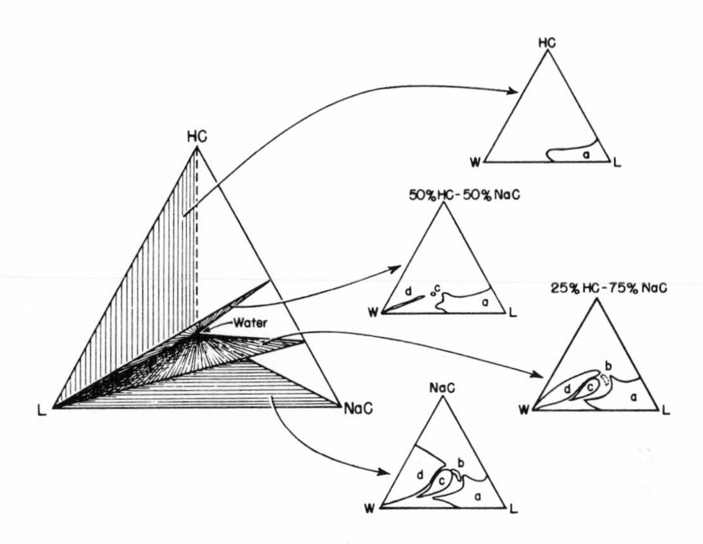

Fig. 63. The quaternary system cholic acid–lecithin–sodium cholate–water. Expressed as wt%. The tetrahedron at the left represents this quaternary system. HC, cholic acid; L, lecithin; NaC, sodium cholate; W, water. Also: a, lamellar liquid crystalline phase; b, cubic liquid crystalline phase; c, hexagonal liquid crystalline phase; d, micellar phase. The shaded areas represent four sections of the quaternary system, taken at different proportions of cholic acid and sodium cholate. These sections are from top to bottom: cholic acid–lecithin–water; 50% cholic acid; 50% sodium cholate–lecithin–water; lecithin–water. Note that the top triangle HC–L–H_2O has only the lamellar liquid crystalline phase (42).

Porter (198) were able to show that a commercial soap (a mixture of sodium and potassium soaps of different long-chain fatty acids) acted as one component except in the nearly dry state. Egg lecithin, a mixture of different phosphatidyl cholines, also acts as a single component except when nearly dry and at low temperatures (7). Further, a mixture of conjugated bile salts also behaves as a single component under certain conditions (4). Some caution should be used in applying the phase rule to these types of mixtures, for only if the mixture of chemically similar molecules behaves as a single component can the phase rule be applied.

A less complete but often useful method of examining phase relations in multicomponent systems is to hold one or more components constant (for instance water or water and NaCl). One may then examine superficially the phase relations of several compounds. For instance, in a six-component system ABCDEF (Fig. 64), one might keep E and F constant in all mixtures. Then the ternary system ABC could be examined in different amounts of D.

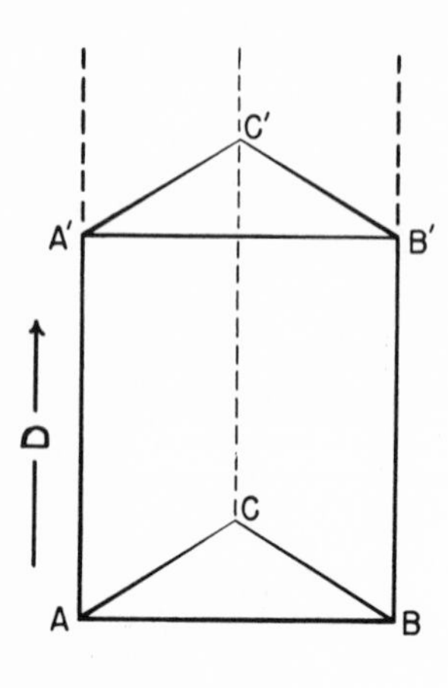

Fig. 64. Method for representing phase relations in a multicomponent system ABCDEF. The phase relations of components AB and C are examined as in increasing amounts of component D in constant amounts of E and F. For further explanation see text.

The representation for this type of complex system would be a regular prism with ABC as the base and the height D. The specific relations can be shown as a series of ternary systems (ABC) to which a constant amount of D has been added (1% D, 2% D, 3% D, . . ., 10% D, etc.), all carried out in a constant amount of E and F. Boundaries of single homogeneous phases may be easily delineated but composition and number of phases in multiphase regions is difficult to determine.

The usefulness of this type of approach is illustrated by results concerning the sodium oleate–sodium taurocholate–1-monoolein–water system as a function of *p*H (Fig. 65). Since hydrogen ion concentration changes sodium oleate partly to oleic acid in the *p*H range studied, sodium oleate really acts (from the point of view of the phase rule) as two components. Thus the system has at least five components. If excess counterion (NaCl) is added, the system becomes even more complex. The results given below were carried out in 98% water (or 0.15 *M* NaCl) at 37°C and give only the composition range of the three components [Na oleate (or oleic acid), Na taurocholate, and 1-monoolein] which form a stable aqueous micellar solution. Only those compositions falling in the zone of vertical shading form stable, clear micellar solutions. All other mixtures separate into two or three other phases (oil, liquid crystals, micellar solution). Since major products of pancreatic lipolysis are fatty acids like oleic acid and monoglycerides like monoolein, and the *p*H of the intestine can vary markedly, one might predict from this model system that *p*H in the intestinal lumen could have a profound effect on the composition of the "micellar" phase of intestinal content.

At *p*H 10, obviously well above intestinal *p*H, the sodium oleate is ionized and the micellar zone is very large. At this *p*H all combinations of sodium taurocholate and sodium oleate form mixed micelles. These micelles

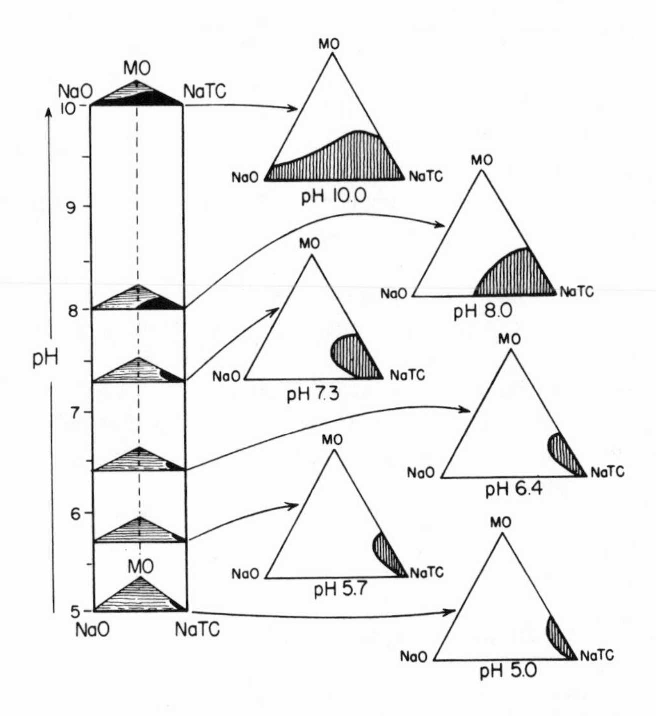

Fig. 65. Multicomponent system sodium oleate (NaO)–sodium taurocholate (NaTC)–1-monoolein (MO)–water as a function of *p*H. 37 °C. Total lipid 2% by weight, total water 98% by weight. Triangles on right are given in mole % 100 moles NaO + NaTC + MO.

solubilize varying quantities of monoolein and in structure are similar to other bile salt micelles discussed previously (Sections IX. C, and X, Fig. 60). As the *p*H is lowered to 8 the micellar zone is reduced and only certain proportions of sodium taurocholate and sodium oleate form micelles. This is due to the formation of more oleic acid than can be solubilized in the mixed bile salt–soap micelle. As *p*H is decreased to 7.3, a more physiological *p*H, the micellar sone is decreased further. At this *p*H it takes quite a large amount of sodium taurocholate to solubilize the oleate present because of the increasing amount of oleic acid present in the system. The micellar zone continues to decrease as the *p*H is lowered. At *p*H 5 very large amounts of sodium taurocholate are needed to solubilize small amounts of oleic acid. Monoolein appears to help the sodium taurocholate in solubilizing the partially ionized oleic acid. Thus, these studies define the combinations of sodium oleate, sodium cholate, and 1-monoolein which will form a micellar solution over the range of *p*H which one would expect to be present in the small intestine. Similar studies were carried out in 0.15 M NaCl. The fixed components in these systems are water and NaCl. The differences in the micellar zone were minor in the presence of 0.15 M NaCl. However, when the total water concentration was increased from 98% to 99.5% (i.e., 0.5% total monoolein plus sodium taurocholate plus sodium oleate) the micellar zone was considerably smaller at every pH. This is simply due to the fact that there are fewer micelles present in these mixtures to solubilize the insoluble components.

Other attempts have been made to study the partition of multilipid–bile salt–aqueous systems into oil and aqueous phases (199–201). These studies are very complex, and discussion of other possible phases such as liquid crystals have been ignored. Despite the difficulty in interpretation, these studies represent the only attempt to define the oil and micellar phases of very complex systems analogous to physiological phenomena occurring during fat absorption.

ACKNOWLEDGMENTS

This work was supported in part by NIH Grants AM-11453, AM-12890, and AM-0525.

I wish to thank the American Oil Chemists' Society, Academic Press, Macmillan Ltd., Elsevier Publishing Co., the American Chemical Society, and the Chemical Societies in Denmark for permission to reproduce tables and figures from their journals.

I wish to acknowledge the helpful support of Drs. F.J. Ingelfinger and R.M. Donaldson and the friendly tutelage of Prof. D.G. Dervichian while I was visiting the Service de Biophysique, Institut Pasteur, from 1963 to 1965.

I also wish to thank Mrs. Connie Castles for her able secretarial help.

REFERENCES

1. A.F. Hofmann and D.M. Small, *Ann. Rev. Med.* **18,** 333 (1967).
2. D.M. Small, M. Bourges, and D.G. Dervichian, *Nature* **211,** 816 (1966).
3. D.M. Small and M. Bourges, *Mol. Crystals* **1,** 541 (1966).
4. D.M. Small, M. Bourges, and D.G. Dervichian, *Biochim. Biophys. Acta* **125,** 563 (1966).
5. M. Bourges, D.M. Small, and D.G. Dervichian, *Biochim. Biophys. Acta* **137,** 157 (1967).
6. M. Bourges, D.M. Small, and D.G. Dervichian, *Biochim. Biophys. Acta* **144,** 189 (1967).
7. D.M. Small, *J. Lipid Res.* **8,** 551 (1967).
8. W.H. Admirand and D.M. Small, *J. Clin. Invest.* **47,** 1043 (1968).

9a. D.M. Small, *New Eng. J. Med.* **277,** 588 (1968).
9b. D.M. Small, *Advan. Intern. Med.* **16,** 243 (1970).
10. A.F. Hofmann, "The Function of Bile Salts in Fat Absorption," Doctoral thesis, University of Lund, Sweden (1964).
11. A.F. Hofmann, *Biochim. Biophys. Acta* **70,** 306 (1963).
12. A.F. Hofmann, *Biochem. J.* **89,** 57 (1963).
13. J.M. Dietschy and M.D. Siperstein, *J. Clin. Invest.* **44,** 1311 (1965).
14. R.H. Palmer, *Science* **148,** 1339 (1965).
15. R.H. Palmer and Z. Ruban, *J. Clin. Invest.* **45,** 1255 (1966).
16. F.G. Zaki, J.B. Carey, Jr., F.W. Hofbauer, and C. Nwokolo, *J. Lab. Clin. Med.* **69,** 737 (1967).
17. P. Holsti, *Nature* **186,** 250 (1960).
18. A. Stalk, *Experientia* **16,** 507 (1960).
19. R.D. Hunt, G.A. Leveille, and H.E. Sauberlich, *Proc. Soc. Exp. Biol. Med.* **113,** 139 (1963).
20. R.D. Hunt, G.A. Leveille, and H.E. Sauberlich, *Proc. Soc. Exp. Biol. Med.* **115,** 277 (1964).
21. G.A. Leveille, R.D. Hunt, and H.E. Sauberlich, *Proc. Soc. Exp. Biol. Med.* **115,** 569 (1964).
22. G.A. Leveille, R.D. Hunt, and H.E. Sauberlich, *Proc. Soc. Exp. Biol. Med.* **115,** 573 (1964).
23. G.A. Leveille, R.D. Hunt, and H.E. Sauberlich, *Proc. Soc. Exp. Biol. Med.* **116,** 92 (1964).
24. R.D. Hunt, *Fed. Proc.* **24,** 431 (1965).
25. N.B. Javitt, *Nature* **210,** 1262 (1966).
26. N.B. Javitt and S. Emerman, *J. Clin. Invest.* **47,** 1002 (1968).
27. "Elsevier Encyclopedia of Organic Chemistry," pp. 167–214, Elsevier, Amsterdam (1940).
28. H. Sobotka, "Chemistry of Steroids," Williams and Wilkins, Baltimore (1938).
29. G.A.D. Haslewood, *Biol. Rev. Cambridge Phil. Soc.* **39,** 537 (1964).
30. G.A.D. Haslewood, in "Bile Salts" (R. Peters and F. G. Young, eds.), Methuen, London (1967).
31. G.A.D. Haslewood, Review, *J. Lipid Res.* **8,** 535 (1967).
32. G.A.D. Haslewood and V. Wootton, *Biochem. J.* **47,** 584 (1950).
33. H. Sobotka, "The Physiological Chemistry of the Bile," Williams and Wilkins, Baltimore (1937).
34. B. Josephson, *Physiol. Rev.* **21,** 463 (1943).
35. P. Ekwall, *Acta Acad. Aboensis Math. Physica,* **17,** 3 (1951).
36. P. Ekwall, *J. Coll. Sci.,* Suppl. 1, 66 (1954).
37. P. Ekwall, K. Fontell, and A. Sten, Proc. 2nd Intl. Cong. Surface Activity, Butterworth, London, p. 357 (1957).
38. K. Fontell, *in* "Surface Chemistry" (P. Ekwall, K. Groth, and v. Runnstrom-Reio, eds.), p. 252, Academic Press, New York (1965).
39. P. Ekwall, Intl. Conf. on Biochem. Probs. of Lipids, *Koninkl. Vlaam. Acad. Wetenschap. Belgie.* 103 (1953).
40. G. A. D. Haslewood, *Physiol. Rev.* **35,** 178 (1955).
41. D.M. Small, *Gastroenterology* **52,** 607 (1967).
42. D.M. Small, *J. Am. Oil Chem. Soc.* **45,** 108 (1968).
43. D.M. Small, *Advan. Chem. Ser.* **84,** 31 (1968).
44a. M.C. Carey and D.M. Small, *J. Coll. Interface Sci.* **31,** 382 (1969).
44b. M.C. Carey and D.M. Small, *Am. J. Med.* **45,** 590 (1970).
45. D.M. Small and W.H. Admirand, *Nature* **221,** 265 (1969).
46. A.F. Hofmann, *Advan. Chem. Ser.* **84,** pp. 53–66 (1968).
47. D.M. Small, *in* "Bile Salt Metabolism" (L. Schiff, J. Carey, and J. Dietschy, eds.),

Charles C Thomas, Springfield, Ill., Chap. 19 (1968).
48. L.J. Bellamy, "The Infrared Spectra of Complex Organic Molecules" (2nd ed.), Wiley, New York (1958).
49. N.B. Colthup, L.H. Daly, and S.E. Wiberly, "Introduction to Infrared and Raman Spectroscopy," Academic Press, New York, London (1964).
50. R.N. Jones and C. Sandorfy, *in* "Technique of Organic Chemistry" (A. Weissberger, ed.), p. 247, Interscience, New York (1956).
51. H.M. Randall, R.G. Fowler, N. Fuson, and J.R. Dangl, "Infrared Determination of Organic Structures," D. Van Nostrand, Princeton, N.J. (1949).
52. R.M. Silverstein and G.C. Bassler, "Spectrometric Identification of Organic Compounds" (2nd ed.), J. Wiley, New York (1967).
53. H.A. Szymanski, "IR Theory and Practice of Infrared Spectroscopy," Plenum Press, New York (1964).
54. I.D.P. Wooten and H.S. Wiggins, *Biochem. J.* **55,** 292 (1953).
55. G. Chihara, K. Matsuo, K. Arimoto, S. Sugano, K. Shimizu, and K. Mashimo, *Chem. Phar. Bull.* **9,** 939 (1961).
56. G. Chihara, K. Matsuo, K. Arimoto, and S. Sugano, *Chem. Pharm. Bull.* **10,** 1190 (1962).
57. G. Chihara, K. Matsuo, A. Mizushima, E. Tanaka, K. Arimoto, and S. Sugano, *Chem. Pharm. Bull.* **10,** 1184 (1962).
58. T. Hoshita, S. Nagayoshi, and T. Kazuno, *J. Biochem.* **54,** 369 (1963).
59. T. Hoshita, S. Nagayoshi, M. Koughi, and T. Kazuno, *J. Biochem.* **56,** 177 (1964).
60. K. Dobriner, E.R. Katzenellenbogen, and R.N. Jones, "Infrared Absorption Spectra of Steroids," Interscience, New York (1953).
61. L.M. Jackman, "Application of Nuclear Magnetic Resonance Spectroscopy in Organic Chemistry," Pergamon Press, New York (1959).
62. J.A. Pople, W.G. Schneider, and H.J. Bernstein, "High-Resolution Nuclear Magnetic Resonance," McGraw-Hill, New York (1959).
63. O. Jarketzky and C.D. Jarketzky, *in* "Methods of Biochemical Analysis," Vol. IX, Interscience, New York (1962).
64. N.S. Bhacca and D.H. Williams, "Applications of NMR Spectroscopy in Organic Chemistry," Holden-Day, San Francisco (1964).
65. D. Chapman and P.D. Magnus, "Introduction to Practical High Resolution Nuclear Magnetic Resonance Spectroscopy," Academic Press, London (1966).
66. J.W. Emsley, J. Feeney, and L.H. Sutcliffe, "High Resolution Nuclear Magnetic Resonance Spectroscopy," Vol. I, Pergamon Press, New York (1965), Vol. II (1966).
67. D.M. Small, S.A. Penkett, and D. Chapman, *Biochim. Biophys. Acta* **176,** 178 (1969).
68. J.N. Shoolery and M.T. Rogers, *J. Am. Chem. Soc.* **80,** 5121 (1958).
69. D. Lavie, S. Greenfield, Y. Kashman, and E. Glotter, *Israel J. Chem.* **5,** 151 (1967).
70. P. Eneroth, B. Gordon, R. Ryhage, and J. Sjövall, *J. Lipid Res.* **7,** 511 (1966).
71. P. Eneroth, B. Gordon, and J. Sjövall, *J. Lipid Res.* **7,** 524 (1966).
72. A. Norton and B. Haner, *Acta Cryst.* **19,** 477 (1965).
73. T.C. Laurent and H. Persson, *Biochim. Biophys. Acta* **106,** 616 (1965).
74. "The Merck Index," 7th ed. (P.G. Stecher, ed.), Merck & Co., Rahway, New Jersey (1960).
75. B. Josephson, *Biochem. J.* **29,** 1484 (1935a).
76. Y. Go and O. Kratky, *Z. Physik.* **B26,** 439 (1934).
77. O. Kratky and G. Giacomello, *Monatsh.* **69,** 427 (1936).
78. V. Caglioti and G. Giacomello, *Gazz. Chem. Ital.* **69,** 245 (1939).
79. G. Giacomello and M. Romeo, *Gazz. Chem. Ital.* **73,** 285 (1943).
80. H. Wieland and H. Sorge, *Z. Physical Chem.* **97,** 1 (1916).
81. G. Giacomello and O. Kratky, *Z. Kristallograph.* **95,** 459 (1936).
82. J. Parsons, W.T. Beher, and G.D. Baker, *Henry Ford Hosp. Med. Bull.* **6,** pt. II, 365 (1958).

83. J. Parsons, W.T. Beher, and G.D. Baker, *Henry Ford Hosp. Med. Bull.* **9,** 54 (1961).
84. J. Parsons, S.T. Wong, W.T. Beher, and G.D. Baker, *Henry Ford Hosp. Med. Bull.* **11,** 23 (1963).
85. J. Parsons, S.T. Wong, W.T. Beher, and G.D. Baker, *Henry Ford Hosp. Med. Bull.* **12,** 87 (1964).
86. N.K. Adam, "The Physics and Chemistry of Surfaces," 3rd ed. (H. Milford, ed.), Oxford Univ. Press, London (1941).
87. G.L. Gaines, Jr., "Insoluble Monolayers at Liquid-Gas Interfaces," Wiley, Interscience, New York (1966).
88. A.W. Adamson, "The Physical Chemistry of Surfaces," Interscience, New York (1960).
89. W.D. Harkins, "The Physical Chemistry of Surface Films," Reinhold, New York (1952).
90. J.T. Davies and E.K. Rideal, "Interfacial Phenomena," Academic Press, New York (1961).
91. I. Langmuir, *J. Am. Chem. Soc.* **39,** 1848 (1917),
92. D.G. Dervichian, "La Technique des Couches Superficielles," Chap. XIII, Techniques de Laboritoire, Masson, Paris (1954).
93. J.W. Gibbs, Collected Works, Vol. 1, 1931, New York (1876).
94. D.J. Crisp, *in* "Surface Chemistry," p. 17, Butterworths, London (1949).
95. E. Otero Aenlle and R.C. Carro, *Anales Fisica Quimica* **52,** 85 (1959).
96. D.G. Dervichian, Comptes Rendus de la 2e Réunion de Chimie Physique, Paris, p. 443, June (1952).
97. D.G. Dervichian, *in* "Progress in Chemistry of Fats and Other Lipids," Vol. 2, p. 193, Pergamon Press, London (1954).
98. D.G. Dervichian and M. Joly, *Nature* **141,** 975 (1938).
99. D.G. Dervichian, *J. Phys.* **7,** 427 (1935).
100. N.L. Gershfeld, "Cohesive forces in monomolecular films at an air-water interface," *Advan. Chem. Ser.* **84,** 115 (1958).
101. H.E. Ries, Jr., *Sci. Am.* March, 2 (1961).
102. P. Ekwall and R. Ekholm, Proceedings of the International Congress on Surface Activity, 2nd ed., London, 1957, p. 23, Butterworths, London (1957).
103. P. Ekwall, R. Ekholm, and A. Norman, *Acta Chem. Scand.* **11,** 693 (1957).
104. P. Ekwall, R. Ekholm, and A. Norman, *Acta Chem. Scand.* **11,** 703 (1957).
105. P. Ekwall, R. Silander, and J. Rydberg, 3rd International Congress Surface Activity, Cologne, Vol. II, B, p. 703 (1960).
106. E. Otero Aenlle, Proceedings of the International Congress on Surface Activity, London, 2nd ed., I, p. 135, Butterworths, London (1957).
107. E. Otero Aenlle and R.C. Carro, *Estos Anales* **B51,** 515 (1955).
108. S.G. Fernandez, A.L. Lopez, and E. Otero Aenlle, *Anales Fisica Quimica* **B60** 137 (1964).
109. A.L. Lopez, S.G. Fernandez, and E. Otero Aenlle, *Koll. Zeit.* **211,** 131 (1966).
110. J.M. Trillo, S.G. Fernandez, and P.S. Pedero, *Anales Fisica Quimica* **B63,** 933 (1967).
111. J. H. Shulman and E.K. Rideal, *Proc. Roy. Soc.* **B112,** 29 (1937).
112. K.D. Dreher, J.H. Schulman, and A.F. Hofmann, *J. Coll. Interf. Sci.* **25,** 71 (1967).
113. P. Desnuelle, J. Molines, and D.G. Dervichian, *Bull. Soc. Chim. France* **18,** 197 (1951).
114. L. de Bernard and D.G. Dervichian, *Bull. Soc. Chim. Biol.* **37,** 843 (1955).
115. L. de Bernard, *Bull. Soc. Chim. Biol.* **40,** 161 (1958).
116. J.B. Leathes, *Lancet* **1,** 853 (1925).
117. G.H.A. Clowes, *in* "Surface Chemistry," *Publ. of Am. Assoc. for Advancement of Science,* No. 21, 1 (1943).
118. D.M. Small, unpublished data.
119. D.J. Crisp, *in* "Surface Chemistry," p. 23, Butterworths, London (1949).

120. J.P. Kratohvil and H.T. Dellicolli, *Can. J. Biochem.* **46**, 945 (1968).
121. P. Ekwall, T. Rosendahl, and A. Sten, *Acta Chem. Scand.* **12**, 1622 (1958).
122. W.S. Bennett, G. Eglinton, and S. Kovac, *Nature* **217**, 776 (1967).
123. P. Ekwall, T. Rosendahl, and N. Lofman, *Acta Chem. Scand.* **11**, 590 (1957).
124. E. Back and B. Steenberg, *Acta Chem. Scand.* **4**, 810 (1950).
125. H. Hammarsten, *Biochem. Z.* 481 (1924).
126. B.A. Josephson, *Biochem. Z.* **263**, 428 (1933).
127. P. Ekwall, E.V. Lindstrom, and K. Setala, *Acta Chem. Scand.* **5**, 990 (1951).
128. R.H. Dowling and D.M. Small, *Gastroenterology* **54**, 1291 (abstract) (1968).
129. R.H. Palmer, P.B. Glickman, and A. Kappas, *J. Clin. Invest.* **41**, 1573 (1962).
130. J.B. Carey, Jr., and G. Williams, *Science* **150**, 620 (1965).
131. W.H. Admirand and C. Trey, *Clin. Res.* **16**, 278 (1968).
132. M.J. Schick, "Nonionic Surfactants," Marcel Dekker, New York (1967).
133. F. Krafft and H. Wiglow, *Ber.* **28**, 2566 (1895).
134. F. Lachampt and R. Perron, *Traite Chim. Organ.* **22**, 837 (1953).
135. M. Demarc and D.G. Dervichian, *Bull. Soc. Chim. France* **12**, 939 (1945).
136. R.H. Palmer, *Proc. U.S. Nat. Acad. Sci.* **58**, 1047 (1967).
137. M. Raison, Proc. Intern. Cong. Surface Activity, London, 2nd ed., p. 374, Butterworths, London (1957).
138. H. Lecuyer and D.G. Dervichian, *Kolloid Z.* **197**, 115 (1964).
139. D. Rudman and F.E. Kendall, *J. Clin. Invest.* **36**, 538 (1957).
140. R.T. Reinke and J.D. Wilson, *Clin. Res.* **15**, 34 (1967).
141. K. Shinoda, T. Nakagawa, B. Tamamushi, and T. Isemura, "Colloidal Surfactants," Acadmic Press, New York (1963).
142. P. Ekwall, K. Fontell, and A. Sten, Proc. Intern. Cong. Surface Activity, London, p. 357, Butterworths, London (1957).
143. A. Norman, *Acta Chem. Scand.* **14**, 1295 (1960).
144. P. DeMoerloose and R. Ruyssen, *J. Pharm. Belg.* **14**, 95 (1959).
145. T. Furusawa, *Fukuoka Acta Med.* **53**, 124 (1962).
146. H. Miyake, T. Murakoshi, and T. Hisatsugu, *Fukuoka Igaku Zassi* **53**, 659 (1962).
147. T. Bates, M. Gibaldi, and J.L. Konig, *Nature* **210**, 1331 (1966).
148. T.R. Bates, Doctoral thesis, Columbia University, New York (1966).
149. M.C. Carey, D. Chapman, M.D. Barratt, and D.M. Small, unpublished data.
150. M.C. Carey, D. Chapman, C.J. Hart, and D.M. Small, unpublished data.
151. S.A. Johnston and J.W. McBain, *Proc. Roy. Soc. London, Ser. A* **181**, 119 (1943).
152. D.M. Small, unpublished observations on the surface tension of bile salt solutions by the Willhelmy plate method.
153. F.P. Woodford, *J. Lipid Res.* **10**, 539 (1969).
154. G. Nemethy and H.A. Scheraga, *J. Phys. Chem.* **66**, 1773 (1962).
155. P. Mukerjee, *Adv. Colloid Interface Sci.* **1**, 241 (1967).
156. B.D. Flockhart, *J. Colloid Sci.* **16**, 484 (1961).
157. E.D. Goddard and G.C. Benson, *Can. J. Chem.* **35**, 986 (1957).
158. M.J. Schick, S.M. Atlas, and F.R. Eirich, *J. Phys. Chem.* **66**, 1326 (1962).
159. M.J. Schick, *J. Phys. Chem.* **68**, 3584 (1964).
160. M.J. Schick, *J. Colloid Sci.* **17**, 801 (1962).
161. P. Mukerjee, *J. Phys. Chem.* **66**, 1375 (1962).
162. G. Stainsby and A.E. Alexander, *Trans. Faraday Soc.* **46**, 587 (1950).
163. J. Tl. G. Overbeek and D. Stigter, *Rec. Trav. Chim.* **755**, 1263 (1956).
164. H.S. Frank and M.W. Evans, *J. Chem. Phys.* **13**, 507 (1945).
165. J.D. Bernal, "The State and Movement of Water in Living Organisms," p. 7ff, Cambridge University Press, London (1964).
166. F. Franks, *Chem. Ind.* **1968**, 560 (May 4, 1968).
167. J.A. Olson and J.S. Herron, *Proc. Intern. Cong. Biochem. 6th, New York,* **1964**, pp. 7, 112 (abstract) (1964).

168. J. Herron and J.A. Olson, personal communication.
169. B. Borgstrom, *Biochim. Biophys. Acta* **106,** 171 (1965).
170. H. Sobotka and N. Czeczowczka, *J. Colloid Sci.* **13,** 188 (1958).
171. A. Rich and D.M. Blow, *Nature* **182,** 423 (1958).
172. D.M. Blow and A. Rich, *J. Am. Chem. Soc.* **82,** 3566 (1960).
173. M.J. Schick, *J. Am. Oil Chem. Soc.* **40,** 680 (1963).
174. M.F. Emerson and A. Holtzer, *J. Phys. Chem.* **71,** 3320 (1967).
175. J. Clifford, *Trans. Faraday Soc.* **60,** 276 (1964).
176. P. Ekwall, K. Setala, and L. Sjoblom, *Acta Chem. Scand.* **5,** 175 (1951).
177. M.C. Carey and D.M. Small, *Gastroenterology* **58,** 1057 (1970).
178. D.B. Neiderhiser and H.P. Roth, *Proc. Soc. Exp. Biol. Med.* **128,** 221 (1968).
179. D.B. Neiderhiser and H.P. Roth, personal communication.
180. A.N. Campbell and N. O. Findlay Smith, "The Phase Rule," 9th ed., Dover, New York (1951).
181. G. Masing, "Ternary Systems," Dover, New York (1944).
182. S. Glasstone and D. Lewis, "Elements of Physical Chemistry," 2nd ed., Macmillan London (1963).
183. R.D. Vold and J.W. McBain, *J. Am. Chem. Soc.* **63,** 1296 (1941).
184. D.M. Small, *in* "Surface Chemistry of Biological Systems" (M. Blank, ed.), Plenum Press, New York p. 55 (1970).
185. D. G. Dervichian, *in* "Progress in Biophysics and Molecular Biology" (J. Butler and H. Huxley, eds.), Ch. 14, Academic Press, New York (1964).
186. F. Rosevear, *J. Am. Oil Chem. Soc.* **31,** 628 (1954).
187. F. Husson, H. Mustacchi, and V. Luzzati, *Acta Crystal.* **13,** 668 (1960).
188. V. Luzzati and F. Husson, *J. Cell. Biol.* **12,** 297 (1962).
189. K.D. Lawson and T.J. Flautt, *Mol. Crystals* **1,** 241 (1966).
190. T.J. Flautt and K.D. Lawson, *Adv. Chem. Series* **63,** 26 (1965).
191. D. Chapman, P. Byrne, and G.G. Shipley, *Proc. Trans. Faraday Soc.* **290,** 115 (1966).
192. D. Chapman, R.M. Williams, and B.D. Ladbrooke, *Chem. Phys. Lipids* **1,** 445 (1967).
193. D. Chapman and N.J. Salsbury, *Trans. Faraday Soc.* **62,** 2607 (1966).
194. N.J. Salsbury and D. Chapman, *Biochim. Biophys. Acta* **163,** 314 (1968).
195. S.A. Penkett, A.G. Flook, and D. Chapman, *Chem. Phys. Lipids* **2,** 273 (1968).
196. D. Champman, *Adv. Chem. Ser.* **63,** 157 (1967).
197. D.M. Small and S. Rapo, *New Eng. J. Med.* **283,** 53 (1970).
198. J.W. McBain, R.D. Vold, and R. Porter, *Ind. Eng. Chem.* **33,** 1049 (1941).
199. B. Borgstrom, *J. Lipid Res.* **8,** 598 (1967).
200. E.B. Feldman and B. Borgstrom, *Biochim. Biophys. Acta* **125,** 136 (1966).
201. E.B. Feldman and B. Borgstrom, *Lipids* **1,** 430 (1966).
202. G. Benzonana, *Biochim. Biophys. Acta* **176,** 836 (1969).
203. D.M. Small, *Fed. Proc.* **29,** 1320 (1970).
204. G. Friedel, *Ann. Physique* **17,** 273 (1922).
205. G.H. Brown and W.G. Shaw, *Chem. Rev.* **56,** 1049 (1957).
206. G.W. Gray, "Molecular Structure and the Properties of Liquid Crystals," p. 114, Academic Press, New York (1962).
207. J. H. Hildebrand and R.L. Scott, "The Solubility of Nonelectrolytes," Dover, New York (1964).

DEFINITIONS

Acid This refers to any substance HA that has the capacity to give a proton.

Aggregation Number (Ag ♯) The aggregation number is the average number of molecules in a micelle, i.e., micellar molecular weight divided by molecular weight of the monomer. The letter *n* is often used to mean Ag ♯. See Section VIII.

Amphiphile A relatively large organic compound that has a water-soluble (hydrophilic) part and an oil-soluble (lipophilic) part. These compounds may be classified into: insoluble nonswelling amphiphiles (amphiphiles that will spread on the surface to form stable monolayers but are insoluble in the bulk); insoluble swelling amphiphiles (compounds that form stable monolayers and are insoluble but swell in water to form lyotropic liquid crystals); and soluble amphiphiles (those compounds that form unstable monolayers and in the bulk form micellar solutions). A more complete discussion is available (42, 203).

Bile Salt In this chapter this term usually means the alkaline metal salt of the bile acid (e.g., Na^+A^-).

Critical Micellar Concentration (CMC) The concentration at which molecular solubility of a detergent is reached and molecular aggregation begins to occur.

Critical Micellar Temperature (CMT) The temperature to which a given mixture of detergent and water must be raised to transform the detergent from a suspension of crystals in water or gel to a clear micellar phase. Since the critical micellar temperature may vary over large concentrations of detergent, the detergent concentration should be specified.

Emulsions Microscopically visible droplets of one liquid suspended in another, stabilized on their surface by an *emulsifier* (an amphiphilic compound).

Krafft Point The critical micellar temperature at the critical micellar concentration of the detergent (see Section VII. D).

Liquid Crystals (Mesophases) In this article we will be speaking of lyotropic liquid crystals, that is, liquid crystals that form in the presence of a solvent (water). These liquid crystals have characteristics of both crystallinity, that is, they have a long-range order in one or two dimensions, and a liquid-like order in the other dimensions. Certain lyotropic liquid crystals have a characteristic birefringence, give specific sharp X-ray spacings, but have some characteristics of liquid—for instance they can be poured or made to flow through tubes [for further explanation see (3) and monographs by Friedel (204), Brown and Shaw (205), and Gray (206)].

Micelle A small aggregate of amphiphilic molecules that forms spontaneously above a certain critical concentration (CMC) and above a certain critical temperature (CMT). The micelle may be made up of any number of monomers (from 2 to 1000) that are in constant equilibrium with a surrounding medium containing only monomers.

Monomer Individual molecules or ions in solution (not aggregated molecules).

Phase A homogeneous, physically distinct, and mechanically separable entity. The typical phases are solid, liquid, and gas. In this review I will be dealing with solid phases (crystalline or amorphous), liquid phases (ideal solutions or micellar solutions), and liquid crystalline phases (e.g., lamellar, hexagonal, cubic, etc.). For a complete discussion of phases and the phase rule, see Campbell and Smith (180).

Solubilization The bringing into solution of substances that are otherwise insoluble in a given medium. In the case of bile salts and detergents, solubilization involves the presence of micelles that incorporate within or upon themselves the otherwise insoluble material. Other types of solubilization are possible (207).

Author Index

Note: numbers in parentheses are reference numbers (keyed to the list at the end of the chapter in which the cited page occurs).

Subject Index